Plant Pest Risk Analysis

Concepts and Application

_____

Plant Pest Risk Analysis

Concepts and Application

_____

Edited by

Christina Devorshak

CAbi
www.cabi.org

CABI is a trading name of CAB International

CABI
Nosworthy Way
Wallingford
Oxfordshire OX10 8DE
UK

Tel: +44 (0)1491 832111
Fax: +44 (0)1491 833508
E-mail: info@cabi.org
Website: www.cabi.org

CABI
38 Chauncey Street
Suite 1002
Boston, MA 02111
USA

Tel: +1 800 552 3083 (toll free)
Tel: +1 (0)617 395 4051
E-mail: cabi-nao@cabi.org

A catalogue record for this book is available from the British Library,
London, UK.

Library of Congress Cataloging-in-Publication Data

Plant pest risk analysis : concepts and application / C. Devorshak, ed.
    p. cm.
  Includes bibliographical references and index.
  ISBN 978-1-78064-036-5 (alk. paper)
  1. Plants, Protection of. 2. Risk assessment. 3. Insect pests--Control.
4. Agricultural pests--Control. I. Devorshak, Christina, 1969-

  SB950.P55 2012
  628.9'6--dc23

                        2012025556

ISBN-13: 978-1-78064-036-5

Commissioning editor: Rachel Cutts
Editorial assistant: Chris Shire
Production editor: Simon Hill

Typeset by SPi, Pondicherry, India.
Printed and bound in the UK by CPI Group (UK) Ltd, Croydon, CR0 4YY.

Contents

_____

Contributors

_____

Chapter 2: Robert Griffin (Robert.L.Griffin@aphis.usda.gov)
Chapter 7: Lottie Erikson (Lottie.Erikson@aphis.usda.gov)
Chapter 10: Robert Griffin
Chapter 11: Christina Devorshak (Christina.Devorshak@aphis.usda.gov) and Alison Neeley (Alison.Neeley@aphis.usda.gov)
Chapter 12: Glenn Fowler (Glenn.Fowler@aphis.usda.gov) and Yu Takeuchi (Yu.Takeuchi@aphis.usda.gov)
Chapter 13: Robert Griffin and Alison Neeley
Chapter 14: Robert Griffin
Chapter 15: Alison Neeley and Christina Devorshak
Chapter 16: Robert Griffin
Chapter 17: Stephanie Bloem (Stephanie.Bloem@aphis.usda.gov) and Kenneth Bloem (Kenneth.Bloem@aphis.usda.gov)
Chapter 18: Anthony Koop (Anthony.Koop@aphis.usda.gov)
Chapter 19: Robert Griffin

Acknowledgements

This text represents the team effort of many individuals and colleagues from the USDA APHIS PPQ CPHST Plant Epidemiology and Risk Analysis Laboratory (PERAL). They have served as authors, contributors, reviewers, editors and sources of information. Between them, they are experts in economics, climate and mapping, entomology, plant pathology, weed science, botany, risk analysis, international trade and international development. They are practising pest risk analysts, working on projects as diverse as developing predictive models for pest spread, to economic models on pest impacts, to developing international guidance on pest risk analysis.

Together, they bring years of experience in pest risk analysis from different backgrounds – including working in the US Department of Agriculture, the International Atomic Energy Agency, the Food and Agriculture Organization of the United Nations, the International Plant Protection Convention and as faculty at universities – with expertise in areas as diverse as international standards development to biological, economic and climate modelling in risk analysis. As a group, they have been active in educating students and plant protection professionals on various aspects of pest risk analysis at universities and professional workshops, both nationally and internationally. Their expertise and contributions to this text were of immeasurable value and I am grateful to them for their time and efforts in contributing chapters to this text.

In addition to the significant efforts of the authors listed above, many others assisted with the production of this text. I am grateful to Robert Griffin for donating not only his own time and expertise, but that of many of his staff in PERAL as writers, reviewers, experts and editors. Lucy Reid,[1] the PERAL librarian, tracked down valuable resources for the writers and always managed to come through for us. Tara Holtz,[1] Alison Neeley, Ashley Jackson,[1] Lottie Erikson, Glenn Fowler and Stephanie Bloem all provided reviews and inputs during various stages of the development of this work.

Much of the material that served as the basis for this book came from PERAL's flagship course, 'Risk Analysis 101' – the materials for that course having been developed over the past six years. In teaching that course, I always learned just as much from the students as I have taught them. It is thanks to their valuable suggestions for improvements to course materials that PERAL has developed one of the best pest risk analysis training courses available – which in turn served as the inspiration and basis for this book.

I must also express sincere appreciation to colleagues in other countries and other plant protection services who are also working in the field of pest risk analysis and regulatory plant protection. My work in the International Plant Protection Convention (IPPC) for five years, then subsequently as a risk analyst in PERAL, has given me unique and valuable opportunities to interact with risk practitioners from all over the world – and more importantly, the chance to learn more about how they conduct pest risk analysis. Over the years, and at countless meetings, I've always been amazed at the passion, commitment and wealth of knowledge so many of them have for this work. And because of these colleagues, I was able to learn about pest risk analysis from some of the best practitioners in the world. We have all learned from each other, and we eagerly continue to learn – such is the nature of the field of pest risk analysis.

Finally, additional acknowledgments go to Roger Magarey,[1] Leslie Newton,[1] Manuel Colunga-Garcia,[2] Dan Borchert,[1] Barney Caton,[1] Ron Sequeira,[1] Bill Smith,[3] Frank Koch,[3] Ross Meentemeyer,[4] Matt Dedmon,[5] Joe Russo,[5] Mike Simon,[6] Gary Lougee[6] and Lisa Ferguson[1] for their assistance with Chapter 12 and to Larry Fowler[1] and Claire Wilson[7] for their assistance with Chapter 18.

Notes

[1] Mailing address and affiliation: USDA-APHIS-PPQ-CPHST, 1730 Varsity Drive, Suite 300, Raleigh, NC, 27606 USA
[2] Michigan State University.
[3] USDA Forest Service.
[4] UNC-Charlotte.
[5] ZedX, Inc.
[6] USDA-APHIS-PPQ-QPAS.
[7] Canadian Food Inspection Agency.

Part I

Pest Risk Analysis Background and History

_____

1 Introduction

Christina Devorshak

1.1. Purpose of this Textbook

Pest risk analysis is an evolving and dynamic field. In the past 20 years, countries all over the world have adopted pest risk analysis as a means to inform regulatory decisions for plant protection. During that time, pest risk analysis has undergone an evolution – pest risk analysts have adapted methods used in other sectors, scientific information to support analyses is more available than ever, and the need for more, better and different types of pest risk analyses continues to grow.

Pest risk analysis informs and guides decisions for regulatory plant protection – from domestic activities such as prioritizing pests for surveillance to making quarantine decisions regarding the importation of products from other countries. The need for analytical support for decision makers has become indispensable. Moreover, regulatory plant protection decisions affecting imported products are required, under international treaties, to be supported by an assessment of risk.

The need for better, faster and more diverse types of pest risk analyses places demands on plant protection services for more analytical support. Despite the importance of pest risk analysis, and the critical need for it, there is very little opportunity for formal education in pest risk analysis. Most pest risk analysts are entomologists, plant pathologists, botanists, weed scientists, ecologists or economists who have learned about risk analysis outside of a traditional educational setting. In fact, most pest risk analysts have probably learned about pest risk analysis through being mentored by colleagues, attending workshops and professional meetings, through specialized professional training, or simply by learning on the job.

In the past several years, plant protection services, as well as institutes of higher education, have recognized the importance of developing curricula in regulatory plant protection, and for pest risk analysis in particular. Courses are being developed in different countries and at different universities to educate students about regulatory plant protection, and to foster a new generation of plant protection specialists. However, there is a noticeable gap in educational materials available to support the development of such courses.

This textbook is intended to fill that gap – it provides a solid foundation in the field of pest risk analysis and its application to regulatory plant protection. The intended audience of this text is upper-level students in agricultural sciences, as well as regulatory plant protection professionals who wish to learn about pest risk analysis: why we do it, how it's done and what it requires. Basic methods for pest risk analysis are addressed in this book – it describes pest risk analysis from the perspective of regulatory science, rather than from a purely academic perspective.

The distinction between an academic approach to risk analysis and the regulatory approach to pest risk analysis is important to understand. When we do pest risk analysis, we conduct risk analysis in a very specific regulatory context where we are legally accountable, and the resulting decisions have significant impacts – on the environment, on trade, and on our stakeholders. We do those analyses typically with limited information for complex biological systems, and with limited time and resources. Many excellent texts address risk and risk analysis in the academic sense, and are useful references for anyone doing risk analysis in general. But there has never been a textbook published on how we can apply pest risk analysis in a regulatory environment – a world where we balance science and academic concepts with real life problems that must be solved in the context of national and international rules, regulations, requirements, laws and agreements. The information contained in this text is unique and has not been recorded previously in any other sources – much of the information is based on the collective wisdom of practising risk analysts and years of experience of solving problems in a regulatory context. Quite simply, there are few references available that cover the whole of pest risk analysis and its roles, applications and methods.

Advanced or specialized methods developed in other fields (e.g. engineering, ecological modelling, toxicology, economics, etc.) may be modified and adapted to do pest risk analysis. However, such methods are beyond the scope of an introductory text on pest risk analysis (readers may wish to consult other texts to learn more about analytical methods used in other fields). Each chapter of this book includes specific references – many of these references are instructive for some of the more specialized methods that are used less frequently in pest risk analysis, including in particular quantitative methods such as probabilistic modelling and Bayesian analysis.

Examples of methods and applications for pest risk analysis are provided throughout the text and many of these examples are based on real analyses performed by plant protection organizations. The examples are provided only to illustrate specific points, but we do not promote any one method, tool, application or technique over another – they all have strengths and weaknesses. Instead, we recommend learning and understanding the most commonly used basic tools and methods for pest risk analysis first – and by understanding the basic tools and methods, the reader should be positioned to read, prepare, review and develop scientifically sound, informative and technically defensible pest risk analyses.

1.2. Format of this Textbook

This textbook provides the fundamental tools and skills for the reader to understand what risk is, and descriptions of different systems and options for conducting systematic analyses in a regulatory plant protection context.

The textbook is organized into five sections:

- **Part I: Pest Risk Analysis Background and History** – provides the reader with a basic introduction to risk, risk analysis and pest risk analysis. It also provides a historical perspective of how pest risk analysis has come to be so critical in regulatory plant protection and provides the international legal framework for pest risk analysis.
- **Part II: Pest Risk Analysis – Components and Applications** – provides the reader with background on key components of pest risk analysis. This includes the use of terminology, what types of information are needed, how we incorporate economic factors into pest risk analysis and, lastly, the applications of pest risk analysis in regulatory plant protection.
- **Part III: Pest Risk Assessment Methods** – provides the reader with an overview of basic methods used to analyse risk. This includes using qualitative and quantitative methods, as well as the use

of economic analysis and geographic information systems.

- **Part IV: Pest Risk Management, Risk Communication and Uncertainty** – provides the reader with a basic introduction to the theory and application of pest risk management. It also discusses the role and importance of effective risk communication (including how it relates to pest risk management), and addresses the role of uncertainty in pest risk analysis.
- **Part V: Special Topics for Pest Risk Analysis** – provides further information on specialized topics within pest risk analysis, including risk analysis for beneficial organisms and weeds, as well as the role of dispute settlement and precaution in the application of pest risk analysis. Lastly it provides a short review of how other organisms, namely invasive alien species and living modified organisms, are addressed through pest risk analysis within the International Plant Protection Convention (IPPC)/sanitary and phytosanitary (SPS) framework.

1.3. Major Themes

1.3.1. The Agreement on the Application of Sanitary and Phytosanitary Measures, the International Plant Protection Convention and International Standards for Phytosanitary Measures

This text approaches pest risk analysis from the perspective of the IPPC and the guidance it provides on pest risk analysis through International Standards for Phytosanitary Measures (or ISPMs). This is also consistent with the requirements laid out in the World Trade Organization Agreement on the Application of Sanitary and Phytosanitary Measures (or SPS Agreement). The roles of the SPS Agreement, the IPPC and ISPMs are described in detail in Part I, but are integral throughout the entire text. It is because of the SPS Agreement and the IPPC that pest risk analysis has become – in the last

20 years – such a critical activity for plant protection organizations all over the world. And it is these Agreements, and the ISPMs, that provide the overall framework for pest risk analysis under which plant protection organizations work.

1.3.2. Terminology in pest risk analysis

Pest risk analysis has its own language, and considerable efforts have been made internationally to develop and use consistent terms related to pest risk analysis. The terminology used throughout this text is consistent with IPPC terminology and *The Glossary of Phytosanitary Terms* (ISPM No. 5). Throughout the different chapters in this text, definitions of key terms are highlighted to frame discussions of various topics, and a Glossary is provided at the end of the text. In addition, Chapter 5 reviews the key terminology used in pest risk analysis. It is important to understand that specific terms have particular meanings within the context of pest risk analysis, and the reader should make a special note of a term whenever it is discussed in detail. Some terms are discussed in several places – this only underscores the importance of understanding the language that is used in pest risk analysis, and the need to be precise in how we communicate about risk.

1.3.3. Uncertainty

Uncertainty is an integral part of pest risk analysis – if we had perfect knowledge of a system, we wouldn't need to do risk analysis because we'd be certain of the outcome. We do risk analysis because we have uncertainty. It occurs throughout any pest risk analysis, from the start (initiation) to the conclusion (risk management). Chapter 16 provides an overview of uncertainty in pest risk analysis, and it is also discussed throughout the chapters in this text in relation to the specific aspects of pest risk analysis. We need to understand and document what our uncertainties are and where they come from, and determine how we can best address our uncertainty in our analyses and in risk management.

1.3.4. Pest risk analysis is multi-disciplinary

Pest risk analysis is a multi-disciplinary field – entomology, plant pathology, botany, weed science, ecology, economics, geography, mathematics, statistics and even social sciences can play important roles in conducting pest risk analysis. Just as no one type of method, process or scheme will fit every type of analysis, there is no one type of expert best suited for every type of analysis. Throughout this text, we reference specialized types of analyses (e.g. weed risk analysis) and specialized tools (e.g. economic analysis and mapping) that require subject matter expertise. Thus, the reader should bear in mind that pest risk analyses are commonly conducted by cross-disciplinary teams, and will often require expertise in several technical and scientific fields.

1.3.5. Pest risk analyses are variable

In the world of pest risk analysis, one size does not fit all, and there is no recipe or 'cookie-cutter' approach to pest risk analysis. Numerous schemes, methods, guidelines and processes have been developed and implemented by different experts and organizations to address the many types of pest risk analysis problems. But, no matter how good a guideline or scheme is, it will never fit every problem or situation.

Therefore, this text does not promote any one method, process or scheme as being the best solution, nor do we provide any formulaic approach to how pest risk analysis should be done. Rather, we believe that those who undertake pest risk analysis should understand the basic concepts of risk, and how to analyse risk – and then provide the foundation and building blocks for the conduct of pest risk analysis. In reading this text, we intend for readers to gain a better understanding of the critical elements of pest risk analysis, the benefits and limitations of various models, and when and how to apply appropriate tools and methods. We don't provide an answer – we provide the reader with the tools to develop their own answers.

1.3.6. Pest risk analysis is a dynamic field

The field of pest risk analysis is a young discipline that has evolved quickly over the past 20 years. The pest risk analysis community has come a long way in creating better models, refining methods and incorporating diverse areas of expertise, but we are still improving. Some organizations are beginning to incorporate more complex mathematical models into their analyses, while others are attempting to streamline and simplify their work. It is virtually certain that the pest risk analyses that are done 20 years from now will look different from the ones we do today, as we continue to learn and evolve. This only underscores the importance of understanding the concepts and building blocks of pest risk analysis, rather than becoming dependent on any one method or procedure. Regardless of how the field of pest risk analysis evolves, understanding the 'basics' will enable understanding of any new models that emerge over time.

1.4. Summary

The focus of this text is on creating a solid foundation and understanding of basic pest risk analysis concepts and methods. We provide the reasoning and theory behind pest risk analysis as well as practical options for conducting pest risk analysis. By the end of this book, the reader should have a working knowledge of:

- The framework for pest risk analysis;
- The language of pest risk analysis;
- Applications for pest risk analysis;
- Methods for pest risk assessment;
- Theory and application of pest risk management;
- Uncertainty in pest risk analysis;
- Risk communication.

2 Basic Concepts in Risk Analysis

Robert Griffin

2.1. Introduction

It is easy to find many different descriptions of risk. A plethora of information is also available online at the many forms of risk analysis, but much less is written about *pest risk analysis*, the label created by the phytosanitary community for its unique form of risk analysis. One reason for this is that the systematic practice of risk analysis in plant quarantine has a relatively short history compared with the application of risk analysis in other disciplines.

A heightened interest by national plant protection organizations (NPPOs) in some form of defined, harmonized risk analysis began with the Uruguay Round negotiations under the General Agreement on Tariffs and Trade (GATT), which were launched at Punta del Este, Uruguay in September, 1986 (FAO, 2000). Among the agreements completed during these negotiations was the Agreement on the Application of Sanitary and Phytosanitary Measures (the SPS Agreement), which was designed to, for the first time, establish internationally agreed disciplines on the measures applied by countries to protect from sanitary (human and animal health and life) and phytosanitary (plant health and life) hazards that may be associated with cross-boundary trade. That is not to say that risk analysis was completely unknown in plant quarantine prior to the SPS Agreement, but rather there was no globally recognized approach or discipline and little motivation to create an international framework which might limit the flexibility that countries had historically enjoyed with decision making on matters of plant quarantine.

The coming into force of the SPS Agreement in 1995 with the transition from the GATT to the World Trade Organization (WTO) with binding dispute settlement mechanisms, brought risk analysis to the forefront for the phytosanitary community. The SPS Agreement specified that phytosanitary measures must be based on either international standards or risk assessment (= risk analysis; discussed in greater detail in Chapter 4). Since the phytosanitary community had no history of standard setting at the time, this meant that essentially all phytosanitary measures must be based on risk analysis.

All plant quarantine authorities whose countries were Members of the WTO suddenly faced the potential for WTO challenges from their trading partners for unjustified phytosanitary measures unless they could produce an SPS compliant risk analysis. The worst trade distorting issues causing tension at the time caused several countries to move quickly to develop risk analysis capabilities while also considering the possibility of launching disputes against their trading partners. In addition, policy makers that historically may not have seen science as a strong influence on their decisions were in a vulnerable position if they did not give sufficient attention and emphasis to the scientific basis for measures applied in trade.

At the same time, all WTO Members recognized the need to develop international standards and identified the priority for risk analysis standards to be developed by the standard setting organizations identified in the SPS Agreement: the Codex Alimentarius for human health (food safety); the Office International des Epizooties (or OIE, now called the World Organization for Animal Health), and the IPPC for plant health. Both of the first two organizations already had a history of standard setting when the Uruguay Round negotiations began; Codex since 1962 and OIE since 1924, but the IPPC had no history of standard setting in 1989, and none of the organizations had standards specifically addressing risk analysis (FAO, 2000).

All of this is to say that pest risk analysis was born of the need to satisfy the expectations of the SPS Agreement. As a result, it has developed under the IPPC and is relatively early in its development and application. This context is crucial to understanding the basic concepts associated with pest risk analysis in order to progress from the fundamentals of risk to the concepts of risk analysis, and finally to the contemporary understanding and practice of pest risk analysis, which is unique in many ways because of the framework within which it is applied and the evolution it has experienced.

2.2. What is Risk?

Risk: the likelihood of an adverse event and the magnitude of the consequences.

Simple enough, or is it? Everyone understands risk in some way or maybe in various ways depending on the situation. Indeed, risk is such an integral part of our everyday life it is largely taken for granted until we are reminded that a particular thing or activity has risk associated with it (Vose, 2000; Yoe, 2012). Differences in interpretation emerge when we look at risk critically; analyse its component elements and compare interpretations as they apply to different situations where we are concerned about *safety*.

Safety however is not measured. Risks are measured. Only when risks are considered against social and political values can safety be judged. This fundamental distinction is at the heart of the SPS Agreement: the idea that every country has the sovereign right to establish their acceptable level of protection (the social and political judgement of safety), but that the process they use to determine and manage the risks to achieve that level of safety will be based on scientific principles and evidence (the measurement = risk assessment).

Measuring risk is an empirical, scientific activity. Judging safety is a normative, political activity. An analogous example is the difference between nutritious and tasty. The component aspects of nutrition are well studied and measurable. An analysis can be completed to determine and compare the nutritive value of foods. Whether or not the same foods are tasty will vary considerably depending on the preferences of individuals.

The acceptable level of risk is a discussion by itself (see Chapter 19 for more on acceptable level of risk). The characterization (or measurement) of risk is the analysis that provides the basis for a judgement regarding the acceptability of a risk. It is not however the only factor that is considered. Perceptions can play an important role in judging safety. Risks that are well known, easily observed and repeated consistently are more comfortable for decision makers. Risks that are cryptic, invisible and surrounded by uncertainty are typically considered to be worse even if analysis shows they are comparable with or less than better known risks.

It is human nature to be more comfortable with risks that we believe we can control (e.g. smoking) even if the risks are known to be high, versus risks that are forced upon us or for which we have little or no influence (e.g. air pollution). In addition, there is a tendency to view risks for which some form of management is possible as being more acceptable than risks which are difficult or impossible to manage. In sum, understanding the relationship between risk (which is measured) and safety (which is judged) is an important first step to appreciating the role of risk analysis in the context of the WTO-IPPC framework for phytosanitary applications. Further, understanding the *perceptions* which underlie our judgements of safety is important to provide perspective on the acceptability or not of risk.

Assume, for example, that a scientific study was conducted on the relationship of cellular phone usage to brain cancer. Assume also that the results show, with a high degree certainty, that one in a million cellular phone users develop brain cancer. The societal judgement in this situation may be to accept the risk of one poor user developing cancer in order for cellular phones to be available for business and daily life to a broad segment of the general population. Cellular phone usage is not likely to stop and may not even decrease as a result of this study, but users might complain to industry and policy makers about the lack of concern for public safety. The political response may be to regulate for limits on cellular phone power or require the industry to provide more shielding.

The cellular phone scenario assumes that the risk is averaged across the population of normal users, but what if the hazard is found to affect primarily teens and children? Suddenly a risk that is the same in every other measurable aspect has become much more menacing. Hence the power of perception and the importance of recognizing there are value judgements imbedded in our ideas about safety. This is a critical distinction to understand with respect to the SPS-IPPC framework within which pest risk analysis operates, and an important factor to consider in *risk communication* (see Chapter 15).

2.3. Hazards and Likelihoods and Risk

In the phytosanitary world, risk is associated with a *hazard*, which may require a regulatory response in order to reduce, avoid or eliminate the risk. In most cases, this hazard will be the *introduction* (entry and establishment) of a harmful plant pest. This would be the adverse event referred to in the definition above. The term hazard implies the existence of a threat with the potential to occur. No hazard; no risk. But on the other hand, a threat may exist that has no possibility to occur, in which case the hazard has no risk. Conversely, a hazard may exist that has a high probability to occur but there are no practical measures that can be taken to reduce, avoid or eliminate the risk, meaning that the risk must be accepted, as is often the case with the natural spread of plant pests.

According to the definition, the risk of any hazard occurring has both a likelihood and consequences. The likelihood is the 'chance' that the hazard will be manifest. This requires a potential pathway for exposure to the adverse event. A series of events must occur, each of which has some probability. If any of the necessary events does not occur, the hazard is not manifest. The events therefore have a multiplicative relationship because any event with a zero probability results in zero hazard.

In a simple example, there is no probability of being struck by lightning on a sunny day. Lightning strikes are clearly a hazard, but without a storm to produce the conditions for lightning, there is no risk even if you are barefoot, outdoors, on a hilltop, leaning against a flagpole. In a phytosanitary context, this means that unless there is a pathway for pest introduction and all the required events have some probability, there is no risk.

Now things get a little more complicated. First, we have introduced the term *probability*. Is there a difference between probability and likelihood? Not really. The term probability is typically associated with numerical descriptions of likelihood and may be considered equivalent to the odds, or the chance of something occurring. In most cases this will be an expression of percentage (e.g. a 50% chance of rain) or a decimal fraction (e.g. a probability of 0.5), recognizing that a probability of 1.0 is equivalent to 100%, which means that the event is expected to occur. But is it certain? That depends on the confidence we have around the estimate. For example, we could have 95% confidence of a 50% chance of rain, meaning that 95 times out of 100 the estimate will be correct, and 5 times out of 100 it will be incorrect. The confidence here is an expression of our uncertainty about the probability.

Next, we need to think about the difference between a probability and *possibility*. In this case, we need to study the SPS Agreement for guidance, in particular the history of interpretations in jurisprudence resulting from dispute settlement. This history is discussed in more detail in Chapter 19, but the key point to recall here is the emphasis on scientific evidence as the basis for our conclusions. Events that may be possible but are not supported by scientific evidence are not considered legitimate events. For example, it may be possible that a pest of petunias will mutate and begin feeding on maize, but unless there is direct scientific evidence that this actually occurs, it cannot be considered to be legitimate even if experts agree that it is 'possible'.

2.4. Consequences and Risk

The consequences side of risk is very different from likelihood. If there are any adverse consequences from the hazard, there will be something to be concerned about. In other words, any consequence is some consequences. The consequences therefore have an additive relationship summing to the degree or level of impact (Yoe, 2012). Consequences are commonly, but not always, expressed in economic terms. For example, the consequences of an automobile accident can be both physical (personal injury) and financial (cost to repair or replace vehicles).

In the phytosanitary world, consequences are almost always expressed in economic terms, even though they may be primarily non-market in nature such as aesthetic, social, environmental or other impacts that are not usually measured in direct market effects. The IPPC substantially

affects our understanding of this component of risk in the phytosanitary context. This is because the IPPC requires all consequences to have economics as a common denominator for measurement and comparison. The IPPC also refers to various analytical techniques that can be used for measuring and converting non-market consequences into economic terms (see also Chapter 7).

Another important difference between the likelihood and consequences is that, in most cases, the prediction of potential consequences will be based on the full impact of the hazard (the worst case scenario), which is not balanced by whatever benefit there may be from taking the risk. This is another point where the role of the SPS Agreement becomes an important factor in shaping the view of risk in the phytosanitary world. A key assumption behind the Agreement is that all WTO Members benefit from fairly applying the disciplines of the Agreement which include, but are not limited to, specific consequences that should be considered (see Chapter 4 for details) but not specific benefits.

2.5. Uncertainty and Risk

Throughout all of this, we also have uncertainty. If we had perfect knowledge; that is to say if we knew with certainty that the hazard would (or would not) occur and we also knew the consequences, then where is the risk? Uncertainty obviously plays a central role in the concept of risk even though it is not explicit in the definition. No uncertainty; no risk (Kaplan, 1993).

Uncertainty takes several forms including incomplete or conflicting information, linguistic imprecision, bias, inappropriate methodologies, incorrect assumptions and more. As a general rule, uncertainty can be categorized as either *variability*, which cannot be changed with more or better information or *error*, which can be changed. It is important to know the source, type and degree of uncertainty as a means to completely understanding the risk and what can be done to affect it.

Another example: Your rich uncle has died and left you his fortune. You know that his estate is worth more thanUS$10 million, but you don't know exactly how much. In this case, you clearly have uncertainty, but do you have risk? No; there is no risk because (i) you are certain of the event; and (ii) it is not an adverse event (no hazard or adverse consequences).

To summarize: hazard, likelihood, consequences and uncertainty form the core elements of risk that we need to consider for risk analysis and later for PRA, but our understanding and use of these concepts in the PRA process is strongly influenced by the SPS Agreement and the IPPC.

2.6. What is Risk Analysis?

Each of the core elements discussed above can be described and their relationship analysed to understand if there is a risk, how bad it may be, how certain we are and what can be done about it. We do this continually in our daily lives, so much so that it is intuitive and probably not considered in explicit terms until or unless we are challenged or we challenge ourselves in the face of doubts, doubters or undesirable outcomes. Otherwise, we continue to evaluate and manage daily risks, gaining experience and gathering more and better information in the process. In other words, risk analysis is something we already know, practise and cultivate, but we probably don't think about it as a systematic process or express our analysis in terms of its component elements.

Some characteristics of risk analysis can be identified at the outset based on our personal, daily experiences. For instance, the more information we have, and the more time we have to gather and consider relevant information, the better we usually feel about our conclusions. The most uncomfortable risk analysis is when we are faced with a quick decision and little information. We normally prefer to have plenty of information and time to consider it fully, even if the ultimate position or action is the same as our initial reaction. This tells us something

about the importance of information and its relationship to uncertainty and our comfort level.

Another characteristic of risk analysis that is obvious from our daily experience is the difference between something new and something we have experienced previously. Even though a particular hazard may be very bad, we tend to be more confident if we have dealt with it before. The more we have been exposed to similar risks, the more confident we are about our conclusions. This is why constant training and retraining is so important for professionals in hazardous occupations such as soldiers, the police, emergency medical personnel and firefighters. As a result, the analysis we do today may not be the same and have the same result as the one we did for a similar situation in the past, nor would we expect it to be identical in the future.

Perhaps the most critical aspect of our personal risk analysis experience has to do with the outcome. First, we will make a judgement on whether or not the risk is acceptable, and if it is judged to be unacceptable, what we will do about it. Ultimately, we will accept, avoid, reduce or eliminate the hazard. Our success in this endeavour provides us with feedback on the effectiveness of our risk analysis and also gives us a point of reference for future decisions. That is to say that our judgement regarding our ability to manage risk plays an important role in our analysis of the risk.

The same is true for a NPPO charged with identifying, analysing and managing phytosanitary risks in the regulatory environment framed by the IPPC and the SPS Agreement. NPPOs share the need for cross-border commerce, recognizing that there are inherent risks which they should identify, understand and manage without being unnecessarily restrictive to their trading partners who are in the same position.

Much like the decisions we take as individuals, regulatory decisions often cannot wait for more or better information and may therefore be uncomfortable. Likewise, routine situations that have a history of acceptable outcomes are in contrast to the many new and unique challenges associated with managing the pest risks associated with hundreds of commodities and thousands of potential pest threats in the face of very little direct information. In this light, risk analysis begins to emerge as a necessary tool and something we may want to understand better.

Is it science, or is it scientific? The SPS Agreement speaks to us about assessments based on scientific principles and evidence that we translate into risk analysis. Thus it is clear that risk analysis requires science, but is it a science itself? This question is often asked and also easily answered. Recall that a defining feature of science is reproducibility. Risk analysis, although it includes heavy doses of science and scientific processes, also incorporates judgements that change over time and based on conditions. This is one important reason why risk analysis is difficult to characterize as a science itself but most experts easily refer to it as a *discipline*.

> Risk analysis: A systematic way of gathering, evaluating, and recording information leading to recommendations for a position or action in response to an identified hazard.

Another simple definition; but does it make sense? Why do we do risk analysis? Ultimately, we aim to identify, understand, and manage the hazards around us so that we reduce, avoid or eliminate the potential for harm. But, before we can manage something, we must be able to measure it.

Kaplan and Garrick (1981) describe three basic questions the analysis will try to answer:

1. What can go wrong?
2. What is the likelihood that it will go wrong?
3. What are the consequences?

The first step – to identify the hazard – is key to beginning the process. Any number of 'inputs' can result in a hazard (or potential hazard) being brought to our attention. Our reaction is to ask, 'What is the risk?' This leads us to whatever information is available or needed to characterize the risk considering the likelihood, consequences and uncertainty associated with the hazard. Throughout this process of characterization, we are also beginning to understand the shortcomings of our information and our ability to process the information, i.e. the uncertainty. At the same time, we are thinking about similar situations we have faced previously, how we dealt with them and the outcome. This is the *risk assessment* phase. The result of this process will be a conclusion regarding the level or acceptability of the risk based on the likelihood and consequences, and considering these in light of the uncertainty.

It is natural for a very high level of uncertainty to elicit a more precautionary conclusion, which brings us to another important aspect of the SPS Agreement that must be understood in the context of *transparency* (see Chapter 4 for a detailed discussion of SPS principles, including transparency). As stated above, the SPS Agreement requires measures to be based on scientific principles and evidence. It is therefore crucial for a risk assessment done in association with SPS measures to distinguish the conclusions based on scientific principles and evidence from those that are influenced by uncertainty.

To demonstrate this point, imagine that scientific evidence has been used to establish that a certain contaminant in drinking water is below the level that would elicit public health problems. Assume also that there are a number of uncertainties around the testing methodology, including in particular whether the test is sensitive enough to reliably detect extremely low levels of contamination. There is also natural variability in sensitivity to the contaminant among the general population. Although the likelihood and consequences of contamination above the threshold are clear from the scientific evidence, the question of whether the water is safe to drink must also consider the uncertainty associated with the evidence. In an SPS environment, the risk analysis should clearly separate the evidence from the uncertainty and characterize the degree to which uncertainty influences the judgement of safety (see also the discussion of precaution in Chapter 19).

Going back to our risk analysis, depending on the urgency of our decision and how comfortable we are with the uncertainty, we eventually decide whether the hazard is acceptable (see also the discussion on the appropriate level of protection in Chapter 19). It may be acceptable because we are confident we can manage it, or it may be that we have no practical way to manage it and must accept it. In either of these two scenarios, no additional risk analysis is needed. But on the other hand, if we judge the risk to be unacceptable, our analysis moves to the next phase, which begins by identifying ways to reduce, avoid or eliminate the hazard. This is the beginning of the *risk management* phase.

2.7. Risk Management

Risk management is where we decide what to do about the hazard. Because we aim to change the risk, this aspect of the analysis is necessarily linked to the risk assessment phase, which allows us to understand the degree to which the hazard is *mitigated* by applying one or another measure. In other words, risk assessment can be done in isolation, but risk management requires a link to risk assessment for the iterative process of evaluating the change in risk due to the application of mitigations (Fig. 2.1).

If the risk assessment is addressing the questions of what can go wrong, what is the likelihood and what are the consequences, we can draw a corollary to risk management and say that risk management is concerned with:

Fig. 2.1. Risk analysis process paradigm.

- What can be done, and what are the options?
- What are the trade-offs (e.g. cost/benefit, etc.)?
- What are the impacts of the risk management decisions?

The primary focus of analysis in risk management is the *efficacy* of mitigations. Efficacy however is not the only aspect of risk management that requires analytical attention. We also want to know if the mitigations are feasible and the impact of applying one compared with another or none at all. These questions of feasibility and impact combine with the analysis of efficacy and uncertainty to form the risk management analysis.

As regards the analysis of efficacy, it is useful to recall that the two core components of risk are the likelihood and consequences. Reducing or eliminating either one can mitigate the risk, although we typically focus primarily on reducing the likelihood, often to the exclusion of any consideration for the role of consequences.

If we imagine the risk of crossing a busy intersection and assume that the likelihood of being struck by an automobile, a bicycle or a pedestrian is exactly the same, then the difference in risk is totally dependent on the consequences. We might decide to 'take our chances' with a pedestrian and perhaps even a bicycle, but we are unlikely to accept the risk with an automobile. The point of this thought exercise is to highlight the role of consequences in risk assessment, but also to note that the consequences of an adverse event should not be discounted or forgotten after risk assessment because consequences can play a role in risk management. The question of whether consequences can be mitigated is at the heart of one of the oldest and best known users of risk analysis in the world: insurance.

Insurance is purchased to mitigate the consequences of an adverse event. Over centuries of development, the insurance industry has advanced some of the most sophisticated and well-tested methodologies for risk analysis, paying particular attention to the relationship of the likelihood of adverse events to the consumer's interest in mitigating the consequences.

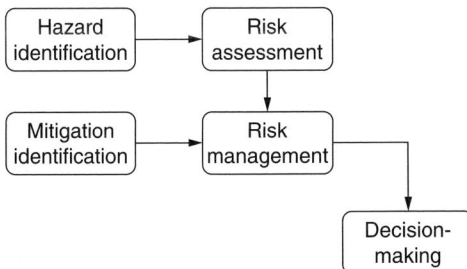

Unfortunately for the phytosanitary community, there are few instances where risk management can focus on mitigating consequences (compensation for crop losses may be one possibility). This means that most of our attention is focused on affecting the likelihood, recalling that likelihood is the result of a series of events with a multiplicative relationship.

Once we have identified a mitigation measure (or measures) deemed to provide adequate protection (the desired level of efficacy), we need to understand whether the measure is *feasible* to apply and the *impacts* of the measure. Two points to note here are that (i) feasibility may involve some form of cost–benefit analysis; and (ii) the impacts of measures is not the same as consequences of the hazard discussed above.

We learned above that the benefit of accepting risk is not an explicit factor to consider in risk analysis as it applies to SPS issues. There is however a role for benefits in risk management as it applies to the analysis of mitigations. The cost–benefit of one measure compared with another is an important aspect of the analysis of feasibility (see Chapter 13 for more information on cost–benefit analysis in risk management). A mitigation that is relatively more expensive and less efficacious than another option that is more effective and costs less will probably be considered less desirable. The less desirable mitigation option based on cost–benefit could however become the more desirable option if the impacts of the measure are also considered.

Impacts in this context refer to the impacts of measures, not pests. For clarity, it is helpful to refer to 'consequences' when the focus is on pest damage and 'impacts' when the focus is on the effects of mitigation measures. For example, the consequences of pest introduction may be a 40% yield loss, whereas the impacts of pesticide spraying for the pest may be the pollution of local surface water.

The results of risk management will be conclusions regarding the most appropriate mitigation measures and a description of the uncertainties associated with the analysis. Recommendations may also be included.

2.8. The Relationship Between Likelihood and Consequences

Up to this point, we have assumed that the likelihood and consequences of a hazard have some relationship that ties them together for risk analysis purposes. This relationship may be expressed as a conceptual formula:

$$\text{Risk} = (\text{Likelihood}) \times (\text{Consequences}) \qquad (2.1)$$

Whether or not this represents a mathematical reality depends on the units and calculations that are used to populate the formula. Putting quantitative questions aside, however, we can see some usefulness for this formula as a general reference to help us remember that there are two core elements of risk whose magnitude and relationship are the business of our analysis.

Formulae such as this are common in risk analysis literature but vary considerably depending on the discipline where they are applied. For instance, the *benefit* aspect of risk will be prominent in most conceptual descriptions of risk analysis in the financial industry. Likewise, the concept of *exposure* will be found in descriptions of risk analysis for toxicology and related studies. The most common formula is:

$$\text{Risk} = (\text{Hazard}) \times (\text{Exposure}) \qquad (2.2)$$

It is particularly useful to be familiar with this concept for two reasons. First, it is likely to be encountered in the world of SPS applications, especially those associated with food safety. Second, it is important to contrast this with the previous concept to understand why the earlier description is more appropriate for phytosanitary applications.

The second formula assumes that the hazard is static and measurable, typically as a dose–response, and that exposure data can be collected and are relatively constant. Think about the cellular phone example discussed earlier, or a poison, or any other static hazard and contrast this to the phytosanitary situation. How often do we have dose–response or exposure data? Not usually. How static are our hazards? Not very static.

Toxins and contaminants (sometimes including microorganisms) in food are typically studied as static hazards, using exposure data which can be useful for analyses associated with food safety issues in the SPS framework. Plant pests and their hosts have complex and dynamic relationships that are affected by a number of external and internal variables, not the least of which is their genetic programming to survive. The result is tremendous variability in cause-and-effect relationships, which make them much less suitable for dose–response and exposure studies. As a result, the risk analysis standards that have been developed and the processes that have been created to support phytosanitary decision making are based on the concept of risk being the product of the likelihood of a hazard multiplied by the magnitude of the consequences of that hazard.

2.9. Why Do Risk Analysis?

The preceding discussion on the conceptual background for risk analysis, its core elements and the influence of the SPS-IPPC framework in shaping our interpretation and application of the concepts, is crucial to understanding phytosanitary applications that are used to justify trade decisions. This raises the question of whether these are the only applications for risk analysis and the only reason we use risk analysis.

It is useful to recall that risk analysis, in one form or another, is part of our everyday life. As such, there are many and various applications, methods and results we use or encounter on an ongoing basis. The same is true for an NPPO. The charge to manage the risk associated with plant pest hazards isn't limited to phytosanitary measures for imports. In fact, every import issue for one country is also an export issue for another. It is easy to imagine how risk analysis capacity would be important for reviewing and possibly challenging the justification for measures put in place by a trading partner for its imports (your exports), or just to provide relevant information to a trading partner to facilitate or improve exports.

WTO obligations under the SPS Agreement make it important for phytosanitary measures to be based on a risk analysis (if they are not based on an international standard), and we know that binding dispute settlement could result from poor or no scientific justification. The benefits of using risk analysis in the WTO context are obviously the ability to justify and defend phytosanitary measures, and to evaluate or possibly challenge the measures of trading partners. It is therefore clear that risk analysis has a key role in both import and export issues. Furthermore, the risk analysis methodology used for these purposes should conform to the principles and processes adopted by the relevant international standard setting organization (the IPPC in the case of phytosanitary measures). This does not however mean that risk analyses need only be done for trade issues or that risk analyses done for other purposes need to meet the same requirements.

Because the general charge for NPPOs is to protect their nation's plant resources from harmful pests, there is ample opportunity for the application of risk analysis beyond the hazards associated with trade. Domestic programmes, regulatory policy, prioritization for surveillance and practically every other facet of national plant protection programmes can benefit from applying risk analysis or aspects of risk analysis. From an internal standpoint, risk analysis can be a very powerful tool for evaluating and prioritizing programmes, allocating resources, identifying research priorities and focusing on technical points of difference with trading partners as well as stakeholders.

Perhaps the most important benefit of risk analysis is the linkages that are created and strengthened as a result of engaging in a scientific exchange and dialogue. Policy makers become more aware of the importance of research, and regulators become more effective through greater coordination with the research community. Trading partners, stakeholders and civil society all benefit from the focus on science rather than politics to meet legitimate protection needs.

2.10. How Do We Do Risk Analysis

Now let's imagine the perfect risk analysis. In a perfect world, we would have plenty of time, information and experience. And if we had a lot of experience, we would probably also have proven methodologies. But what would the product look like, and how would it be used?

We can begin by describing what risk analysis *is not*. We already know it is not a science. We also know that uncertainty is a key element of risk, so a risk analysis will not be certain. And, despite the emphasis on scientific information, the process also includes judgements that can change over time or under different conditions, so we would expect risk analyses to be dynamic, not stable or static. Finally, we know that the results of risk analysis provide conclusions and may have recommendations, but do not offer a solution or make a decision. Risk analysis may be one of several inputs considered by policy makers in making a decision.

Risk analysis *is* important to precisely identify hazards and characterize both the likelihood and the magnitude of the consequences of the hazards using scientific principles and evidence. A good risk analysis recognizes uncertainty and describes both the nature and sources of uncertainty. It may also identify and prioritize research needs to reduce uncertainty. Finally, a good risk analysis will summarize conclusions and may recommend options.

A very transparent risk analysis will provide good documentation of information sources, explain methodologies, describe analyses, identify assumptions and uncertainties, and be written for optimal understanding by the intended audience. The completion of a risk analysis represents a snapshot in time of the information, judgements, analysis and conclusions associated with the question being addressed.

Observers sometimes assume that two analyses using the same information should not only have the same result, but also have the same appearance. Another common assumption is that a risk analysis done in the past continues to be as valid in the future. In both cases, the risk analyses in question may be very well done, but we need to recall that risk analysis includes judgements and assumptions that can greatly affect both the appearance and the results, and that these factors can be dynamic. The keys to distinguishing a good risk analysis from a poor one are transparency and consistency. A good risk analysis clearly explains the methods, assumptions and uncertainty, and consistently applies the criteria used in the analysis to arrive at conclusions. Another analysis of the same information can legitimately differ, but can be just as well done if it is likewise transparent and consistent.

Vose (2005) identifies five steps in the risk assessment process:

1. Identifying the risk that is to be analysed and potentially controlled.
2. Developing a qualitative description of the risk (what might happen, why, factors that affect the risk, etc.).
3. Developing a semi-quantitative or quantitative analysis of the risk and associated risk management options.
4. Implementing the approved risk management strategy.
5. Communicating the decision and its basis to stakeholders.

In plant health, we often focus primarily on qualitative and semi-quantitative descriptions of risk for routine risk analyses while quantitative analyses are usually used for special cases. Further, risk communication – while important – is not necessarily a requirement under the SPS-IPPC framework.

2.11. Terminology

Like many technical fields, risk analysis has its own jargon that needs to be understood for one to be fluent in the work. A multitude of references are available to identify and describe key terms and concepts, but unfortunately, the application of risk analysis in the framework of the SPS Agreement and the IPPC provides some special linguistic challenges. A clear understanding of the terminology, especially where there are conflicts, inconsistencies and ambiguities, is an important first step to being comfortable with the concepts and the practice of risk analysis in this framework. Chapter 5 provides a more detailed discussion of IPPC and phytosanitary terminology, however, a few key points are highlighted here as well.

The first and primary point of confusion comes from the use of the term *risk assessment* in the SPS Agreement (WTO, 1994). Nowhere does the agreement refer to *risk analysis*, yet all three of the standard setting organizations identified by the SPS Agreement refer to the overall process as risk analysis, and designate risk assessment as the sub-process for characterizing the likelihood and consequences of the hazard. On this point, it is noteworthy that most sources consider risk analysis to be composed of three sub-elements: risk assessment, risk management and risk communication. Although it may be argued whether or not risk communication is actually an analytical process, nearly everyone agrees that it is an essential element of risk analysis.

The confusion between risk assessment and risk analysis gets deeper when the discussion is limited to risk analysis for phytosanitary purposes. This is because the IPPC uses the term *pest risk analysis* to refer to its specific form of risk analysis (IPPC, 2004, 2010). OIE and Codex simply refer to risk analysis (see also Chapter 4).

The SPS Agreement also refers to the appropriate level of protection (sometimes represented by the acronym ALP or ALOP). Most of the Agreement uses this term except it is noted that some WTO Members refer to the concept it represents as the acceptable level of risk (ALR) (WTO, 1994). These terms are sometimes used interchangeably based on the assumption that they represent opposite aspects of the same concept.

The next biggest challenge comes from the reference in the SPS Agreement to *pest* or *disease*. This is inconsistent with the IPPC, which includes diseases in its definition of a *pest*. In other words, the term pest is used broadly by the IPPC to include arthropods, pathogens and weeds.

The SPS Agreement clearly states that the consequences of introducing a pest or disease are *biological and economic*, however consequences for IPPC purposes, are only economic. Likewise, the SPS Agreement recognizes two forms of discrimination: *discrimination* between trading partners is one form; discrimination between national producers and trading partners is another. The latter is known as *national treatment* for SPS purposes, a term that is not used by the IPPC.

2.12. Summary

Risk analysis came to the forefront for plant protection organizations with the adoption of the SPS Agreement in 1995. Risk – the probability of an adverse event, and the magnitude of the consequences – and uncertainty (which is a part of risk) can be measured and described. We use risk analysis – a systematic way of gathering information, analysing information and making recommendations – as means to communicate about risk with stakeholders. In analysing risk, we need to identify adverse events, make judgements regarding how likely those events are to happen, the magnitude of the consequences and our level of uncertainty.

We must be ever mindful that risk analysis should be science-based, defensible and transparent. Some of the best guidance in this respect comes from Morgan and Henrion (1990), who provide the Ten Commandments for good policy analysis:

1. Do your homework with the literature, experts and users.
2. Let the problem drive the analysis.
3. Make the analysis as simple as possible, but no simpler.
4. Identify all significant assumptions.
5. Be explicit about decision criteria and policy strategies.
6. Be explicit about uncertainties.
7. Perform systematic sensitivity and uncertainty analysis.
8. Iteratively refine the problem statement and the analysis.
9. Document clearly and completely.
10. Expose the work to peer review.

References

FAO (2000) *Multilateral Trade Negotiations in Agriculture: a Resource Manual*. FAO, Rome, http://www.fao.org/DOCREP/003/X7351E/X7351e00.htm#TopOfPage, accessed 20 December 2011.

IPPC (2004) International Standards for Phytosanitary Measures, Publication No. 11: *Pest Risk Analysis for Quarantine Pests Including Analysis of Environmental Risks and Living Modified Organisms*. Secretariat of the International Plant Protection Convention (IPPC), Food and Agriculture Organization of the United Nations, Rome.

IPPC (2010) International Standards for Phytosanitary Measures, Publication No. 5: *Glossary of Phytosanitary Terms*. Secretariat of the International Plant Protection Convention (IPPC), Food and Agriculture Organization of the United Nations, Rome.

Kaplan, S. (1993) The general theory of quantitative risk assessment – Its role in the regulation of agricultural pests. In: NAPPO. *International Approaches to Plant Pest Risk Analysis: Proceedings of the APHIS/NAPPO International Workshop on the Identification, Assessment, and Management of Risks due to Exotic Agricultural Pests*, Alexandria, Virginia, 23–25 October 1991. Ottawa, Canada. Bulletin (North American Plant Protection Organization) No. 11.

Kaplan, S. and Garrick, B.J. (1981) On the quantitative definition of risk. *Risk Analysis* 1 (1), 11–27.

Morgan, M.G. and Henrion, M. (1990) *Uncertainty: a Guide to Dealing with Uncertainty in Quantitative Risk and Policy Analysis*. Cambridge University Press, New York.

Vose, D. (2005) *Risk Analysis: a Quantitative Guide*. John Wiley & Sons Ltd, Chichester, West Sussex.

WTO (1994) *The WTO Agreement on the Application of Sanitary and Phytosanitary Measures*. World Trade Organization, Geneva.

Yoe, C. (2012) *Principles of Risk Analysis: Decision Making Under Uncertainty*. CRC Press, Boca Raton, Florida.

3 History of Plant Quarantine and the Use of Risk Analysis

Christina Devorshak

3.1. Introduction

We should have ever in mind that quarantine must be based on biological principles established through scientific research, that inspection and regulation are makeshift, although necessary, measures forced upon us by the failure to prevent the original invasion of these pests, and that it is our paramount duty to contribute out of our abundant opportunity to fundamental investigations of means of preventing the introduction and spread of insects and plant diseases.

Those words were written nearly 100 years ago by W.A. Orton (Orton, 1914) in his discussion of problems facing plant quarantine, but the sentiment still holds true to this day. Orton went on to call for specific domestic actions and for greater international cooperation in preventing the spread of pests.

Although pest risk analysis is a relatively new discipline, the story of quarantine and plant protection begins more than 600 years ago. This chapter will highlight four key components that together have brought us to where we are today in analysing the risks associated with the introduction and spread of pests: enacting national and international laws and regulations; understanding and analysing scientific information as the basis for those laws;

international cooperation to ensure success of those laws; and finally, establishing risk and risk analysis as the basis for decision making.

3.2. Origin of the Concept of Quarantine

The concept of quarantine is not new – it dates back hundreds of years to the Middle Ages, at a time when the bubonic plague, or Black Death, was spreading across Asia and into Europe. As with diseases and pests today, the bubonic plague spread through human movement, largely related to human migration and the trade in goods coming from Asia and arriving in Europe. Italy was a major stop on the trading routes, and it is from the Italian word for 'forty' or *quaranta* that the term quarantine is derived (Gensini *et al.*, 2004).

In 1377, the seaport of Ragusa (then part of the Venetian Republic) enacted laws that required travellers to remain in isolation (or quarantine) for a period of 40 days to prevent introducing plague to the city. This requirement was actively enforced, and those who violated the law were subject to fines or other penalties. Subsequent to that first law, other countries also enacted similar laws aimed at protecting their citizens from the spread of a deadly disease (Gensini *et al.*, 2004).

As time went on, the laws established, in effect, the first quarantine stations – places where individuals could be isolated from the rest of the population in order to assess whether they were carrying the disease (Gensini *et al.*, 2004). What is remarkable about these actions and laws is that, at that time, there was a very poor understanding of disease transmission and infectious agents. None the less, those measures applied more than 600 years ago are not much different than the same types of measures we still use today to protect ourselves from the spread of harmful diseases and organisms.

3.3. Quarantines, Phytosanitary Measures and Plant Health

3.3.1. Early plant protection laws – 1600s–1800s

Today, the term 'quarantine' is often used in plant health to refer generically to any phytosanitary measure. However, strictly speaking, quarantine (e.g. isolating an organism to determine if it is carrying pests) is one type of measure (IPPC, 2010). Other types of measures include laws and regulations, inspection, treatments, surveillance, certification or other activities aimed at managing pests (IPPC, 2010). In the case of plants, the first phytosanitary measures designed to protect plant health came about 300 years after the first quarantine laws were enacted to protect human health.

One of the first plant protection laws, enacted in France in 1660, was aimed at controlling the spread of wheat stem rust (*Puccinia graminis* Pers.), an important disease of wheat. Although the specifics of the disease were poorly understood, farmers deduced that barberry plants (an alternate host for the fungus that causes wheat stem rust) growing in close proximity to wheat made the disease worse. Therefore, the law at that time allowed for the destruction of barberry plants. Other places followed suit – Rhode Island and Massachusetts in the United States passing similar laws in the 1700s (Roelfs, 2011), and several states in

Germany (Ebbels, 2003) passing laws to control barberry in the 1800s.

3.3.2. Science and early regulatory plant protection

Beginning in the late 1800s, countries began putting in place specific laws to prevent the entry or spread of exotic pests. This period of time coincided with a growth in scientific knowledge about plant pests – prior to the 1800s, decisions were largely based on only a crude understanding of how pests affected plant health. However, the 1800s began a time of scientific exploration, particularly in the field of natural biology. And as our understanding of organisms (particularly plant pests) grew, so did our ability to analyse those organisms and base control actions for those organisms on the best scientific information available. Interestingly, most of these laws came about due to the introduction and/or spread of just a few key pests like the grape phylloxera (*Daktulosphaira vitifoliae* (Fitch)) and Colorado potato beetle (*Leptinotarsa decimlineata* (Say)).

In 1874–1875, the Colorado potato beetle, considered one of the most destructive pest of potatoes at the time, was introduced into Germany, in particular near major ports of entry. Swift action was taken to eradicate the localized populations and, in 1875, Germany enacted a decree prohibiting the importation of potatoes, and materials associated with potatoes (e.g. packing materials, sacks, etc.) (Mathys and Baker, 1980). Thus, this legislation was one of the first that established a link between a serious pest and controlling the pathways (e.g. the materials associated with potatoes) by which the pest can move from one place to another.

The grape phylloxera, a tiny aphid-like insect, feeds on the roots and leaves of grapevines. It had been introduced to France on grape cuttings from the United States in 1859, and over the next 20 years spread throughout the grape-producing regions of Europe. By the late 1800s the grape phylloxera was causing major devastation to vineyards all over Europe, and in France, the losses to the wine industry were considered massive. The grape phylloxera was also introduced into Victoria and New South Wales, Australia in 1872 on vine cuttings from France. It soon became a major pest in vineyards in Australia, and in 1877 strict legislation was enacted to control the pest (Maynard *et al.*, 2004).

As a result of the devastating losses to the wine industry, countries became aware of the need to prevent the spread of this pest, and in 1878, the first international agreement for cooperation in preventing of a spread of a pest was adopted. The International Convention on Measures to be Taken against *Phylloxera vastatrix* (the name the pest was originally assigned in Europe) was signed by several European countries. The agreement included several provisions that are echoed in later treaties for plant protection, including prohibiting the movement of certain material, the exchange of information about the pest, requirements for inspection, and allowance for the written assurance that materials in trade were free of the pest (Ebbels, 2003).

The story of the control of grape phylloxera did not end there however. European countries were desperate to find effective controls for the pest, and the solution came from the very place the pest had originated – the United States. An American entomologist named C.V. Riley was instrumental in demonstrating that the pest affecting grapes in France was the same species as a pest of grapes in the United States. Through scientific investigation, C.V. Riley established that certain varieties of grapevine grown in the United States were resistant to the pest. Beginning in 1871, resistant varieties were shipped to France and, by 1872, over 400,000 cuttings had been shipped to revitalize the French wine industry (Sorensen *et al.*, 2008).

The spirit of international cooperation in controlling this pest continued over the next several years. Riley's French colleague, J.E. Planchon – the principle investigator into the grape phylloxera in France – visited Riley in the US in 1873 and continued scientific investigations into the pest and the apparent resistance of US varieties of grapevines. Further investigations continued over the next several years, with scientists travelling back and forth between countries, primarily researching which roots stocks were most resistant. Riley's initial contributions were ultimately recognized when, in 1889, he was awarded the Legion of Honour by the French government (the country's highest recognition) in honour of his work (Sorensen *et al.*, 2008).

Other pests were the subject of national and international regulations during that time as well. In 1877, four American states (Kansas, Missouri, Minnesota and Nebraska) enacted legislation aimed at controlling the spread of the Rocky Mountain locust (*Melanoplus spretus* (Walsh)), grasshoppers that were documented to swarm in devastating numbers (Norin, 1915). The purpose of the legislation was not to prevent the migration since the available scientific information indicated that migration could not be prevented. Therefore, based on the state of knowledge, and the best available science, the legislation was aimed at controlling the pest locally so that the outbreaks would be reduced in subsequent years.

Around the end of the 1800s, we begin to see that plant protection legislation was taking into account the biology of the pests and the state of scientific knowledge at that time, as in the case above. Another example is that DeBary established, based on scientific evidence, that barberry served as an alternate host for the fungus that caused wheat stem rust (the first pest subject to plant protection laws highlighted earlier). This finally established the scientific basis for the regulations that had been enacted more than 200 years prior. Subsequent to his discovery and beginning in 1869 through the 1920s, several European countries and US states enacted laws for the eradication of barberry as a means of controlling wheat stem rust (Ebbels, 2003).

3.3.3. The first broad national phytosanitary laws

As a result of several serious pests being introduced to different countries in the late

1800s, it soon became apparent that broader action was needed to prevent future pests from causing similar damage. For European countries, the San Jose scale (*Quadraspidiotus perniciosus* (Comstock)) was of great concern (Mathys and Baker, 1980). It was a serious pest of fruit trees and there were concerns it could be carried on nursery stock, and establish in Europe much like the grape phylloxera had several years before. Countries came to understand that prevention was the best strategy and that regulating on a pest by pest basis would not be as effective.

Several countries all over the world began to enact broad laws for plant protection and establish national plant protection services responsible for exercising those laws. In 1887, Great Britain passed the Destructive Insects Act and established a Board of Agriculture. The act was revised in 1907 to become the Destructive Insects and Pests Act that addressed both insects and pathogens (Ebbels, 2003). Similarly, in 1899, the Netherlands established a national plant protection service whose purpose was to prevent the introduction of new pests. France and Germany also passed similar laws during that time period, restricting the movement of nursery stock and fresh fruits.

Other countries, in particular colonies or former colonies of European countries, also began passing quarantine laws during that time. The Cape of Good Hope, in what is now South Africa, passed 'An Act to Regulate the Introduction into This Colony of Articles and Things, By reason of Disease or Otherwise, Might be Injurious to the Interests Thereof' in 1876, and Australia passed its first federal plant quarantine law in 1907 (Norin, 1915). Canada initially passed the San Jose Scale Act in 1898 after the pest was introduced into California, and then later passed the Destructive Insects and Pests Act in 1910, granting government the power to take appropriate actions to prevent the entry and spread of pests. Interestingly, one of the restrictions in the act was to limit importations into Canada to certain seasons of the year, presumably as a measure to reduce the risk of pest establishment.

3.3.4. Risk enters the picture – The US Plant Quarantine Act of 1912

Over the next 20 years, other countries would follow suit, establishing national plant protection services and enacting broad laws and regulations (as opposed to the species-specific laws enacted in the 1800s) aimed at preventing the introduction and spread of plant pests. The USA passed the Plant Quarantine Act of 1912 after several more serious pests had been introduced (Castonguay, 2010). This act granted the Secretary of Agriculture the authority to restrict and control the importation of plant pests, in particular by restricting the entry of nursery stock and other types of plant products that could carry pests (Weber, 1930).

The act included provisions for various types of phytosanitary actions including phytosanitary inspections (for both foreign and domestic products) of different commodity classes (e.g. nursery stock, fruits and vegetables for consumption), certification, phytosanitary treatments and other key functions for protecting plant health.

Over the next several years, different quarantine regulations were promulgated under the act. These included both domestic and foreign quarantines aimed at particular pests or commodities or commodity classes. The first broad quarantine enacted under the Plant Quarantine Act came in 1919 under 'Quarantine 37', which related specifically to nursery stock (Weber, 1930).

It is worth noting that at the time the original act was passed in 1912, special concessions were granted to nurserymen who were resistant to having undue restrictions placed on the importation of nursery stock. What was finally agreed under Quarantine 37 (or 'Q-37' as it is commonly known) was that importation of nursery stock would be allowed, if the material was exported from countries having official inspection and certification systems. Under these circumstances, import permits could be issued, and inspection and certification were a condition of entry. Material from countries lacking inspection and certification systems was prohibited, except for experimental or scientific purposes, and then only under permit (Weber, 1930).

The passage of Q-37 was the subject of considerable controversy. G.A. Weber (1930) noted:

> Few if any acts of the Department of Agriculture have aroused so much discussion and so much adverse criticism and condemnation on the one hand and commendation on the other as the promulgation by the Secretary of Agriculture on November 18, 1918, effective June 1, 1919, of 'Nursery stock, plant, and seed quarantine No. 37,' the purpose of which is 'to reduce to the utmost the risk of introducing dangerous plant pests with plant importations'.

Nursery stock was understood to be 'risky' material in terms of its ability to carry unwanted pests, so controls on the importation of nursery stock were deemed necessary by scientists and government officials. However, the industry groups resented the hindrances on trade, noting in particular that the restrictions were in effect an embargo on trade, and that other countries may retaliate against US exports as a result of restrictions on their exports to the USA. In essence, the industry groups were concerned that other countries would view the regulations as unjustified and disguised barriers to trade – a discussion that still goes on today.

This seeming tension between protecting a country from the spread of pests, while not implementing undue restrictions on trade remains with us today and is the reason for establishing one of the major international agreements governing phytosanitary measures – namely the Agreement on the Application of Sanitary and Phytosanitary Measures. This Agreement is covered in detail in Chapter 4.

What is important to understand here is that the debate over Q-37 demonstrated, for the first time, the need to balance the risks associated with the movement of plant material and for measures to manage those risks against the need for establishing rules for trade that were fair, predictable and – most importantly – based on science.

Another major regulation under the Plant Quarantine Act was that of 'Quarantine 56' (or Q-56) regulating the importation of fresh fruits and vegetables for consumption. Materials regulated under Q-56 are subject to prohibitions unless specifically permitted. Ironically, the Q-56 regulations were, and still are, more restrictive than the Q-37 regulations, even though the risk associated with materials for consumption is generally much lower than for nursery stock. This apparent contradiction in the regulations compared with the level of risk remains under debate – efforts to modify Q-37 regulations to address the level of risk associated with nursery stock are still controversial, 100 years after the original law was enacted.

3.4. International Cooperation – Finding Solutions Through Mutual Interests

The introduction of grape phylloxera into Europe, and the introduction of other pests into the USA, played a major role in highlighting the need for international cooperation – both in scientific research and in the development of internationally coordinated plant protection regulations. In the late 1800s, scientists all over the world recognized the need to exchange scientific information about pests, particularly with respect to control and prevention of those pests (Castonguay, 2010; MacLeod et al., 2010).

Various levels of cooperation were already taking place on a pest by pest basis, as evidenced by the cooperative efforts between the USA and France to control grape phylloxera. But larger efforts were needed, and scientists in many countries embraced the cause. In 1891, a Swedish botanist named Jacob Eriksson called attention to the need for international cooperation to prevent the spread of pests at the International Congress for Agriculture and Forestry meeting at The Hague. He continued his efforts, and in 1903 he presented the cause again to the International Congress, meeting in Rome (Ebbels, 2003).

Around the same time, an American named David Lubin was petitioning governments in Europe for the formation of an international organization that would gather and disseminate information on agricultural issues. Because of the efforts of these two men, and wide support from scientists and scientific organizations, the International Institute of Agriculture was established in Rome in 1905, under the patronage of the King of Italy. One of the objectives of the Institute was the better control of plant diseases (Orton, 1914; Castonguay, 2010).

Scientists from different countries were very supportive of improving international cooperation. The American Phytopathological Society recognized the importance of this movement at a symposium on international phytopathology in 1912 and passed a resolution stating in part:

> Resolved, That the American Phytopathological Society, appreciating the fact that plant diseases do not heed national limits or geographical boundaries and also the evident limitations imposed upon investigations when restricted by national bounds, respectfully recommend that administrators of research institutions, whether state or national, as well as individual investigators recognize the importance of establishing closer international relations.

Although scientists led early efforts at international cooperation, there was a clear need for governments to be on board as well. Professor Cuboni suggested that the General Assembly of the International Institute of Agriculture adopt several recommendations with regard to phytosanitary inspections for moving plants in trade that stated:

> The general assembly recommends that the governments adhering to the institute:
>
> **1.** Organize, if they have not already done so, a government service of phytopathological inspection and control, especially for nurseries and establishments trading in living plants intended for reproduction.
> **2.** Enact that all consignments of plants intended for reproduction be accompanied by a certificate similar to that required by the Berne phylloxera convention to be

> delivered by the government inspector, certifying that said plant comes from a nursery subject to his control and free from dangerous cryptogamic or entomological disease.

The recommendations further called for an international agreement for the protection of agriculture against pests. The call for cooperation at the inter-governmental level was met when countries adopted the International Convention for the Protection of Plants, originally in 1914 and revised in 1929. However, the agreement was only weakly supported – world events from the 1920s through the Second World War kept countries from fully supporting the new international treaty for plant protection.

After the Second World War, there were renewed efforts for cooperation in international plant protection at the inter-governmental level. With the formation of the United Nations (UN) after the war, the Food and Agriculture Organization (FAO) of the UN was established in Rome, Italy – replacing the International Institute of Agriculture. Member countries of the FAO began drafting a new plant protection agreement – and in 1951 adopted the IPPC. This new agreement superseded all previous plant protection agreements (including the first international agreement for cooperation from 1878, International Convention on Measures to be Taken against *Phylloxera vastatrix*) (Castonguay, 2010). The IPPC was subsequently revised in 1979 and again in 1997 – it is covered in more detail in Chapter 4.

3.5. Modern Laws – Risk Analysis Becomes a Legal Obligation

Most major trading countries had enacted national plant protection laws by the first two decades of the 20th century. Risk and risk analysis were not explicit in those laws, though the management of risk was implicit because the laws were intended to prevent entry and spread of pests. However, several key events beginning after the Second World War would change the state of plant protection, and bring risk analysis to the foreground.

After the Second World War, countries were keen to begin economic recovery, and promoting increased trade between countries was seen as the best option for economic stability, and therefore lasting peace. Thus, in 1947 countries met in Seattle to negotiate what would become the General Agreements on Tariffs and Trade or the GATT (GATT, 1947).

The GATT was primarily aimed at reducing tariffs, duties and other barriers to trade. Agriculture, and trade in agricultural products were not explicitly mentioned, and the potential risks associated with such trade received little attention. However, it was recognized that there could be legitimate barriers to trade, and exceptions were noted in Article XX.b, which put in place measures designed to protect human, animal and plant health from the introduction and spread of diseases.

The effect of the GATT was to liberalize trade – countries could import and export a greater variety of goods, to and from many different origins. As the variety of products moving in trade increased and origins of products became more diverse, the potential risks for moving pests with these products also increased. This underscored the need for countries to be able to put in place protective measures to prevent the entry of pests. However, there were concerns that countries would use quarantine barriers (as noted in Article XX.b of the GATT) to protect domestic markets and prevent trade – that is, using quarantine measures arbitrarily and without justification.

In the 1980s, countries recognized the need to revise the GATT, and in 1986, undertook negotiations (called 'the Uruguay Round' of negotiations) that lasted almost 10 years. Unlike previous negotiations to the GATT, agriculture and trade in agricultural products were central to discussions – in particular with regard to market access, subsidies and SPS measures (or measures to protect human, animal or plant life or health). At that time, countries recognized that the SPS concepts inherent in Article XX.b (*necessary to protect human, animal*

or plant life or health) needed greater guidance and discipline to prevent abuse (e.g. using quarantine measures as disguised barriers to trade). As a result, the Agreement on the Application of Sanitary and Phytosanitary Measures (or SPS Agreement) was negotiated as part of the revision of the GATT (FAO, 2000).

One of the major pillars of the SPS Agreement is that measures to protect human, animal or plant life or health must be based on scientific information, and must be technically justified. The technical justification could come in the form of an international standard, guideline or recommendation written by an international standard setting body (discussed in Chapter 4) OR in the form of a risk assessment. When the SPS Agreement entered into force in 1995, risk assessment as the basis for national laws and regulations for plant protection became an international obligation for all Member countries.

The SPS Agreement also identified standard setting bodies for human, animal and plant health – and as the IPPC already existed, it was identified as the standard setting body for plant health. However, the 1951 and 1979 texts of the IPPC did not include provision for the development of standards. Thus, the IPPC was revised to modernize the concepts and to align it with the new expectations placed upon it by the SPS Agreement. The 'New Revised Text of the IPPC' was adopted in 1997 – the revision established a Secretariat, a Commission on Phytosanitary Measures and a process for the development of international standards (FAO, 1997, 2000). The 1997 text of the IPPC remains in force today. More detail on the history of the development of the SPS agreement will be provided in Chapter 4.

3.6. Adoption of ISPM No. 2: *Guidelines for Pest Risk Analysis* (Currently *Framework for Pest Risk Analysis*)

It was clear that the SPS Agreement would create obligations for all Member countries who wanted to participate in trade – phytosanitary requirements had to be based on international standards or an assessment of risk. In 1991, an international meeting was held, 'International Approaches to Plant Pest Risk Analysis' (NAPPO, 1993). The objectives of the meeting were to improve approaches to pest risk analysis and to foster discussions to facilitate the development of a document that would ultimately form the basis for an international approach to pest risk analysis (i.e. an international standard). The meeting was spearheaded by regional plant protection organizations (see section below on regional plant protection organizations) and organized jointly by the Animal and Plant Health Inspection Service of the US Department of Agriculture (APHIS-USDA) and the North American Plant Protection Organization (NAPPO), and provided the first steps towards an international standard for risk analysis.

The IPPC had only just begun developing international standards, and one of the first standards developed was general guidance on risk analysis, demonstrating the high profile that risk analysis now had in plant protection. The original ISPM No. 2 (*Guidelines for Pest Risk Analysis*, revised in 2007 and now called *Framework for Pest Risk Analysis*) was endorsed by FAO in 1995 and published in 1996, just after the SPS Agreement entered into force, and even before the IPPC revision was completed.

3.6.1. ISPM No. 2 adopted – countries align their national procedures with international standards

Before the SPS Agreement, there was not a requirement for formal risk analysis as the basis for measures. Countries used various methods – mostly to assess risk for their own purposes, rather than as a means of providing justifications to trading partners. Pest risk analysis was slowly beginning to emerge as a discipline in its own right, and a few countries had implemented procedures for at least rudimentary analyses in the late 1980s.

In some cases, decisions were made based on past history and established relationships with trading partners, or on various types of 'decision sheets' that provided a minimum of information on pests, hosts and options for managing pests on different commodities. Very little specific guidance existed for countries on how to analyse risk associated with the movement of pests on commodities in trade, as now required by the SPS Agreement.

Immediately prior to the negotiation of the SPS Agreement, countries were developing their own national guidance on risk analysis for plant protection. By the early 1990s, most major trading countries had implemented the use of risk analysis as the basis for decision making but there was a lack of consensus or harmonization (see Chapter 4 for more on harmonization) on the specific steps that should be included in pest risk analysis. At the time, countries understood that consensus and harmonization would be essential, especially under the new global trading framework.

Subsequent to the adoption of ISPM No. 2, countries sought to align their national procedures to the requirements laid out in the newly agreed standard. Although the approaches were consistent with the standard, there was considerable variation in methods used. Countries like the USA, Canada, New Zealand and Australia implemented commodity-based approaches to risk analysis. Their guidelines were written so that a commodity could be analysed for the types of pests it might carry, depending on the specific origin and the level of processing of the commodity. Pests that are considered to be likely to follow the commodity in trade are analysed further to determine whether risk management is necessary to manage those pests.

Other countries have implemented pest-specific approaches to risk analysis. The European and Mediterranean Plant Protection Organization (EPPO) is a regional organization for plant protection in Europe. It has developed a pest risk analysis scheme that provides very specific guidance on analysing the risk associated with a single pest – taking into account any of the pathways that pest may move on (rather than examining a specific commodity or single pathway carrying several types of pests).

Although these represent two very different approaches to pest risk analysis, there are some common elements. Regardless of the approach (e.g. pest-specific or commodity) the vast majority of pest risk analyses currently rely, at least in part if not entirely, on 'qualitative' descriptions and ratings of likelihood and consequences in the analyses. Simply put, a 'qualitative' rating is a non-numerical, non-quantitative method to describe the relative level of a particular element of the analysis. For instance, a pest may be described to have a 'low' likelihood or probability of being detected, or 'high' environmental consequences. Such qualitative descriptions are a legitimate means of estimating risk, particularly when quantitative data are missing. Uncertainty is also described in qualitative terms, using ratings such as high, medium and low. Later chapters in this text will cover different methods for conducting risk analyses in depth.

3.6.2. A note about regional organizations and pest risk analysis

The IPPC recognizes several regional organizations for plant protection (or RPPOs) – RPPOs are inter-governmental organizations that function as coordinating bodies for NPPOs on a regional level. The functions of RPPOs included in the text of the IPPC (1997) include:

- Coordination and participation in activities among their NPPOs in order to promote and achieve the objectives of the IPPC.
- Cooperation among regions for promoting harmonized phytosanitary measures.
- Gathering and disseminating information, in particular in relation with the IPPC.
- Cooperation with the IPPC in developing and implementing international standards for phytosanitary measures.

Each RPPO has its own activities and programme, and most of the RPPOs develop their own regional standards for phytosanitary measures (or RSPMs). These RSPMs apply only to countries that are Members of the RPPO that adopts a given standard. Because of the importance of pest risk analysis, several RPPOs have also developed guidance on aspects of pest risk analysis in the form of RSPMs – some of these RSPMs pre-date the first IPPC standards on PRA, and were used in the development of ISPM No. 2. Other RSPMs on PRA came later, and are based on ISPM No. 2.

There are currently nine recognized RPPOs:

- Asia and Pacific Plant Protection Commission (APPPC; established in 1956);
- Comunidad Andina (CA; 1969);
- Comite de Sanidad Vegetal del Cono Sur (COSAVE; 1980);
- Caribbean Plant Protection Commission (CPPC; 1967);
- European and Mediterranean Plant Protection Organization (EPPO; 1951);
- Inter-African Phytosanitary Council (IAPSC; 1954);
- North American Plant Protection Organization (NAPPO; 1976);
- Organismo Internacional Regional de Sanidad Agropecuaria (OIRSA; 1947);
- Pacific Plant Protection Organization (PPPO; 1995).

3.7. Summary

Although pest risk analysis is a relatively new discipline, countries have been managing the risks associated with pests for hundreds of years. In the past 150 years, countries began to cooperate internationally in managing pest risk with the aim to prevent the introduction and spread of important pests. At the same time, countries began to adopt laws and regulations at the national level to protect themselves from new pests. More recently, international treaties, like the SPS Agreement and the IPPC have been adopted to provide a

framework for how countries should implement national laws and regulations related
to managing the risks associated with foreign pests. Those agreements include obligations for countries to technically justify

their measures through risk analysis. Thus,
risk analysis has become the basis for
national laws and regulations aimed at preventing the introduction and spread of new
pests.

References

Castonguay, S. (2010) Creating an agricultural world order: regional plant protection problems and international phytopathology, 1878–1939. *Agricultural History Society* 84, 46–73.

Ebbels, D.L. (2003) *Principles of Plant Health and Quarantine.* CAB International, Wallingford.

FAO (1997) *New Revised Text of the International Plant Protection Convention.* Food and Agriculture
Organization of the United Nations, Rome.

FAO (2000) *Multilateral Trade Negotiations in Agriculture: a Resource Manual.* Food and Agriculture
Organization of the United Nations, Rome, Available online at: http://www.fao.org/DOCREP/003/
X7351E/X7351e00.htm#TopOfPage, accessed 20 December 2011.

GATT (1947) *General Agreement on Tariffs and Trade.* GATT, Geneva. Available online at http://www.wto.org/
english/docs_e/legal_e/gatt47_e.pdf, accessed 20 December 2011.

Gensini, G.F., Yacoub, M.H. and Conti, A.A. (2004) The concept of quarantine in history: from plague to SARS.
Journal of Infection 49, 257–261.

IPPC (2007) International Standards for Phytosanitary Measures, Publication No. 2: *Framework for Pest Risk
Analysis.* Secretariat of the International Plant Protection Convention (IPPC), Food and Agriculture
Organization of the United Nations, Rome.

IPPC (2010) International Standards for Phytosanitary Measures, Publication No. 5: *Glossary of Phytosanitary
Terms.* Secretariat of the International Plant Protection Convention (IPPC), Food and Agriculture
Organization of the United Nations, Rome.

MacLeod, A., Pautasso, M., Jeger, M.J. and Haines-Young, R. (2010) Evolution of the international regulation
of plant pests and challenges for future plant health. *Food Security* 2, 49–70.

Mathys, G. and Baker, E.A. (1980) An appraisal of the effectiveness of quarantines. *Annual Review of
Phytopathology* 18, 85–101.

Maynard, G.V., Hamilton, J.G. and Grimshaw, J.F. (2004) Quarantine – phytosanitary, sanitary and incursion
management: an Australian entomological perspective. *Australian Journal of Entomology* 43, 318–328.

NAPPO (1993) *International Approaches to Plant Pest Risk Analysis: Proceedings of the APHIS/NAPPO
International Workshop on the Identification, Assessment, and Management of Risks due to Exotic
Agricultural Pests,* Alexandria, Virginia, 23–25 October 1991. Ottawa, Canada. Bulletin (North American
Plant Protection Organization) No. 11.

Norin, C.A. (1915) Agricultural quarantine and Inspection in the United States. Thesis. Oregon Agricultural
College, Corvallis, Oregon, USA.

Orton, W.A. (1914) Plant quarantine problems. *Journal of Economic Entomology* 7, 109–116.

Roelfs, A.P. (2011) Epidemiology in North America. In: Bakum, J. (ed.) *The Cereal Rusts Volume I and II* (original eds Bushnell, W. and Roelfs, A.P. 1984). Elsevier, Amsterdam. Available online at: http://www.
globalrust.org/traction, accessed 5 January 2011.

Sorensen, W.C., Smith, E.H., Smith, J. and Carton, Y. (2008) Charles V. Riley, France, and Phylloxera. *American
Entomologist* 54 (3), 134–149.

Weber, G.A. (1930) *The Plant Quarantine and Control Administration: its History, Activities and Organization.*
Brookings Institution, Washington, DC), No. 59.

WTO (1994) *The WTO Agreement on the Application of Sanitary and Phytosanitary Measures.* World Trade
Organization, Geneva.

4 International Legal and Regulatory Framework for Risk Analysis

Christina Devorshak

4.1. Introduction

Every country has an organization that serves as the NPPO, charged with protecting the plant resources, both natural and cultivated, of its territories. Since 2000, most countries' NPPOs have worked under an international framework of treaties that govern and guide the measures that countries should use to protect their plant resources from the introduction and spread of pests and diseases. 'Measures', in this book, refers to any laws, regulations, requirements, actions or procedures that have the purpose of reducing risks associated with the introduction and spread of pests. The two most important treaties are the Agreement on the Application of Sanitary and Phytosanitary Measures and the IPPC. Both of these treaties, and their role in protecting plant health, are discussed below.

4.2. History of the Development of the SPS Agreement

4.2.1. The Uruguay Round

After the end of the Second World War, countries all over the world were in a state of economic turmoil. At the time, world leaders established the 'GATT' in order to liberalize global trade and encourage economic growth. The GATT (1947) entered into force in 1948, and had as its objective:

> Recognizing that their relations in the field of trade and economic endeavour should be conducted with a view to raising standards of living, ensuring full employment and a large and steadily growing volume of real income and effective demand, developing the full use of the resources of the world and expanding the production and exchange of goods.

> Being desirous of contributing to these objectives by entering into reciprocal and mutually advantageous arrangements directed to the substantial reduction of tariffs and other barriers to trade and to the elimination of discriminatory treatment in international commerce.

They felt that only through economic stability, established through a strong system of global trade, would countries be able to achieve lasting peace. The GATT created a framework of rules for countries to follow – these rules were primarily related to the types of tariffs and subsidies countries could apply to imported and exported goods. At that time, agricultural products were not given any special treatment in the GATT, except for one clause that related to protecting health. That clause recognized that there could be legitimate barriers to trade, and exceptions were noted in Article XX.b. With respect to sanitary and phytosanitary measures, Article XX.b of the GATT stated:

> Subject to the requirement that such measures are not applied in a manner which would constitute a means of arbitrary or unjustifiable discrimination between countries where the same conditions prevail, or a disguised restriction on international trade, nothing in this Agreement shall be construed to prevent the adoption or enforcement by any contracting party of measures:
>
> (b) necessary to protect human, animal or plant life or health.

Article XX.b made provisions for countries to put in place measures designed to protect human, animal and plant health from the introduction and spread of diseases. It is from Article XX.b of the GATT that the SPS Agreement evolved.

Beginning in the 1980s, countries agreed that a new round of negotiations on international trade was needed to modernize the global trading system and to further reform economic policies related to international trade. That round of negotiation, called the Uruguay Round, resulted in the establishment in 1995 of the WTO and its agreements. The WTO was established in Geneva, Switzerland and comprises several agreements, all related to international trade. These agreements encompass trade in goods, services and intellectual property. In the case of goods, this includes commodities like agricultural products, as well as other products that are imported and exported between countries (FAO, 2000).

4.2.2. Agreement on Agriculture

One of the new features of the WTO compared with the previous GATT agreement was the Agreement on Agriculture (recall that agriculture as a sector was not specifically mentioned in the GATT). The WTO's Agreement on Agriculture basically states that countries should strive to minimize

trade distorting economic practices. These practices include subsidizing domestic agriculture and assigning high tariffs to imported agricultural products. Both subsidies and tariffs are may be regarded as protectionist practices designed to protect domestic markets from cheaper imported products. Protectionist practices are viewed to be trade distorting because the overall effect is to artificially drive up prices of consumer goods. Reducing tariffs and subsidies, according to the WTO, means that all countries will be able to take advantage of a more fair and predictable trading environment – a 'level playing field' so that all countries can benefit from free trade (FAO, 2000).

Although countries agreed that a more liberalized trade environment would be beneficial, there remained concerns that – in the absence of protectionist measures in the form of tariffs and subsidies – countries might use other types of measures to protect domestic markets. Specifically, countries were concerned that quarantine measures (measures designed to protect human, animal and plant health) or other types of technical barriers might be used to protect domestic markets. Because of this concern, countries renegotiated Article XX.b of the GATT and established the SPS Agreement to provide structure and guidance to countries in the application of measures to protect human, animal and plant life or health (or sanitary and phytosanitary measures) so that such measures were not used as disguised barriers to trade.

4.2.3. Agreement on the Application of Sanitary and Phytosanitary Measures

The Agreement on the Application of Sanitary and Phytosanitary Measures provides guidance to countries on how sanitary and phytosanitary (SPS) measures should be applied. Such measures, including quarantine measures, are designed to protect human, animal and plant life or health from the introduction and spread of pests and diseases (Table 4.1). In the case of human health, these measures also apply to contaminants in food.

Table 4.1. Definition of an SPS Measure. (From WTO, 1994; Devorshak, 2007.)

To protect:	From:
Human or animal life	Risks arising from additives, contaminants, toxins or disease-causing organisms in their food, beverages, feedstuffs
Human life	Plant- or animal-carried diseases (zoonoses)
Animal or plant life	Pests, diseases, or disease-causing organisms
A country	Damage caused by the entry, establishment or spread of pests

The SPS Agreement has been in force for most countries since 1995 (and in force for all countries since 2000). The purpose of the SPS Agreement is to facilitate trade while allowing countries to put in place measures to protect health. The Agreement requires, however, that such measures are technically justified, meaning that there is a scientific and technical basis for requiring a protective measure. Simply put, it means that there must be scientific and technical evidence that a potential risk to human, animal or plant health exists, and that the measure to manage that risk is applied in such a way as to not overly interfere with trade. The requirements for using scientific evidence as the basis for measures are laid out in the various Articles of the SPS Agreement (described below). The text of the SPS Agreement can be seen on the WTO website: www.wto.org.

4.2.4. The SPS Committee

The SPS Agreement is administered by a body called the SPS Committee. The SPS Committee is composed of 'Members', or countries that are signatory to the WTO. There are currently 191 Members of the WTO, and each Member has an equal vote

in the Committee. The SPS Committee serves as a forum for trade concerns arising from SPS issues to be raised, discussed and resolved between Members. It also serves as a venue for Members to meet many of the obligations of the SPS Agreement (including transparency and notification discussed below).

4.3. Key SPS Provisions

The SPS Agreement includes several key provisions, contained in Articles that govern the application of SPS measures by Member countries. A summary of key provisions is provided here (WTO, 1994), but the entire SPS Agreement can be found on the WTO website (see www.wto.org). The key provisions are:

4.3.1. Sovereignty

The principle of sovereignty states that all Members retain the right to protect human, animal and plant life or health from the introduction of pests and diseases into their territories, and that no other Member or organization may interfere with that right (Article 2). The principle of sovereignty means that a Member may put in place an SPS measure without interference by any other Member, so long as that measure is justified and supported by scientific evidence.

4.3.2. Transparency (and notification)

A key principle to the functioning of the SPS Agreement is the principle of transparency (Articles 5 and 7 and Annex B of the SPS Agreement). Transparency means that, if an importing country Member puts in place an SPS measure, they should be able to provide other Members with the justification for that measure. It also means that an importing country Member should provide other Members with any requirements they are implementing on imported goods within a 'reasonable period of time'. The transparency provisions of the SPS Agreement are a central requirement and essential to the overall function and discipline of the SPS Agreement. All of the other provisions, obligations and requirements of the SPS Agreement are hinged on the application of transparency by all Members, primarily because if Members are not fully informed of other Members' justifications and import requirements they are unable to function together impartially and fairly.

In order to meet transparency requirements of the SPS Agreement, Members must establish two different types of authorities in their countries: the Notification Authority and the Enquiry Point. The notification authority is an office or individual in the national government of each Member that is responsible for the notification process (discussed below). The Enquiry Point is an office or individual in the national government of each Member that is responsible for responding to enquiries from other trading partners related to national SPS measures (such as import requirements on specific commodities). In principle, the Enquiry Point for a Member should be able to answer such questions, or direct enquiries to the relevant authority in its country.

Another aspect of transparency is the notification process of the SPS Committee. The notification process requires Members to notify, through the Notification Authority of their country, the SPS Committee (i.e. other Members) of any substantial changes or new requirements that could affect their trading partners. This may include changes in regulations, requirements, specific procedures or laws related to SPS measures.

The notification itself should include a description of the proposed changes to national regulations that will have a substantial effect on trade. Members should notify the SPS Committee of such changes at least 60 days prior to any new requirement entering into force. During this time period, other Members may comment on the proposed changes. In addition, the 60-day time period should allow for other Members to prepare for any changes in requirements on products they may be exporting to the notifying Member.

As a final note, the SPS Agreement does not define certain terms, such as 'reasonable period of time'. However, the concepts may be tested through the SPS Committee, or if necessary through a process of dispute settlement (discussed in Chapter 19).

4.3.3. Scientific justification

The basic tenet of the SPS Agreement is that SPS measures should be scientifically justified (Article 2 of the Agreement). This means that there must be scientific evidence that indicates there is a risk to human, animal or plant health and that any related measure to protect health must also be based on scientific evidence. If there is no scientific evidence indicating that there is a risk, then an SPS measure is not considered justified.

The SPS Agreement does not set a requirement for the minimum amount or quality of information needed to make a judgement. Rather, the Agreement states that measures should be based on *available* information. A lack of information is not a justification for implementing a measure such as a precautionary measure (see Chapter 19 for more information on precaution), or for delaying judgement about what type of measure should be implemented. Disputes have arisen when Members have tried to use lack of information as a justification for SPS measures, however the WTO has routinely rejected this point, and affirmed that measures must be based only on *available* information (see Chapter 19 for more information on disputes). Measures can be scientifically justified in one of two ways: by conducting a scientific risk assessment to determine what measures are appropriate (see Section 3.4 on risk assessment) or by basing measures on international standards (see Section 3.5 on harmonization).

4.3.4. Risk assessment

The method that is used to collect, analyse and present scientific information to justify an SPS measure is called 'risk assessment' in the SPS Agreement (Article 5 of the Agreement). This article defines what risk assessment is, and what kind of information should be considered in conducting a risk assessment. However, specific methods for risk assessment are not defined by the SPS Agreement itself, but rather by relevant international organizations (discussed later in this chapter).

Risk assessment in the SPS Agreement is defined as:

> the evaluation of the likelihood of entry, establishment or spread of a pest or disease within the territory of an importing Member according to the sanitary or phytosanitary measures which might be applied and of the associated potential biological and economic consequences; or the evaluation of the potential for adverse effects on human or animal health arising from the presence of additives, contaminants, toxins or disease-causing organisms in food, beverages or feedstuffs.

The SPS Agreement states that a 'risk assessment' must be done if a country wishes to put in place SPS measures. The risk assessment should take into account and consider, in the case of plant health, both biological and economic evidence. Relevant economic factors that should be considered in risk assessment include: the potential damage in terms of loss of production or sales as result of the introduction of a pest; the costs of control or eradication of a pest; and the relative cost effectiveness of risk management options. Likewise, in the assessment of risk, relevant production practices, inspection, sampling and testing methods, the prevalence of pests, the existence of pest free areas and other ecological and environmental conditions should be considered.

The SPS Agreement is clear that the obligation lies on the importing country to demonstrate that there is a risk and to analyse that risk to justify an SPS measure; however, there is a balance of obligations for both importing and exporting countries in the provision of evidence and scientific information. Furthermore, while it is the importing country that is ultimately responsible for any decisions that are made as a result of a risk assessment, the risk assessment itself may be performed by anyone (including, in some cases, other countries or universities).

Although the SPS Agreement uses the term 'risk assessment', the term 'risk analysis' is more commonly used outside of the SPS Agreement (see Chapter 2 and Chapter 5 for discussions on this point). In this book, we will be using the term risk analysis, comprising the three parts – risk assessment, management and communication. In plant protection, we also use the term 'pest risk analysis' to refer to the process of analysing risks of pest of plants specifically.

4.3.5. Harmonization

Article 3 of the SPS Agreement covers the principle of 'harmonization'. It states that, where possible, Members should base their SPS measures on international standards, guidelines and recommendations. The SPS Agreement is unique in that it actually identifies other organizations in the SPS Agreement that are responsible for developing international standards. These organizations (discussed in Section 4.4 of this chapter) are the Codex Alimentarius for food safety and human health, the World Organization for Animal Health (formerly the OIE) for animal health and the IPPC for plant health.

As mentioned above, SPS measures must be scientifically justified either through a scientific risk assessment or by basing the measures on international standards. If a country chooses to base its measures on an international standard, it cannot be challenged within the WTO context (i.e. subject to dispute) because such measures have been agreed through the standard. International standards provide the scientific justification of measures, because the standards themselves are based on scientific evidence and an analysis of risk. Moreover, basing measures on international standards creates a more predictable and fair trading environment for trading partners since international standards are agreed upon by member countries.

Although the SPS Agreement encourages countries to base their measures on international standards, it also allows countries to base their measures on a risk assessment (in SPS terms) – in cases where a Member wishes to achieve a higher level of protection than a standard provides for, or in cases where an international standard does not exist. In the case of phytosanitary standards, there are relatively few standards that address specific pests or commodities. Thus, it is this allowance – that countries may use risk assessment as an alternative to standards – which has led to the importance of risk assessment as the basis for measures aimed at protecting plants.

4.3.6. Regionalization

The SPS Agreement recognizes that pests and diseases follow ecological and climatic conditions, but do not typically follow political borders. Article 6 of the SPS Agreement allows countries to define entire countries, parts of countries, regions or even groups of countries as areas that can be recognized as free of particular pests or diseases. This concept of 'area freedom' is a critical component of many risk management programmes. It means that an importing country can recognize an exporting country as free of a pest or disease and therefore it should not require additional measures to manage the risk associated with that pest or disease.

4.3.7. Equivalence

Another important principle of the SPS Agreement is the recognition of equivalence (Article 4). The recognition of equivalence means that an importing country should recognize as equivalent an alternative measure that achieves the same level of protection as a measure they might require. For instance, a country may require an imported fruit to be fumigated for a particular pest. If an exporting country can demonstrate that an alternative to fumigation (e.g. heat

treatment) is as effective as fumigation, then the importing country should accept that measure as equivalent. In this case, the country proposing the alternative treatment (typically the exporting country) must be able to demonstrate equivalence, based on scientific evidence. Equivalence is such a critical concept in the SPS-IPPC framework that there is a standard devoted specifically to recognition of equivalence as a phytosanitary measure, ISPM No. 24 *Guidelines for the Determination and Recognition of Equivalence of Phytosanitary Measures* (IPPC, 2005).

4.3.8. Dispute settlement

The Agreement includes a provision for settlement of disputes between Members (Article 11). Typically, disputes arise over the use and application of risk assessment or the interpretation of scientific information. If the dispute cannot be resolved through mediation, it may be resolved through a legally binding dispute settlement process. Chapter 19 includes a more detailed discussion of dispute settlement and the findings of several important cases.

4.3.9. Provisional measures

Although Article 2 requires that measures be scientifically justified, Article 5 (risk assessment) of the SPS Agreement allows countries to provisionally adopt measures in the event of a lack of information or in cases of emergencies. If a country invokes provisional measures, however, it is then obligated to actively seek out scientific information within a reasonable period of time to conduct a risk assessment and they must then review the measure in light of any evidence they subsequently gather. This requirement is extremely important – if a country implements provisional measures, but fails to actively seek out information to conduct a risk assessment, the measure may be challenged by any affected trading partners and the country implementing the measure may be subject to dispute settlement. However, the benefit of provisional measures is that they allow countries to implement measures to protect health, while providing them time to conduct a risk analysis.

4.3.10. 'Appropriate level of protection'/consistency

The concept of consistency and appropriate level of protection (ALOP) is also addressed in Article 5 along with risk assessment. While the concept of 'appropriate level of protection' (or acceptable level of risk) is very difficult to define in practice, in theory it means that countries should strive to apply measures to reach a given level of protection – consistently across sectors and for different pests and diseases. This may be done by defining a given level of harm (typically economic harm) that would be considered unacceptable if a given pest or disease was to be introduced into a country (see also Chapter 9, Section 9.4.3 for more on how countries may define the ALOP). Alternatively, and more commonly, the appropriate level of protection is defined by the history of decision making within a country so that new requirements are consistent with previous requirements. The purpose of defining the appropriate level of protection, and for achieving consistency, is to create a more predictable, fair and equitable trading environment.

4.3.11. Least trade restrictive (minimal impact)

According to Article 2 of the SPS Agreement, Members should apply SPS measures only to the extent necessary to reach the appropriate level of protection. The appropriate level of protection can only be determined by the importing country based on risk assessment. Members should refrain from requiring measures that are excessive or unduly interfere with trade. If a less stringent

measure can achieve the necessary level of protection, then the Member should not require a more stringent measure.

4.3.12. Non-discrimination/ national treatment

The SPS Agreement requires that, where similar conditions prevail, Members shall be treated equally under the principle of non-discrimination. This is often referred to as the 'most favoured nation' principle – meaning that every trading partner should be treated as 'most favoured'. Furthermore, the SPS Agreement requires that trading partners should be treated equally compared with domestic producers under the principle of 'national treatment'. Meaning that if a country does not require their domestic producers to adhere to specific measures they cannot require their trading partners adhere to those same measures. Although the principles may be difficult to apply in practice, especially where domestic producers would prefer preferential treatment, the intent of these principles is to ensure a fair and predictable trading environment for all countries.

4.4. International Standard Setting Bodies

As mentioned above, there are three organizations identified in the SPS Agreement as internationally recognized standard setting bodies.

4.4.1. Codex Alimentarius Commission

The Codex Alimentarius Commission was established in 1960 and is jointly administered through the WHO and FAO. The Codex Alimentarius Commission is the international standard setting body for food safety and human health. It comprises several different committees, each of which deals with particular issues related to food safety, such as microbiological contaminants or pesticide residues. The Codex Alimentarius Commission develops international standards for food safety that provide guidelines to countries on the levels of contaminants, toxins or other types of additives that would pose an unacceptable risk in food, beverages and feedstuffs (FAO, 2000).

4.4.2. World Organization for Animal Health (formerly the Office Internationale des Epizooties or OIE)

The World Organization for Animal Health is the oldest of the standard setting bodies, having been established in 1924 to address animal disease outbreaks. The World Organization for Animal Health develops standards aimed at managing the risks associated with animal diseases. In addition to developing international standards, the World Organization for Animal Health also performs an important function in notifying member of disease outbreaks in other countries. The purpose of this notification process (distinct from the SPS notification process) is to ensure that prompt action can be taken to prevent or slow the spread of animal diseases (and in particular those diseases that may also affect humans, such as some strains of influenza) (FAO, 2000).

4.4.3. International Plant Protection Convention (IPPC)

The IPPC is identified in the SPS Agreement as the standard setting body for plant health. Of the three standard setting bodies, the IPPC has the most recent history of developing international standards. Although it was originally adopted as an international treaty in 1951, its history of standard setting dates back only to 1995. The IPPC is unique among the standard setting bodies in that it exists as an international treaty in its own right – alongside the SPS Agreement – as a plant protection agreement that includes

provisions for international trade (FAO, 2000). The IPPC is described in more detail below.

4.5. The International Plant Protection Convention in Detail

The IPPC originally existed as a plant protection agreement, first adopted in 1951. It has since been revised twice, once in 1979 and more recently in 1997. The most recent revision in 1997 resulted in the 'New Revised Text of the International Plant Protection Convention' (IPPC, 1997). The 1997 text of the IPPC was a direct result of the IPPC being identified in the SPS Agreement as the standard setting body for plant health. Prior to 1997, the IPPC existed solely as a treaty, but did not have any mechanism for developing and adopting international standards. Therefore, countries re-negotiated the text and established both a Secretariat (to administer the Convention) and a Commission (the Commission on Phytosanitary Measures composed of Member countries or 'contracting parties'), which is the body that governs the work of the IPPC and is responsible for adopting newly developed standards.

4.5.1. Purpose of the IPPC

The purpose of the IPPC is to protect plants and plant health from the introduction and spread of pests. It encourages international cooperation in preventing the spread of pests across borders and between countries. Although the IPPC clearly makes provisions for international trade, it is primarily a plant protection agreement.

4.5.2. Scope of the IPPC

The scope of the IPPC applies to all plants, both wild and cultivated, and includes everything from forests to agricultural crops. Furthermore, the scope of the IPPC applies to anything that is capable of introducing and spreading pests. This may include agricultural commodities (e.g. fruits and vegetables), live plants for planting, seeds, plant products, handicrafts, solid wood packing material and other types of wood, conveyances, containers, ships and shipping materials, and even such materials as used cars and machinery or marble tiles. Essentially, any material that can carry or harbour a pest may be subject to measures defined and promoted under the IPPC. Lastly, under the IPPC, the term 'pest' refers to any organism that is injurious to plants or plant health. This includes plant pathogens (fungi, viruses, bacteria and nematodes), arthropods (insects and mites), other types of invertebrates (molluscs) and weeds.

4.5.3. Key provisions of the IPPC

Like the SPS Agreement, the IPPC includes several provisions that address the rights and obligations of member countries. The rights and obligations of countries that are signatory to the IPPC are very similar to those identified in the SPS Agreement and include scientific justification, sovereignty, harmonization, least trade restrictive measures, consistency, regionalization, equivalence, appropriate level of protection, non-discrimination/national treatment, provisional measures, dispute resolution and transparency (FAO, 1997). Risk analysis, or 'pest risk analysis' in IPPC terms, is also one of the key provisions of the IPPC and is described in detail below.

4.5.4. Relationship to other international organizations (SPS, FAO, CBD)

The scope of the IPPC overlaps to some extent with other international agreements. As described above, the SPS Agreement identifies the IPPC as the internationally recognized standard setting body for plant health. As such, the IPPC and WTO have a clear relationship, where the work of the IPPC supports the aims of the SPS Agreement. The SPS Committee looks to the IPPC for advice and information on technical issues related to plant health, and WTO Members are expected to participate in the IPPC standard setting process (including the development and implementation of standards).

Besides the WTO, the IPPC also cooperates with other international organizations, particularly where there are overlaps in scope. The Convention on Biological Diversity (CBD) is an international agreement that has the purpose of promoting the conservation and sustainable use of biological diversity. Part of the work of the CBD is directed at managing the risks associated with 'invasive species' or introduced species that pose a risk to biological diversity and the environment.

Beginning in 2000, the IPPC and CBD recognized that there was a clear overlap in scope between the two agreements (MacLeod et al., 2010; Schrader et al., 2010). The IPPC is concerned with introduced organisms that have the potential to harm plant health in particular, while the CBD is concerned with harm to any organism or environment (not limited to plant health). At that time, the two organizations recognized that the IPPC already had structures and mechanisms in place (including international standards) aimed at managing risks associated with invasive species that have the potential to harm plant health. Therefore, the two organizations established a joint work programme to enhance cooperation and coordination in achieving common goals.

A subsidiary agreement of the CBD is the Cartagena Protocol on Biosafety. The Protocol addresses potential risks to biodiversity associated with 'living modified organisms' or LMOs (also known as genetically modified organisms) intended for release into the environment. LMOs include everything from maize seed to genetically modified animals like salmon. As part of the coordinated work programme, the CBD recognizes that the IPPC and its standards also address potential risks of LMOs to plant health.

4.6. Types of Standards Produced by the IPPC

International standards for phytosanitary measures (ISPMs) provide the basis for national regulatory measures implemented by countries that are contracting parties to the IPPC. These standards provide guidance on all aspects of phytosanitary measures, including phytosanitary certification, implementation of pest free areas and related measures, pest eradication, pest risk analysis (including pest risk management), risk management for different types of pests or commodities, requirements for import and export of commodities and other legal and regulatory measures. The number of adopted and revised standards continues to change. The most current list of standards can be seen on the IPPC website (http://www.ippc.int).

There are four general types of standards: reference, pest risk analysis, concept and specific.

4.6.1. Reference standards

Reference standards are those standards that provide guidance on the Convention and the use of specific terms. The two reference standards are:

- ISPM No. 1 – *Phytosanitary Principles for the Protection of Plants and the Application of Phytosanitary Measures in International Trade* (IPPC, 2010a).
- ISPM No. 5 – *Glossary of Phytosanitary Terms* (IPPC, 2010b).

4.6.2. Pest risk analysis standards

Another category of standards are the pest risk analysis standards. There are currently four different standards that are specifically related to pest risk analysis (including risk management):

- ISPM No. 2 – *Framework for Pest Risk Analysis* (IPPC, 2007).

- ISPM No. 11 – *Guidelines for Pest Risk Analysis for Quarantine Pests, Including Analysis of Environmental Risks and Living Modified Organisms* (IPPC, 2004a).
- ISPM No. 14 – *The Use of Integrated Measures in a Systems Approach for Pest Risk Management* (IPPC, 2002).
- ISPM No. 21 – *Pest Risk Analysis for Regulated Non-quarantine Pests* (IPPC, 2004b).

4.6.3. Concept standards

Concept standards include those standards that describe key conceptual activities undertaken by NPPOs. This includes a wide range of activities ranging from phytosanitary certification to the implementation of pest free areas or pest eradication programmes, to phytosanitary treatments, to the design and operation of post-entry quarantine facilities. These standards provide the fundamental building blocks to all the activities that an NPPO must undertake to protect plant health. Examples of concept standards are ISPM No. 4 (*Guidelines for Pest Free Areas*, IPPC, 1996) or ISPM No. 8 (*Determination of Pest Status in an Area*, IPPC, 1998).

4.6.4. Specific standards

Specific standards address specific pests or commodities. ISPM No. 15 (*Guidelines for Regulating Wood Packaging Material in International Trade*, IPPC, 2009) is an example of a commodity-specific standard that provides guidance on measures that should be taken to reduce risk associated with wood packing material. There are also several standards that address fruit fly pests (Diptera: Tephritidae), such as ISPM No. 26: *Establishment of Pest Free Areas for Fruit Flies (Tephritidae)* (IPPC, 2006). These pest specific standards provide guidance on how countries should undertake a range of measures related to managing risks associated with this very important group of pests, including treatments, pest free areas and areas of low pest prevalence.

4.7. Role of Risk Analysis in the IPPC

Like the SPS Agreement, the primary goal of the IPPC is managing risks to plant health. At the same time, the IPPC creates the obligation for contracting parties to technically justify measures aimed at protecting plant health (e.g. 'quarantine' or phytosanitary measures). Measures are considered technically justified if the measure is based either on an international standard or on a risk analysis. The term used in the IPPC, slightly different from the other international standard setting bodies, is 'pest risk analysis'. The IPPC defines pest risk analysis as 'the process of evaluating biological or other scientific and economic evidence to determine whether an organism is a pest, whether it should be regulated, and the strength of any phytosanitary measures to be taken against it'.

The importance of PRA can be seen in the text of the IPPC itself (FAO, 1997). The requirement for PRA and technical justification appears in several different articles, including:

- The Preamble;
- Article I (Purpose and responsibility);
- Article IV (General provisions relating to the organizational arrangements for national plant protection);
- Article VI (Regulated pests);
- Article VII (Requirements in relation to imports);
- Article VIII (International cooperation).

These articles lay out rights, obligations and responsibilities for contracting parties. In particular, the articles require importing countries to conduct pest risk analysis, to provide technical justifications to trading partners, and for both importing and exporting countries to share information and cooperate for the purpose of conducting pest risk analysis (see also Chapter 6 for more information on this topic).

Specific guidance on how countries should meet their obligations for pest risk analysis is provided in the various standards related to pest risk analysis. These include ISPM No. 2 (originally adopted in 1996 and revised in 2007; IPPC, 2007), which provides a general overview on how pest risk analysis should be conducted, and ISPM No. 11 (IPPC, 2004a). These standards outline when and how countries should conduct pest risk analysis, providing general guidelines without being prescriptive as to how those guidelines should be implemented.

According to ISPM No. 2 (IPPC, 2007):

'Pest risk analysis (PRA) provides the rationale for phytosanitary measures for a specified PRA area. It evaluates scientific evidence to determine whether an organism is a pest. If so, the analysis evaluates the probability of introduction and spread of the pest and the magnitude of potential economic consequences in a defined area, using biological or other scientific and economic evidence. If the risk is deemed unacceptable, the analysis may continue by suggesting management options that can reduce the risk to an acceptable level. Subsequently, pest risk management options may be used to establish phyto sanitary regulations.'

Broken down into its components, this means pest risk analysis considers biological and economic information for a pest, estimates the likelihood (or probability) that a pest may be introduced, estimates the consequences of a pest introduction and summarizes the conclusions. It also provides options for managing or reducing the risk.

4.8. Additional Work of the IPPC

The IPPC carries out a range of activities in support of the overall work programme of the IPPC and in relation to the implementation of standards. This section describes the work of the various bodies of the IPPC and other aspects of the work programme besides standard setting.

4.8.1. Commission on Phytosanitary Measures

The work to be conducted by the IPPC is decided by the Commission on Phytosanitary Measures (CPM). The CPM is composed of the Member countries, or contracting parties to the IPPC. At its annual meeting, the CPM adopts standards (developed by expert working groups and reviewed by a Standards Committee), as well as makes decisions on globally important plant health issues. It designates the work that should be carried out by the Secretariat each year, including activities related to developing new international standards, information exchange and technical assistance.

The CPM is represented by the Bureau of the CPM, comprising the Chairperson and two Vice-Chairpersons, and a representative from each of the FAO regions. The Chairperson and Vice-Chairperson positions rotate between regions.

4.8.2. Standards Committee

One of the most important bodies of the IPPC is the Standards Committee (SC). The SC comprises phytosanitary experts representing all of the FAO regions. It meets annually, and provides technical oversight to the standard setting programme. It is responsible for reviewing draft international standards prior to the standards being adopted by the CPM to ensure the drafts are technically correct and consistent with international standards. After the SC reviews the draft standards, the standards are distributed to countries for comment. The comments provided by countries are then considered and/or incorporated into the drafts by the SC and the standards are then sent to the CPM for adoption.

4.8.3. Other subsidiary bodies of the IPPC

There are several other types of bodies that help execute the work programme that is designated by the CPM. These include technical panels and expert working groups that write the initial drafts of standards which are sent to the SC for its review. There is also the Strategic Planning and Technical Assistance group that provides suggestions to the CPM, Bureau and Secretariat on the work programme of the IPPC. Lastly, there is

the Subsidiary Body on Dispute Settlement, which, in the event of a dispute between contracting parties, provides oversight to the dispute settlement process in the IPPC.

4.8.4. Information exchange

The IPPC calls on countries to cooperate in the exchange of information for the purpose of preventing the introduction and spread of pests. This includes providing information on the structure of the NPPO, the laws and regulations of the NPPO, lists of regulated pests and information on pests that occur in the country. Importantly, all countries share an obligation to provide information to other countries specifically for the conduct of pest risk analysis.

In order to facilitate the exchange of information, the IPPC hosts the International Phytosanitary Portal (http://www.ippc.int). Through the International Phytosanitary Portal, countries can both provide and access national information identified above. Likewise, the International Phytosanitary Portal provides important information related specifically to the IPPC including draft and adopted international standards, as well as other important plant protection information.

4.8.5. Technical assistance

The Convention itself includes provision for technical assistance, and the CPM includes technical assistance as part of the work programme of the IPPC. Technical assistance provided through the IPPC is to assist countries in meeting their obligations under the IPPC and to help those countries in developing their NPPO. This includes providing countries assistance in information exchange, developing key functions of the NPPO (e.g. surveillance, certification, administration, etc.) and implementation of international standards at the national level.

More information on all of these topics is available online at the International Phytosanitary Portal.

4.9. Summary

The SPS Agreement and the IPPC provide the international framework for safe and fair trade – on the one hand liberalizing trade to promote economic welfare for all countries, but at the same time creating a structure around how countries should implement measure to protect their territories from the introduction of pests and diseases. While the SPS Agreement is a trade agreement that makes provisions for protecting health (in our case, plant health), the IPPC is a plant protection agreement that makes provisions for safe trade. Both agreements contain several key provisions that give countries rights and obligations. The agreements require that measures must be technically justified (through risk assessment) or harmonized (based on international standards). The principles of transparency, equivalence, non-discrimination (and national treatment), regionalization, appropriate level of protection (and consistency), and least trade-restrictive measures further underscore the need for risk analysis within the framework of the SPS Agreement and the IPPC.

References

Devorshak, C. (2007) Area-wide integrated pest management programmes and agricultural trade: challenges and opportunities for regulatory plant protection. In: Vreysen, M.J.B., Robinson, A.S and Hendrichs, J. (eds) *Area-wide Control of Insect Pests from Research to Field Implementation*. Springer, Dordrecht, pp. 407–415.

FAO (1997) *New Revised Text of the International Plant Protection Convention*. Food and Agriculture Organization of the United Nations, Rome.

FAO (2000) *Multilateral Trade Negotiations in Agriculture: A Resource Manual*. Food and Agriculture Organization of the United Nations, Rome. Available online at: http://www.fao.org/DOCREP/003/X7351E/X7351e00.htm#TopOfPage, accessed 20 December 2011.

GATT (1947) *General Agreement on Tariffs and Trade*. Geneva. Available online atline at http://www.wto.org/english/docs_e/legal_e/gatt47_e.pdf, accessed 20 December 2011.

IPPC (1996) International Standards for Phytosanitary Measures, Publication No. 4: *Requirements for the Establishment of Pest Free Areas*. Secretariat of the International Plant Protection Convention (IPPC), Food and Agriculture Organization of the United Nations, Rome.

IPPC (1997) *New Revised Text of the International Plant Protection Convention*. Food and Agriculture Organization of the United Nations, Rome.

IPPC (1998) International Standards for Phytosanitary Measures, Publication No. 8: *Determination of Pest Status in an Area*. Secretariat of the International Plant Protection Convention (IPPC), Food and Agriculture Organization of the United Nations, Rome.

IPPC (2002) International Standards for Phytosanitary Measures, Publication No. 14: *The Use of Integrated Measures in a Systems Approach for Pest Risk Management*. Secretariat of the International Plant Protection Convention (IPPC), Food and Agriculture Organization of the United Nations, Rome.

IPPC (2004a) International Standards for Phytosanitary Measures, Publication No. 11: *Pest Risk Analysis for Quarantine Pests Including Analysis of Environmental Risks and Living Modified Organisms*. Secretariat of the International Plant Protection Convention (IPPC), Food and Agriculture Organization of the United Nations, Rome, Italy.

IPPC (2004b) International Standards for Phytosanitary Measures, Publication No. 21: *Pest Risk Analysis for Regulated Non Quarantine Pests*. Secretariat of the International Plant Protection Convention (IPPC), Food and Agriculture Organization of the United Nations, Rome.

IPPC (2005) International Standards for Phytosanitary Measures, Publication No. 24: *Guidelines for the Determination and Recognition of Equivalence of Phytosanitary Measures*. Secretariat of the International Plant Protection Convention (IPPC), Food and Agriculture Organization of the United Nations, Rome.

IPPC (2006) International Standards for Phytosanitary Measures, Publication No. 26: *Establishment of Pest Free Areas for Fruit Flies (Tephritidae)*. Secretariat of the International Plant Protection Convention (IPPC), Food and Agriculture Organization of the United Nations, Rome.

IPPC (2007) International Standards for Phytosanitary Measures, Publication No. 2: *Framework for Pest Risk Analysis*. Secretariat of the International Plant Protection Convention (IPPC), Food and Agriculture Organization of the United Nations, Rome.

IPPC (2009) International Standards for Phytosanitary Measures, Publication No. 15: *Regulation of Wood Packaging Material in International Trade*. Secretariat of the International Plant Protection Convention (IPPC), Food and Agriculture Organization of the United Nations, Rome.

IPPC (2010a) International Standards for Phytosanitary Measures, Publication No. 1: *Phytosanitary Principles for the Protection of Plants and the Application of Phytosanitary Measures in International Trade*. Secretariat of the International Plant Protection Convention (IPPC), Food and Agriculture Organization of the United Nations, Rome.

IPPC (2010b) International Standards for Phytosanitary Measures, Publication No. 5: *Glossary of Phytosanitary Terms*. Secretariat of the International Plant Protection Convention (IPPC), Food and Agriculture Organization of the United Nations, Rome.

MacLeod, A., Pautasso, M., Jeger, M.J. and Haines-Young, R. (2010) Evolution of the international regulation of plant pests and challenges for future plant health. *Food Security* 2, 49–70.

Schrader, G., Unger, J.G. and Starfinger, U. (2010) Invasive alien plants in plant health: a review of the past ten years. *EPPO Bulletin*, 40, 239–247.

WTO (1994) *The WTO Agreement on the Application of Sanitary and Phytosanitary Measures*. World Trade Organization, Geneva, Switzerland.

Part II

Pest Risk Analysis – Components and Applications

———————————

5 Terminology Used in Pest Risk Analysis

Christina Devorshak

5.1. Introduction – Why Are Definitions Needed?

One of the key benefits of harmonization is the development and use of a common language to address phytosanitary issues. As an indication of the importance of developing this common language, the IPPC devotes an entire standard to terminology: ISPM No. 5: *Glossary of Phytosanitary Terms* (IPPC, 2010). This standard is updated annually to incorporate new terms and serves as the one and only reference standard for terms used in phytosanitary work, including for pest risk analysis. The terminology used in this text follows the terminology in the *Glossary of Phytosanitary Terms* (ISPM No. 5), and definitions are often re-iterated in the text of this book to emphasize their importance.

Defining terms may seem to be a rather mundane matter, and the work can indeed be rather tedious. However, the importance lies in defining not only the terms, but in describing the concepts encompassed within the terms. When the IPPC was being revised in the 1990s, several fundamental concepts were debated by phytosanitary experts including the concepts of quarantine pest and pest risk analysis. And while many of the bigger concepts have been addressed, there still remain differences in

opinion and interpretation in many terms used in plant protection.

This chapter will highlight some of the most important terms and their related concepts; it will also discuss some terms and concepts that are still under debate, or other terms that may be encountered but for which harmonization is lacking.

5.2. 'Quarantine Pest' – the First Concept that Required Definition

The first, most fundamental concept is that of a 'quarantine pest'. The IPPC defines a quarantine pest as 'a pest of potential economic importance to the area endangered thereby and not yet present there, or present but not widely distributed and being officially controlled'. This definition features several key elements.

5.2.1. 'A pest'

'A Pest' (as defined by the IPPC) is 'any species, strain or biotype of plant, animal or pathogenic agent injurious to plants or plant products'. Thus, a pest, in the broadest sense, is an organism that is capable of harming plants.

5.2.2. 'Of potential economic importance'

'Of potential economic importance' introduces the concept that there must be some potential for economic harm associated with the pest. This is a fundamental concept – in the IPPC and the SPS Agreement, and is also reflected in requirements for pest risk analysis. The potential for economic harm is so important that additional guidance on this concept was produced both in ISPM No. 5 (as a supplement) and in ISPM No. 11 (IPPC, 2004) on how economic importance can be considered. One of the most notable clarifications provided in the supplemental guidance is that social effects as well as effects on the environment can be considered to be economic effects, and that methods exist to quantify economic harm resulting from environmental harm, in addition to monetary damage to commercial crops. More information on economic impacts is provided in Chapter 7.

5.2.3. 'To the area endangered thereby'

'To the area endangered thereby' introduces the concept that a specific area should be defined when discussing potential economic harm from a pest. An area, as defined by the IPPC is 'an officially defined country, part of a country or all or parts of several countries', therefore, the specific area is required to be clearly defined (explained in more detail later in this chapter). A pest may not be able to establish in the entire pest risk analysis area. This concept is reflected in requirements for pest risk analysis – the endangered area is the area where a pest can establish, and the pest risk analysis area is the area in relation to which the pest risk analysis is being conducted.

5.2.4. 'Not yet present there or present but not widely distributed and being officially controlled'

The portion of the statement 'not yet present there' introduces the concept that

in order for a country to consider a pest to be a 'quarantine pest' (and therefore subject to measures), it should not occur in that country. This relates to the principles of technical justification and national treatment in the SPS Agreement and the IPPC (previously discussed in Chapter 4). Measures against a pest are not technically justified if a pest already occurs in a country and is not being officially controlled. The portion of the statement 'or present but not widely distributed and being officially controlled' introduces the concept that in some cases, a pest may have a very limited distribution in a country, and the NPPO establishes measures to contain that pest (see definition for 'official control'), then measures can be justified and considered non-discriminatory.

These elements are absolutely essential to understand the concept of a 'quarantine pest', and play a central role in the conduct of pest risk analysis. Note that the definition of 'quarantine pest' does not make a distinction over taxon, meaning that the term applies to all relevant taxa including arthropods, other invertebrates, pathogens and weeds. The decision to undertake a pest risk analysis is typically based on a pest meeting the criteria for being a quarantine pest. In some instances, pest risk analysis can provide the evidence to support the status of a quarantine pest. Pest risk analyses are a means of providing evidence for the likelihood of establishment (related to 'endangered area') and the potential to cause negative impacts (related to 'potential economic harm'). Finally, a pest risk analysis can provide the basis for the determination of the strength of measures to be taken against it.

5.2.5. Terms related to quarantine pest

Terms related to 'quarantine pest' include:

- Pest;
- Regulated pest;
- Regulated non-quarantine pest;
- Contaminating pest;
- Non-quarantine pest.

These terms are inter-related. The definition of 'pest' is integrated into the definitions of all of the other terms; and the concept of whether the pest should be considered to be 'regulated' is integrated into regulated pest, quarantine pest and regulated non-quarantine pest. A 'regulated pest' includes both quarantine pests (defined above) and 'regulated non-quarantine pest'.

An RNQP is 'a non-quarantine pest whose presence in plants for planting affects the intended use of those plants with an economically unacceptable impact and which is therefore regulated within the territory of the importing contracting party'. RNQPs are a special case – these are pests that already occur in a country (non-quarantine pest), but may be regulated only on commodities such as nursery stock (on plants intended to be planted) if they are capable of causing economic harm to that commodity. In practice, RNQPs are dealt with infrequently and will not be covered to any great extent in this text.

Contaminating pests, also sometimes called 'hitchhiking pests' are pests that may not infest a commodity directly, but which 'hitchhike' or contaminate a commodity. A non-quarantine pest is a type of pest that does not meet the criteria for being a quarantine pest for an area.

For the purposes of this book, the reader should understand the relationship between a pest, a regulated pest, a quarantine pest and a non-quarantine pest (see Fig. 5.1).

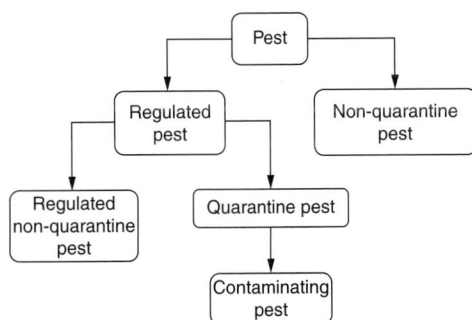

Fig. 5.1. Relationship of terms referring to 'pest' from the *Glossary of Phytosanitary Terms*. (From ISPM No. 5, IPPC, 2010.)

5.3. 'Pest Risk Analysis' – a Key Concept Requiring Definition

As we have just seen, the definition of quarantine pest includes the key elements of economic importance, presence and endangered area. These elements are also integrated into the definition of pest risk analysis – another major concept that required harmonization and definition. According to the IPPC, 'pest risk analysis' is 'the process of evaluating biological or other scientific and economic evidence to determine whether an organism is a pest, whether it should be regulated, and the strength of any phytosanitary measures to be taken against it'.

Prior to harmonizing the terminology, and thus the concepts, there was an array of terms used in the field of pest risk analysis often to convey the same meaning (see Box 5.1). There was little agreement on basic concepts, and terms varied significantly in their meaning and application – and the lack of consensus on what encompassed pest risk analysis was reflected by the lack of discipline in terminology.

The overarching principle behind risk analysis is 'technically justified', defined as 'justified on the basis of conclusions reached by using an appropriate pest risk analysis or, where applicable, another comparable examination and evaluation of available scientific information'. Recall that according to the IPPC, measures must be technically justified, or based on international standards, providing further impetus to defining the term and concept of pest risk analysis.

We see several key elements in the definition of pest risk analysis that can be examined in more detail.

5.3.1. 'The process of evaluating biological or other scientific and economic evidence'

'The process of evaluating biological or other scientific and economic evidence' indicates that pest risk analysis should be based on evidence, and that economic evidence in particular should be evaluated.

Box 5.1. Diverse terminology used in pest risk analysis showing lack of consensus over terms and concepts prior to the adoption of the *Glossary of Phytosanitary Terms* and ISPM No. 2.

Assessing the risk of artificial spread	Pest and pathogen risk analysis
Assessment of quarantine benefits	Pest assessment
Benefit–cost assessment	Pest introduction status
Bio-economic assessment of quarantine options	Pest population risk assessment
Biological analysis	Pest risk analysis
Biological assessment	Pest risk assessment
Biological hazards	Pest risk evaluation
Biological impact of the disease	Quarantine risk assessment
Biological risk assessment	Quarantine risk assessment analysis
Biological risks	Risk assessment
Commodity (carrier) risk assessment	Risk assessment analysis
Controlling new pest risk	Risk reduction measures
Economic analysis of the risk	Risk mitigation
Economic risk assessment	Risks are identified
Evaluation of potential disease risks	Scientific analysis of the biological risk
Evaluation of the risk of pest dissemination	Scientific and objective risk assessment
Identify areas of biological hazards	Scientific risk assessment
Managing the risks	Significant hazard of introducing pests
Organism status evaluation	Statistical risk assessment
Overall risk evaluation	Technical analysis of risk of diseases and pests

Adapted from: NAPPO, 1993

5.3.2. 'To determine whether an organism is a pest, whether it should be regulated'

'To determine whether an organism is a pest, whether it should be regulated' indicates that we should use relevant evidence (biological evidence) to decide if an organism is harmful to plants, and if it meets the definition of a quarantine pest (or a regulated non-quarantine pest – see ISPM No. 5 (IPPC, 2010) for more information). In other words, does the organism satisfy the definition for a regulated pest? The need to examine economic evidence is thus clarified since the definition for quarantine pests requires that the organism has the potential to cause economic harm.

5.3.3. 'And the strength of any phytosanitary measures to be taken against it'

'And the strength of any phytosanitary measures to be taken against it' indicates that there should be a clear linkage between the level of risk a pest may present (based on the evidence) and the level of risk management that may be applied.

5.3.4. Terms related to pest risk analysis

There are several other definitions related to pest risk analysis:

- Pest risk;
- Pest risk assessment;
- Pest risk management;
- PRA area;
- Pathway.

As we learned earlier, risk analysis is typically used as the broad term to refer to the practice of evaluating risk, managing risk and communicating about risk. Similarly, pest risk analysis includes both pest risk assessment and pest risk management, and the term 'pest risk' is used to describe a hazard in a phytosanitary context. Accordingly:

- 'pest risk' is 'the probability of introduction and spread of a pest and the magnitude of the associated potential economic consequences';
- 'pest risk assessment' is 'Evaluation of the probability of the introduction and spread of a pest and the magnitude of the associated potential economic consequences';

- 'pest risk management' is 'evaluation and selection of options to reduce the risk of introduction and spread of a pest'.

The 'PRA area' is the area that is under consideration in the pest risk analysis – this may be an entire country, a region, a part of a country or a group of countries.

The concept of a 'pathway' is important in pest risk analysis. It refers to 'any means that allows entry or spread of a pest'. Thus, in conducting a pest risk analysis, the first step is usually to define the relevant pathway(s) by which the pest can enter and/or spread. Depending on the issue, the pest risk analysis may be conducted on groups of pests that follow a single well-defined pathway (e.g. a commodity) or may examine a single pest that is capable of following one or more pathways. Risk communication is notably absent from the definition in the *Glossary of Phytosanitary Terms* and the ISPMs for pest risk analysis; this differs somewhat from the conventional view of risk analysis. Although communication is an important component in risk analysis, it is only briefly mentioned in ISPM No. 2 (IPPC, 2007). This is because of the nature of most pest risk analyses – being used to justify measures for products moving in international trade. As such, the pest risk analysis document is primarily used as a communication tool to inform decision makers and to justify measures to trading partners. Risk communication in the broadest sense is not part of the obligations a country must satisfy under either the IPPC or the SPS agreement. Thus it is a component of pest risk analysis, but is not explicitly described in the standards.

There is slightly different terminology between the IPPC and the SPS Agreement, but both agreements are consistent in their requirements and expectations. Likewise, the terminology developed for risk analysis under both Codex Alimentarius and the World Animal Health Organization differs slightly from what is now seen in the IPPC; however the major concepts, methods and practices are very similar. Table 5.1 provides a comparison of risk analysis terms currently used in the different international organizations. Refer also to Chapter 2 for a discussion on the variations on the risk models applied for static versus dynamic hazards.

5.4. What are 'Measures'?

Another group of terms that are important to understand, particularly when discussing pest risk management, are related to 'measures'. The broadest term that can be used is a 'phytosanitary measure', which is 'any legislation, regulation or official procedure having the purpose to prevent the introduction and/or spread of quarantine pests, or to limit the economic impact of regulated non-quarantine pests'. 'Phytosanitary measure' is therefore an umbrella term that includes several concepts related to managing pest risk. These concepts are each defined in their own right and include:

- *Phytosanitary action*: An official operation, such as inspection, testing, surveillance or treatment, undertaken to implement phytosanitary measures.
- *Phytosanitary legislation*: Basic laws granting legal authority to an NPPO from which phytosanitary regulations may be drafted.
- *Phytosanitary procedure*: Any official method for implementing phytosanitary measures including the performance of inspections, tests, surveillance or treatments in connection with regulated pests.
- *Phytosanitary requirement*: Official rule to prevent the introduction and/or spread of quarantine pests, or to limit the economic impact of regulated non-quarantine pests, including establishment of procedures for phytosanitary certification.
- *Provisional measure*: A phytosanitary regulation or procedure established without full technical justification owing to the current lack of adequate information. A provisional measure is subjected to periodic review and full technical justification as soon as possible.

- *Emergency measure*: A phytosanitary measure established as a matter of urgency in a new or unexpected phytosanitary situation. An emergency measure may or may not be a provisional measure.

What is important to understand about these related terms is that they collectively describe the laws and regulations, requirements, actions and procedures that NPPOs use every day to manage pest risk. Phytosanitary measure, the most general term, is inclusive of all of these other terms – and the more specific terms are used to refer to specific concepts. Figure 5.2 shows the relationship between the different types of measures, and in relation to international agreements and standards. When in doubt, 'phytosanitary measure' is always a correct term to use.

5.4.1. What's so special about provisional measures?

Note that provisional measures (and emergency measures) are special cases compared with other types of measures. Recall that the definition of pest risk analysis includes 'to determine…the strength of any measures to be taken'. This means, by default, that any measure should be technically justified, or based on a pest risk analysis. However, negotiators of both the SPS Agreement and the IPPC understood that countries may need to implement measures in the absence of a pest risk analysis – either due to a lack of information, or due to time-sensitive issues (such as in the case of an emergency) and therefore made allowances for countries to implement so-called provisional measures.

There is often confusion about the relationship between provisional measures and precautionary measures – in both cases, decisions are made in the absence of adequate information. However, the definition of 'provisional measure' states that it is 'subject to periodic review and full justification as soon as possible'. This creates the obligation for the country putting in place a provisional measure to actively seek out information for it to determine whether the measure is justified – usually by conducting a pest risk analysis. If the measure is not eventually (e.g. 'as soon as possible') supported by evidence and a pest risk analysis, then the measure should be amended accordingly.

In contrast, measures that are implemented as 'precautionary measures' are implemented in the absence of adequate information, but without creating a burden on the country implementing the measure to seek out information. This approach goes against the requirements of the SPS Agreement, which requires that measures be technically justified. When a precautionary approach is taken, or 'precautionary

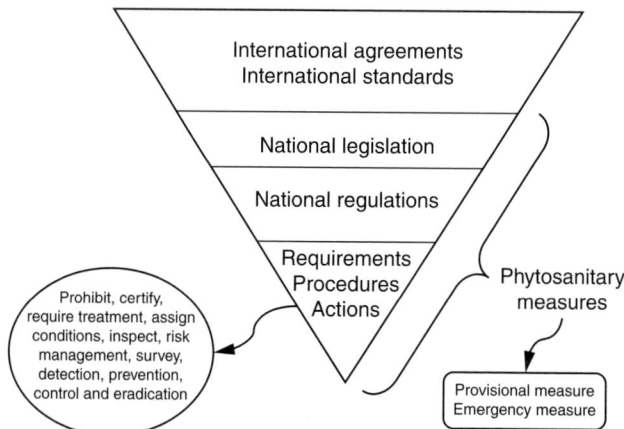

Fig. 5.2. Relationship between different types of measures defined by the IPPC.

measures' are implemented, a country opens itself up to challenge through the WTO dispute settlement process. The sections on dispute settlement and precaution (in Chapter 19) provide more detailed discussions of this point.

5.4.2. What about 'official'? How does that relate to NPPOs?

Running through all the definitions related to measures is the concept of 'official'. In this sense, official has a very specific meaning and refers to 'established, authorized or performed by a National Plant Protection Organization' where a National Plant Protection Organization (or NPPO) is an 'official service established by a government to discharge the functions specified by the IPPC'. Where a definition includes 'official', this means that the NPPO must be involved in the concept being described. For example, 'phytosanitary procedure' refers to 'any official method'– this means that the NPPO must either conduct the procedure itself, or the procedure must be under the explicit authorization of the NPPO.

5.5. What About Introduction and Spread?

We have seen the concepts surrounding 'pests', 'pest risk analysis' and 'measures' are central to understanding the IPPC, and indeed pest risk analysis in general. The next set of related terms form, along with 'quarantine pest', the backbone of the definition of pest risk analysis, and are therefore crucial to understand. Specifically, pest risk analysis, and the IPPC itself, are concerned with 'introduction' and 'spread' of pests. The term 'introduction', under the IPPC, specifically refers to 'The entry of a pest resulting in its establishment'. This definition contains two additional terms:

- *Entry:* Movement of a pest into an area where it is not yet present, or present but not widely distributed and being officially controlled;

- *Establishment*: Perpetuation, for the foreseeable future, of a pest within an area after entry.

Another important term is 'spread', which is the 'expansion of the geographical distribution of a pest within an area'. In this case, spread can refer to the movement of a pest from one area to another (e.g. between countries) as well as the movement of a pest within an area (e.g. spreading within a country after establishment).

As stated above, pest risk analysis is concerned with both the introduction (including entry and establishment) and spread of a quarantine pest. First, looking at all the definitions of these terms, we see that they are all inter-related. Second, we see that in defining these terms, the definition of pest risk analysis is quite rigorous.

Based on all the definitions together, we can now see that a pest risk analysis must address all the above concepts – namely it must evaluate whether a pest meets all the criteria – that it is a quarantine pest that has the potential to enter, establish and spread. Pests that do not meet all the criteria (e.g. a pest that is not capable of establishing in the endangered area) should not be subject to measures since the pest would not be likely to cause negative consequences.

It is hard to overstate the importance of understanding how all of these terms are related to each other, and how they relate collectively to the practice of pest risk analysis. All these terms will be featured repeatedly in discussions throughout this book and we recommend that readers ensure they have a working knowledge of them.

5.6. Types of Areas

Concepts related to 'areas' feature repeatedly in the IPPC, as well as in pest risk analysis. The broadest term is 'area', which is 'an officially defined country, part of a country or all or parts of several countries'. It is important to recognize that this definition is very broad – it can refer to limited geographic areas (e.g. a specific growing area within an individual state) all the way up to large regions encompassing several countries (e.g. Europe). Pest risk analyses should define what areas are being addressed in the pest risk analysis, and areas that are subject to pest risk management; therefore, it is crucial to understand the distinctions being made when specific terms related to 'areas' are being used. Other similar terms include:

- *Endangered area:* An area where ecological factors favour the establishment of a pest whose presence in the area will result in economically important loss.
- *PRA area:* Area in relation to which a pest risk analysis is conducted.
- *Pest free area:* An area in which a specific pest does not occur as demonstrated by scientific evidence and in which, where appropriate, this condition is being officially maintained.
- *Area of low pest prevalence:* An area, whether all of a country, part of a country or all or parts of several countries, as identified by the competent authorities, in which a specific pest occurs at low levels and which is subject to effective surveillance, control or eradication measures.
- *Controlled area*: A regulated area which an NPPO has determined to be the minimum area necessary to prevent spread of a pest from a quarantine area.
- *Regulated area*: An area into which, within which and/or from which plants, plant products and other regulated articles are subjected to phytosanitary regulations or procedures in order to prevent the introduction and/or spread of quarantine pests or to limit the economic impact of regulated non-quarantine pests.
- *Quarantine area*: An area within which a quarantine pest is present and is being officially controlled.
- *Pest status (in an area)*: Presence or absence, at the present time, of a pest in an area, including where appropriate its distribution, as officially determined using expert judgement on the basis of current and historical pest records and other information.

In addition to these terms that relate to pest risk analysis (including, in particular, pest risk management), there are several other terms that are used by NPPOs to describe other 'areas' that come under regulation or other measures. These include:

- Protected area;
- Place of production;
  - ◦ Pest free place of production;
  - ◦ Pest free production site;
- Country of origin;
- Field;
- Habitat;
- Point of entry.

5.7. Pathways, Plants, Commodities, Conveyances

Throughout this book, we will discuss the ways pests can move (pathways), the things pests are moving on and how to analyse risk associated with pest movement. Of course, in order to define how pests move, we need to define the things that pests are moving on. The IPPC is concerned with organisms that are injurious to plants and plant products. And not only do we wish to protect plants and plant products from pests, we need to understand how plants and plant products can serve as pathways for pests.

According to the IPPC, 'plants' are 'Living plants and parts thereof, including seeds and germplasm'. This is a very broad definition – the key feature is that it refers to *living* plants (or fresh parts of living plants), and is therefore distinguished from other types of 'plant products'. Some specific types of 'plants' that are further defined in the *Glossary of Phytosanitary Terms* include:

- Seeds;
- Germplasm;
- Plants for planting;
- Plants *in vitro*;
- Bulbs and tubers;
- Cut flowers and branches;
- Fruits and vegetables.

'Plant products' are 'unmanufactured material of plant origin (including grain) and those manufactured products that, by their nature or that of their processing, may create a risk for the introduction and spread of pests'. 'Plant products' differ from 'plants' in that plant products are typically not a fresh commodity (e.g. flowers or fruit), and may be processed in some way (e.g. bamboo baskets, wood handicrafts, potpourri, grain, etc.). Some other examples of plant products defined in the Glossary are:

- Grain
- Stored product
- Wood
- Wood packaging material
- Round wood
- Sawn wood
- Raw wood
- Processed wood material
- Bark-free wood
- Debarked wood
- Dunnage

There are several specific definitions under the category of 'wood'. This is because wood, and the wood products identified above, represent a special category of products. Wood and wood products may be traded (e.g. logs sold to mills, wooden handicrafts, etc.), or wood products may also be manufactured into pallets, crates and other packing material used in shipping. Depending on the level of processing (e.g. raw, sawn, round, processed, bark-free) and the specific use of the wood, it can present unique risks. In fact, managing risks associated with wood packing material is the subject of its own standard, ISPM No. 15: *Guidelines for Regulating Wood Packaging Material in International Trade.*

There are other important terms used when discussing trade in particular. Article I of the IPPC states that 'where appropriate, the provisions of this Convention may be deemed by contracting parties to extend, in addition to plants and plant products, to storage places, packaging, conveyances, containers, soil and any other organism, object or material capable of harbouring or spreading plant pests, particularly where international transportation is involved.' Essentially, the IPPC, its provisions and the application of pest risk analysis applies to any means by which quarantine pests can move from one place to another, and these means are not limited only to plants.

Figure 5.3 shows several related terms that are used when referring in particular to trade in plants and plant products. The term 'regulated article' is used to convey this concept and is defined as 'any plant, plant product, storage place, packaging, conveyance, container, soil and any other organism, object or material capable of harbouring or spreading pests, deemed to require phytosanitary measures, particularly where international transportation is involved.' In addition to 'plants' and 'plant products' (and their related terms), there are other types of regulated articles that are further defined in the Glossary of Phytosanitary Terms, including:

- Commodity
- Commodity class
- Consignment
- Lot
- Packaging.

The term 'commodity' is used throughout this book, so readers should be familiar with this definition as well. A 'commodity' is 'a type of plant, plant product, or other article being moved for trade or other purpose'. Consignments and lots refer to specific quantities of commodities being moved in trade; generally, a consignment may be comprised of one or more lots.

5.8. Terms surrounding invasive species and environmental concerns

The term 'invasive species' is used frequently in the literature to refer to various types of unwanted organisms. Like 'risk analysis', there are many variations on the theme, leading to confusion and lack of agreement over terms and concepts. The phytosanitary community has sought to provide structure and discipline around many of these same concepts by defining specific terms that are used by and between NPPOs. Thus, ecologists may speak of invasive species, alien species, invasive alien species, alien invasive species, invaders, etc., while the phytosanitary community speaks about 'pests' and 'quarantine pests'.

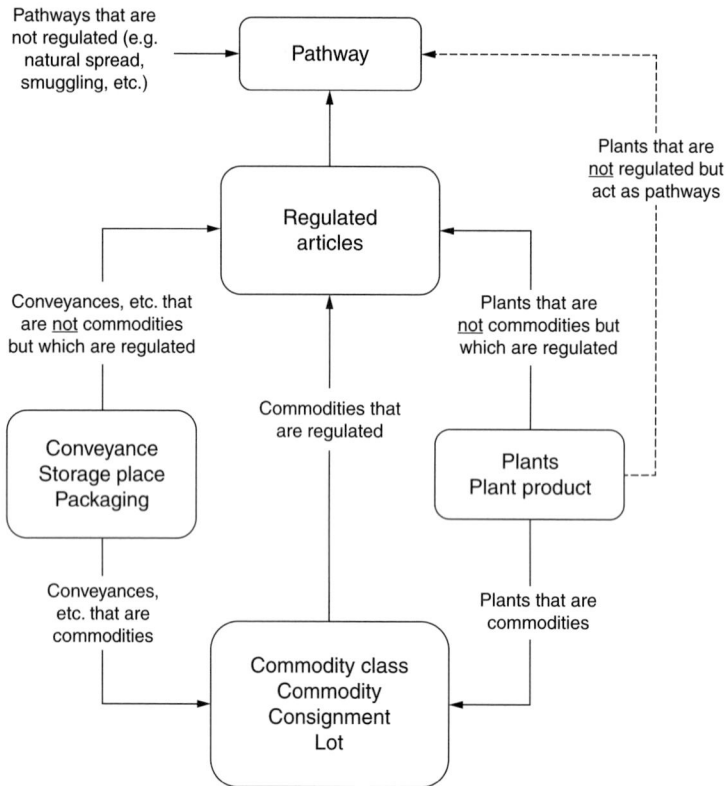

Fig. 5.3. Relationship between key terms related to pathway, plants, regulated articles and plants as defined in the *Glossary of Phytosanitary Terms*. (From ISPM No. 5, IPPC, 2010.)

Another international convention, the CBD is concerned with, among other things, invasive species (CBD, 1992). The CBD and the IPPC overlap in scope to the extent that invasive species (under the CBD) meet the definition of 'quarantine pest' under the IPPC. Most countries that are signatory to the IPPC are also signatory to the CBD. This creates the potential for there to be confusion over different concepts and terms between the two agreements, as well as the potential for conflicting obligations. Recognizing that additional guidance was needed, the IPPC and CBD agreed to recognize either other's contributions – to that end, the IPPC added an Appendix (Appendix 1) to ISPM No. 5 (IPPC, 2010) on interpreting similar terms and concepts between the two agreements (see also Chapter 20 for more on the CBD and invasive species, and how these relate to the IPPC).

In some cases, identical terms are used but have been defined very differently. Some of the terminology appearing in the CBD includes:

- Alien species;
- Invasive alien species;
- Introduction;
- Intentional introduction;
- Unintentional introduction;
- Establishment;
- Risk analysis.

5.8.1. Introduction, quarantine pests and alien species

A notable divergence in concepts surrounds the term 'introduction'. In the IPPC, introduction is the result of both entry and establishment of a pest, where the term 'entry' refers to the movement of a pest into a new area. This movement can be through any means – human-mediated or through natural spread. The IPPC is primarily concerned with preventing the introduction of quarantine pests (pests that are not yet present in an area) – through pest risk analysis and the application of phytosanitary measures.

The CBD definition of introduction is 'The movement by human agency, indirect or direct, of an alien species outside of its natural range (past or present). This movement can be either within a country or between countries or areas beyond national jurisdiction.'

The CBD definition of 'alien species' is 'a species, subspecies or lower taxon, introduced outside its natural past or present distribution; includes any part, gametes, seeds, eggs or propagules of such species that might survive and subsequently reproduce'.

Thus, an alien species is considered an 'alien species' only after it has been introduced, according to the CBD definition. In contrast, the IPPC considers a pest a 'quarantine pest' before it is introduced, and therefore subject to measures to prevent introduction (according to the IPPC definition). Under the CBD definition of 'introduction', it is seemingly implied that the alien species must first be introduced to be considered alien (and thus subject to regulation), and that further movement of the organism (either through spread or additional introductions) leads to the additional introductions of the alien species. The CBD definition is also specific in its reference to 'movement by human agency', whereas the IPPC is concerned with the movement of pests by any means (human or natural). Although it is not explicit, interpretations of the CBD imply that an initial introduction (in the IPPC sense) would equate to 'introduction' in the CBD sense as well, although this is not necessarily clear based solely on the relevant CBD definitions.

Another key point to understand with respect to introduction is the role of 'establishment'. The IPPC definition includes 'perpetuation, for the foreseeable future'. The CBD definition of establishment is 'the process of an alien species in a new habitat successfully producing viable offspring with a likelihood of continued survival'. There

are two differences between these definitions. First, 'establishment' in the IPPC is an end-point, but in the CBD it is a process. Second, in the IPPC, establishment includes the concept that the pest will survive in the new area 'for the foreseeable future'; in the CBD, this concept is absent – the process of establishment leads to a population that may or may not survive into the future.

5.8.2. Risk analysis (CBD) and pest risk analysis (IPPC)

Earlier in this chapter, we discussed the definitions and concepts surrounding the term 'pest risk analysis', highlighting that the definition provides structure to the concept of pest risk analysis. Pest risk analysis examines likelihood of entry and establishment of a pest, spread of the pest, and evaluates consequences (in particular economic consequences).

The CBD definition of risk analysis is '1) the assessment of the consequences of the introduction and of the likelihood of establishment of an alien species using science-based information (i.e., risk assessment), and 2) the identification of measures that can be implemented to reduce or manage these risks (i.e., risk management), taking into account socio-economic and cultural considerations'.

Again we see some differences in how similar terms can be defined to convey different concepts. The IPPC is explicit that consequences should be economic consequences. It goes on to define what economic consequences are and how they can be measured. And although environmental and social effects can be difficult to quantify, the IPPC includes environmental and social consequences under the umbrella of economic consequences. Chapters 7 and 11 provide information and guidance on how economic considerations are included in pest risk analyses. In addition, pest risk analysis is typically done before a pest is introduced, in order to determine whether it should be regulated (as a quarantine pest) and the strength of measures to be applied against it.

By contrast, the CBD definition of risk analysis refers to 'assessment of consequences' but does specify what consequences should be considered. In addition, it refers to 'probability of establishment' – it is not clear whether an alien species must first be introduced (with or without establishment) before the process of a risk analysis would be triggered. It examines the 'probability of establishment' (possibly post-introduction), but does explicitly include evaluating the 'likelihood of introduction' (pre-introduction).

Earlier in this section, we provided a brief explanation of 'pathway', stating that it is any means that allows the introduction or spread of a pest. The process of pest risk analysis under the IPPC considers that there must first be a viable pathway for a pest to be able to enter and establish (i.e. be introduced), spread and have consequences. Presumably because the CBD definition of 'introduction' includes 'by human agency', the pathway component is at least partially defined and perhaps this accounts for why likelihood of introduction is not explicitly mentioned in the CBD definition of risk analysis.

The IPPC and CBD differ slightly in how they address 'invasive' organisms. The CBD has a separate definition for invasive alien species, which is 'an alien species whose introduction and/or spread threaten biological diversity'. The IPPC does not make a distinction with regard to invasive characteristics – if an organism meets the definition of a quarantine pest, it is subject to regulation. And although 'invasiveness' is not explicitly defined, invasive characteristics are addressed in pest risk analysis by analysing specific aspects of likelihood of entry, establishment and spread.

5.9. Summary

It is easy to see that there can be considerable variation in how terms are used and defined, and how concepts are applied between even closely related fields. The IPPC and CBD share some common aims (preventing harmful organisms from moving from one place to another), but their respective languages and approaches are quite different. This divergence in terms and concepts can prove problematic – countries integrate terms from both agreements into national laws and regulations, and they use terms from both agreements to communicate with other countries. This only serves to highlight the importance of harmonization, international standards and the development of a common terminology used by countries in communicating about risk.

References

CBD (1992) *Convention on Biological Diversity. Text and Annexes*. Secretariat of the Convention on Biological Diversity/UNEP, Montreal.

FAO (2000) *Multilateral Trade Negotiations in Agriculture: A Resource Manual*. Food and Agriculture Organization of the United Nations, Rome. Available online at http://www.fao.org/DOCREP/003/ X7351E/X7351e00.htm#TopOfPage, accessed 20 December 2011.

IPPC (2004) International Standards for Phytosanitary Measures, Publication No. 11: *Pest Risk Analysis for Quarantine Pests Including Analysis of Environmental Risks and Living Modified Organisms*. Secretariat of the International Plant Protection Convention (IPPC), Food and Agriculture Organization of the United Nations, Rome.

IPPC (2010) International Standards for Phytosanitary Measures, Publication No. 5: *Glossary of Phytosanitary Terms*. Secretariat of the International Plant Protection Convention (IPPC), Food and Agriculture Organization of the United Nations, Rome.

NAPPO (1993) *International Approaches to Plant Pest Risk Analysis Proceedings of the APHIS/NAPPO International Workshop on the Identification, Assessment, and Management of Risks Due to Exotic Agricultural Pests*, Alexandria, Virginia, 23–25 October 1991. Ottawa, Canada. Bulletin (North American Plant Protection Organization) no. 11.

WTO (1994) *The WTO Agreement on the Application of Sanitary and Phytosanitary Measures*. World Trade Organization, Geneva.

6 Information and Pest Risk Analysis

Christina Devorshak

6.1. Introduction

Pest risk analysis requires resources – one of the most important resources is information. Information and evidence are the most basic components of virtually every pest risk analysis, and this chapter will provide an overview of:

- International requirements related to information and evidence.
- What kind of information is needed to conduct pest risk analysis.
- What are sources of information.
- What to do when information is lacking.
- Good practices for gathering and using information.

6.2. International Requirements Related to Information

In earlier chapters, we discussed how – under both the SPS Agreement and the IPPC – countries are obligated to base measures on pest risk analysis (also referred to as risk assessment under the SPS Agreement). Both agreements discuss requirements for information and evidence in relation to risk analysis.

Article 5 of the SPS Agreement addresses risk assessment. Article 5.2 of the SPS Agreement states:

In the assessment of risks, Members shall take into account available scientific evidence; relevant processes and production methods; relevant inspection, sampling and testing methods; prevalence of specific diseases or pests; existence of pest- or disease-free areas; relevant ecological and environmental conditions; and quarantine or other treatment (WTO, 1994).

This article clearly creates the obligation that countries should use scientific and technical information in conducting risk analysis, and then it further defines some specific types of information that should be considered.

The IPPC mirrors the requirements found in the SPS Agreement with regard to evidence-based risk analysis. Pest risk analysis requires that biological, scientific and economic evidence is evaluated. Furthermore, the IPPC includes several provisions related to the exchange of information between NPPOs (FAO, 1997). Article IV (General Provisions Relating to the Organizational Arrangements for National Plant Protection) and Article VII (Requirements in Relation to Imports) include requirements for the provision of information on pest status, surveillance and for the distribution of information on regulated pests, both nationally and between trading partners. Article VIII of the IPPC (FAO, 1997) is explicit regarding requirements for information exchange, stating:

1. The contracting parties shall cooperate with one another to the fullest practicable extent in achieving the aims of this Convention, and shall in particular:

(a) cooperate in the exchange of information on plant pests, particularly the reporting of the occurrence, outbreak or spread of pests that may be of immediate or potential danger, in accordance with such procedures as may be established by the Commission;

(b) participate, in so far as is practicable, in any special campaigns for combating pests that may seriously threaten crop production and need international action to meet the emergencies; and

(c) cooperate, to the extent practicable, in providing technical and biological information necessary for pest risk analysis.

Each contracting party shall designate a contact point for the exchange of information connected with the implementation of this Convention.

6.3. Official Contact Points and Official Information

The requirement for establishing an 'Official Contact Point' under the IPPC is important because it means that every country should have in a place a single point of contact that is responsible for providing official information. Official information is specific information that comes from the NPPO of a given country and is distinguished from scientific evidence or other types of evidence considered in a pest risk analysis. The IPPC (FAO, 1997; IPPC, 2010a) identifies certain types of information countries are required to report to other contracting parties, to the IPPC Secretariat and/or to RPPOs:

- IPPC Official Contact Points (Article VIII.2);
- Official pest report (Article VIII.1a);
- Description of the NPPO (Article IV.4);
- Legislation (Article VII.2b);
- Entry points (Article VII.2d);
- List of regulated pests (Article VII.2i);
- Emergency actions (Article VII.6).

Official information also includes information on pest status, regulated pests, quarantines, pest free areas and other phytosanitary measures that a country has in place. Scientific and other technical information may be included in official information. However, official information is provided solely by the NPPO, through the official contact point – usually to other NPPOs, but also to relevant international or regional organizations (e.g. the IPPC, SPS Committee or RPPOs) (Devorshak and Griffin, 2002).

Official information plays an important role in pest risk analysis. One of the primary questions addressed in all pest risk analyses is the presence or absence of a pest in an area, or the pest status. Information obtained through surveillance provides the basis for determining pest status, and this information should be reported by NPPOs (FAO, 1997; Devorshak and Griffin, 2002). Likewise, NPPOs should determine which pests are considered regulated, and provide that information to trading partners on request, particularly for the purposes of pest risk analysis. Other official information may play a role in pest risk analysis, including the declaration of pest free areas or areas of low pest prevalence, and information on any official programmes an NPPO has in place to manage, contain or eradicate pests (FAO, 1997; Devorshak and Griffin, 2002).

6.4. Guidance from International Standards on Information

Given the importance of information and evidence, it is not surprising to learn that several international standards address information and the exchange of information between NPPOs in detail. ISPM No. 1 (*Phytosanitary Principles for the Protection of Plants and the Application of Phytosanitary Measures in International Trade*, IPPC, 2010a) discusses information under the basic principles of transparency, technical justification and cooperation. It is also included under the specific principles of pest risk analysis, pest reporting and information exchange (FAO, 1997).

Pest status has such an important role in pest risk analysis in particular that it is the subject of its own standard, ISPM No. 8: *Determination of Pest Status in an Area*. This standard describes how countries should list pests that occur (or do not occur) in their territories, including present, absent or other types of status (e.g. present, under eradication or absent, pest eradicated, etc.). The status of a pest in an area is what will determine whether a pest is considered a quarantine pest, and whether it should be considered in a pest risk analysis. The standard also includes a description of the types and quality of information that can be used to describe pest status (see Table 6.1, reproduced from ISPM No. 8) (IPPC, 1998).

Likewise, ISPM No. 6 (*Guidelines for Surveillance*) provides guidance on how countries should conduct surveys for pests, and, importantly, how the results of those surveys should be reported. All countries use information obtained from surveys, as well as pest status, to decide whether pests need to be analysed in pest risk analyses (FAO, 1997). Therefore, these standards provide essential guidance on information, in particular official information, which is critical for conducting pest risk analyses (FAO, 1997).

Information may be qualitative (descriptive and/or with descriptive ratings) or quantitative (numerical, statistical or monetary). In practice, we often make use of both kinds of information in a pest risk analysis, regardless of whether the pest risk analysis itself is qualitative or quantitative

Table 6.1. Information sources used in preparing pest risk analyses. (See also ISPM No. 8: *Determination of Pest Status in an Area* (IPPC, 1998) for more information.)

Information source (Adapted from ISPM No. 8 (IPPC, 1998))

1. Collector/ identifiers	2. Technical identification	3. Location and date	4. Recording/ publication
a. Taxonomic specialist	a. Discriminating biochemical or molecular diagnosis (if available)	a. Delimiting or detection surveys	a. NPPO record/RPPO publication (where refereed)
b. Professional specialist, diagnostician	b. Specimen or culture maintained in official collection, taxonomic description by specialist	b. Other field or production surveys	b. Scientific or technical journal refereed
c. Scientist	c. Specimen in general collection	c. Casual or incidental field observation, possibly with no defined location/ date	c. Official historical record
d. Technician	d. Description and photo	d. Observation with/in products or by-products; interception	d. Scientific or technical journal non-refereed
e. Expert amateur	e. Visual description only	e. Precise location and date not known	e. Specialist amateur publication
f. Non-specialist	f. Method of identification not known		f. Unpublished scientific or technical document
g. Collector or identifier not known			g. Non-technical publication; periodical/ newspaper
			h. Personal communication; unpublished

↑ Level of reliability from highest to lowest ↓

(see Chapters 9 and 10 for more information on qualitative and quantitative pest risk analyses). The two driving factors that typically determine whether an analysis uses qualitative or quantitative information are necessity and availability.

In cases where little or no quantitative information is available, then a quantitative pest risk analysis may not be practical or possible. In many cases, a qualitative pest risk analysis is sufficient – quantitative data may not be necessary, but could be used and incorporated into the qualitative analysis. Pest risk analyses are rarely purely qualitative or purely quantitative – they are usually a combination of different methods and techniques. It stands to reason, therefore, that most of the time we use a mix of qualitative and quantitative information.

It is important to understand that some information can be more or less reliable, depending on the source of information. In general, data and information from refereed, reputable scientific journals are considered to be the highest quality and most reliable. Information from official NPPO records – such as official pest reports, interception records and distribution records – is also considered to be reliable information. Identifications/diagnoses performed by specialists are more reliable, for instance, than identifications done by amateur collectors. Information that comes from sources that are not peer-reviewed, or information which cannot be independently verified is not considered reliable and should be used with caution. Even some commonly used sources of information on the internet may not be reliable – therefore verifying information and using original sources is essential in the information-gathering stage. No matter the information source, the information used should be clearly cited, and any conclusions that are drawn should be clearly linked to the available evidence.

6.5. What Types of Information are Needed for Pest Risk Analysis

The type of pest risk analysis will usually dictate what types of information are needed – and the information can be qualitative or quantitative. Table 6.2 provides a list of general types of information that may be needed, according to the subject of the analysis. The most basic unit of most pest risk analyses is usually a specific pest. It is therefore not surprising that the most common types of information we need to do pest risk analysis cover basic biological information about pests.

Information needed can include how to identify the pest (or diagnostic information), life history, ecology, host range, climatic conditions the pest requires, its global distribution and any other information about how the pest lives, feeds, reproduces or disperses. We also need information on what type and level of damage the pest causes (both to crops and wild flora) within its native range, as well as in areas where it has been introduced. Depending on the type of pest, we may also want to gather information on pathways the pest is known to follow – including interception records from ports of entry, if available (FAO, 1997; Devorshak and Griffin, 2002). Additional guidance on information for pest risk analysis has been considered by EPPO – they have a regional standard *Checklist of Information Required for Pest Risk Analysis (PRA)* (EPPO, 1998).

If we are conducting a pest risk analysis for the import or export of a commodity, we also need information on how the commodity is produced, processed, handled and shipped. In some cases, a country will consider the unmitigated risk (or the risk without any pre- or post-harvest practices) for a commodity and then compare the unmitigated risk to the level of risk associated with the commodity taking into account any of the handling practices (IPPC, 2004). This is because many pre- and post-harvest practices will have an effect on pest prevalence in the commodity, even though the processing is not directly intended to manage pest risk and the effects may be difficult to measure precisely. For instance, it is a common industry practice to wash mango fruits after harvest, primarily to remove sap (which may cause staining) from the skin of the mango. Although the purpose of washing the fruit is not done to reduce pest risk, it will likely dislodge most external pests and

Table 6.2. Types of information needed according to subject.

Information type	Subject			
	Invertebrates[a]	Plant Pathogens[b]	Weeds	Commodities or other pathways
Identification, taxonomy, diagnostics, morphology	✓	✓	✓	✓
Life history/development	✓	✓	✓	✓
Phenology (seasonality, time of growth, etc.)	✓	✓	✓	✓
Ecology	✓	✓	✓	✓
Global distribution	✓	✓	✓	✓
Climate data (including production areas)	✓	✓	✓	✓
Host range/alternate hosts	✓	✓	±	–
Dispersal mechanisms/ ability to spread naturally or through human assisted means/vectors	✓	✓	✓	–
Economic and other impacts	✓	✓	✓	–
Control measures/ management	✓	✓	✓	✓
Production areas and practices[c]	–	–	–	✓
Harvest practices[c]	–	–	–	✓
Post-harvest practices[c]	–	–	–	✓
Degree of processing[c]	–	–	–	✓
Packaging and shipping[c]	–	–	–	✓
Known pathways (including interception records)	✓	✓	✓	–
Inspection, detection and surveillance methods	✓	✓	✓	✓

[a]Invertebrates includes arthropods and molluscs.
[b]Plant pathogens includes viruses, viroids, bacteria, fungi, chromistans, phytoplasmas and nematodes.
[c]Includes effects on pests that may be associated with the commodity.
±Parasitic plants may have specific hosts however most weeds are free-living.

this practice may be taken into account in the pest risk analysis.

6.6. Resources – Where Do They Come From and Where Do We Find Them?

Now that we have an idea of what kind of information we need to gather, we can start thinking about where to gather the information. Access to a university, government or museum library is essential for risk analysts to be able to conduct their research. In addition, access to the internet – with online journals and countless databases – is vital.

Those two resources will provide the majority of information needed to conduct science-based pest risk analyses.

The information itself may originate from a variety of sources – scientific journals may be published by scientific or other professional societies (e.g. Entomological Society of America journals); NPPOs or RPPOs may publish reports or pest records; universities may publish scientific information or extension data sheets; industry groups may publish information on crop production practices. Bear in mind that the NPPOs of trading partners (e.g. an exporting country) may be the best sources of information with regard to pest distribution or industry practices. Information and data may be assembled and provided by:

- Industry groups – farmers, co-ops, councils, etc.;
- Scientific societies;
- NPPOs and RPPOs, including trading partners;
- Universities, institutes and academia
- Extension services;
- Science museums;
- International organizations (e.g. CABI, FAO, UNCTAD, CGIAR centres, etc.);
- Subject matter experts.

Some specific resources provided by such organizations that can be used for pest risk analysis (and are usually available via a library or the internet) include:

- Technical and scientific books on specific topics;
- Trade/economic reports for commodities or countries;
- Scientific and other peer-reviewed journals – hard copy and electronic journals;
- Government reports (NPPO records, reports or analyses);
- Compendia – prepared by scientific organizations or available through other sources (e.g. CABI Crop Protection Compendium);
- Journal indexing services (e.g. CAB Abstracts, Web of Science, AGRICOLA, etc.);
- Information from RPPOs (e.g. pest reports, pest data sheets, etc.);
- Various databases on pests or commodities/internet databases (e.g. ScaleNET);
- General or specific search engines (e.g. Google Scholar, etc.).

6.7. Information Gaps – What Are the Options?

In spite of the huge number, and wide availability, of information resources, it is not uncommon for there to be relatively little published information on a pest. If one considers that the universe of pests includes insects and other arthropods, plants (e.g. weeds) and pathogens – and that these pests can come from any region of the world, it is not surprising to learn that many pests have not been researched in depth. In some cases, we only have a species name, geographic location(s) where the pest has been found, and perhaps a limited description of its host range. In the case of new or undescribed species, even that level of information may be lacking. And even for pests that are well understood, there may still be information gaps.

Recall from Chapters 2 and 4 that under the SPS Agreement there is no qualification on the amount of information needed in order to conduct a pest risk analysis (WTO, 1994). And while the SPS Agreement allows countries to put in place provisional measures in the absence of a pest risk analysis, they are still obligated to gather information and conduct an analysis as soon as possible. A lack of information is not, in itself, sufficient justification for NOT conducting a risk analysis. In addition, we also learned that one of the benefits of conducting a pest risk analysis is that it may serve to identify where we have gaps in knowledge and where further research may be the most useful. In practical terms, it means that we have to have options to deal with situations where information may be lacking.

A common strategy for dealing with lack of information is to extrapolate from one situation to another – that is, we may make certain assumptions in the analysis. We can use information from an analogous situation (e.g. a pest, a pathway, etc.) that we understand well, and for which we have evidence, and apply that knowledge to a situation for which we may have relatively little information (see Box 6.1 for examples). However, when we use this approach, any and all of the assumptions we make must be clearly documented. This ensures transparency, highlights our uncertainty and documents information gaps that could be addressed if and when new information becomes available.

6.8. What About Expert Judgement?

Another option for dealing with information gaps is to use expert judgement. Expert judgement is the expression of opinion, based on

> **Box 6.1.** Examples of extrapolating information.
>
> **1.** We are working on a species of bark beetle, and we decide to examine what we know about other species of bark beetle (provided that we believe that all species of bark beetle in that taxon behave similarly) and use that information as well.
> **2.** We are estimating whether a pathogen could survive in an area, based on where it exists in its current distribution. We may extrapolate that the pathogen could survive in climate zones similar to where it already exists, but we do not know that unequivocally – we have made an assumption based on available evidence.
>
> In practice, this is a relatively common approach to working in an information-poor environment. However, it is absolutely essential that the analyst documents that they have made certain assumptions along the way.

knowledge and experience. As risk analysts, we may use our own judgements in an analysis, or we may also solicit the opinion of specialists and experts (Burgman *et al.*, 2006). Indeed, one can argue that the use of expert judgement in risk analysis is unavoidable (Gray *et al.*, 1991) – if we had perfect knowledge of a hazard, we probably wouldn't be doing an analysis! ISPM No. 11 notes that expert judgement can be used, but should be clearly noted in the analysis.

Throughout an analysis, we usually have to make a variety of judgements – this helps us cope with uncertainty and lack of data, but may add to the overall uncertainty of the analysis. Experts may be able to provide subjective judgements for:

- The nature of a particular hazard;
- The potential for a particular event to occur;
- The potential impacts associated with a hazard;
- The nature of a system (e.g. provide a conceptual model for the hazard to be realized);
- The level of uncertainty associated with any of the components of the analysis (Burgman *et al.* 2006).

They may make these judgements based on their own experience, indirect evidence or extrapolating from analogous situations when specific data are lacking. However, we must understand that opinions obtained through expert judgement are subjective. An example for applying expert judgement would be if we lack data on the possible impacts of a potential weed we are evaluating and decide to consult an expert. That expert would draw on experience, and provide an opinion of the impacts we might expect. As with the examples above, it is imperative that the use of expert judgement in an analysis is explicit and clearly documented.

There are other, more complex methods for dealing with lack of information, including for conducting quantitative analyses. Such methods include the application of Bayesian Belief Networks (BBNs), analytical hierarchy process (AHP) and other qualitative, semi-quantitative and quantitative methods (Dambacher *et al.*, 2007). Generally, these more complex methods are used in special situations – they may be resource-intense and time-consuming to employ and require specific expertise to apply that are beyond the scope of this text.

6.9. Summary – Good Practices for Gathering and Using Information for Pest Risk Analysis

As risk analysts, we must be mindful that our work is always subject to scrutiny – from scientists, trading partners, industry groups, stakeholders and other interested parties. Because evidence and information form the backbone of any analysis, we should do our utmost to ensure that we gather the best information possible in order to support our analysis. And not only do we want to have the best information we can find, we want to use that information accurately and transparently. The rules below outline good practices for gathering, analysing and using information and evidence in pest risk analyses (Holtz, 2011, Rules to live by for preparing defendable risk assessments – experiences learned from two 'highly influential' risk assessments. Plant Epidemiology and Risk Analysis Laboratory, USDA-APHIS-PPQ, PowerPoint presentation).

1. Always use the highest quality information available – remember that peer reviewed journals published by reputable groups are usually the best sources.

2. Avoid citing information that you are not able to validate first hand – if using a review article, compendium or database, it is best to go back to the original sources used and use the original sources directly.

3. All scientific or technical statements should be supported by appropriate scientific or technical references.

Statements and conclusions in the analysis must be supported by appropriate evidence.

Avoid making statements or conclusions that go beyond what the original evidence can support.

4. Ensure that all references are properly cited. It is also highly advisable to retain copies of references for record-keeping. If necessary, the original references can be consulted again if questions are raised about the analysis.

5. If possible, use peer review for your analyses – either (or both) internal and external reviews ensure accuracy and scientific rigour, and may catch errors, mistakes in judgements or oversights before the pest risk analysis is finalized.

References

Burgman, M., Fidler, F., McBride, M., Walshe, T. and Wintle, B. (2006) Eliciting expert judgment: literature review. Presentation, the University of Melbourne, Melbourne.

Dambacher, J.M., Shenton, W., Hayes, K.R., Hart, B.T. and Barry, S. (2007) Qualitative modeling and Bayesian Network Analysis for risk-based biosecurity decision making in complex Systems. Presentation. The University of Melbourne, Melbourne.

Devorshak, C. and Griffin, R. (2002) The role and relationship of official information. In: Hallman, G.J. and Schwalbe, C.P. (eds) *Invasive Arthropods in Agriculture: Problems and Solutions.* Science, Enfield, NH, pp. 51–70.

EPPO (1998) *Guidelines on Pest Risk Analysis Check-list of Information Required for Pest Risk Analysis (PRA).* PM 5/1(1). EPPO, Paris.

FAO (1997) *New Revised Text of the International Plant Protection Convention.* Food and Agriculture Organization of the United Nations, Rome.

Gray, G.M., Allen, J.C., Burmaster, D.E., Gage, S.H., Hammitt, J.K., Kaplan, S., Keeney, R.L., Morse, J.G., Warner North, D., Nyrop, J.P., Alina Stahevitch, O. and Williams, R. (1991) Principles for conduct of pest risk analyses: report of an expert workshop. *Risk Analysis* 18, 6, 773–780.

IPPC (1997) International Standards for Phytosanitary Measures, Publication No. 6: *Guidelines for Surveillance.* Secretariat of the International Plant Protection Convention (IPPC), Food and Agriculture Organization of the United Nations, Rome.

IPPC (1998) International Standards for Phytosanitary Measures, Publication No. 8: *Determination of Pest Status in An Area.* Secretariat of the International Plant Protection Convention (IPPC), Food and Agriculture Organization of the United Nations, Rome.

IPPC (2004) International Standards for Phytosanitary Measures, Publication No. 11: *Pest Risk Analysis for Quarantine Pests Including Analysis of Environmental Risks and Living Modified Organisms.* Secretariat of the International Plant Protection Convention (IPPC), Food and Agriculture Organization of the United Nations, Rome.

IPPC (2007) International Standards for Phytosanitary Measures, Publication No. 2: *Framework for Pest Risk Analysis.* Secretariat of the International Plant Protection Convention (IPPC), Food and Agriculture Organization of the United Nations, Rome.

IPPC (2010a) International Standards for Phytosanitary Measures, Publication No. 1: *Phytosanitary Principles for the Protection of Plants and the Application of Phytosanitary Measures in International Trade.* Secretariat of the International Plant Protection Convention (IPPC), Food and Agriculture Organization of the United Nations, Rome.

IPPC (2010b) International Standards for Phytosanitary Measures, Publication No. 5: *Glossary of Phytosanitary Terms.* Secretariat of the International Plant Protection Convention (IPPC), Food and Agriculture Organization of the United Nations, Rome.

7 Economic Analysis in Pest Risk Analysis

Lottie Erikson

7.1. Introduction

The introduction of non-indigenous (non-native) species around the world has had a tremendous impact on both the global economy and on the world's ecosystems. Some of these introductions were intentional while others occurred as the by-product of international trade and human movement. In many cases, the introduction of non-indigenous species has had beneficial impacts, such as the introduction of various species of plants for landscape restoration, biological pest control, sports, domestic pests and, perhaps most significantly, for food crops and livestock (Pimentel et al., 2000). For example, more than 98% of the crops and livestock that make up the USA's $300 billion agriculture sector are introduced species (e.g. corn, wheat, rice, cattle, poultry, etc.) (Pimentel et al., 2001; US BEA, 2011).

In other cases, however, the introduction of non-indigenous species has resulted in major economic and environmental losses. According to Olson (2006) invasive species are one of the leading causes of global ecological change, and they 'have been implicated in over 42 percent of the world-wide vertebrate extinctions with an identifiable cause'. Further, it is estimated that 40% of the threatened and endangered species are at risk due to pressures from invasive species (Stein and Flack, 1996; Olson, 2006).

Assessing economic impacts is a critical part of pest risk analysis, and a requirement under both the SPS Agreement and the IPPC. The purpose of this chapter is to review existing guidance for economic analysis in pest risk assessment found in international agreements and standards, to consider how this guidance corresponds to accepted approaches to economic analysis, and to attempt to draw some general conclusions. The purpose is not to present a primer on economic analysis, but sufficient background or references are presented to clarify points discussed in this chapter.

The IPPC *Glossary of Phytosanitary Terms* (ISPM No. 5, IPPC, 2010) defines pest risk assessment as evaluation of the probability of the introduction and spread of a pest and the magnitude of the associated potential economic consequences. A key parameter in the assessment of pest risk is economic consequences and a key issue is what constitutes an appropriate measure of economic consequences in a pest risk assessment – not only what should be measured, but how it should be measured.

The answer to this question is not entirely straightforward and depends on a number of factors. A key factor is the nature of the question that the economic analysis in a pest risk assessment is attempting to answer. Is the economic analysis simply attempting to determine whether a phytosanitary measure is technically justified under the SPS Agreement? Or is the analysis attempting to determine whether the phytosanitary measure also meets economic efficiency criteria and maximizes welfare of all affected parties?

Guidance found in international agreements and standards is somewhat ambiguous regarding which questions should be asked and answered by economic analysis in pest risk analysis, and the guidance has been interpreted in different ways by different countries. Issues relating to consistent treatment of trading partners and appropriate level of protection/acceptable level of risk[1] may be emphasized by some countries when choosing an approach to economic analysis. Theoretical considerations, appropriateness of economic modelling techniques to particular situations and availability of resources including time, data and modelling expertise are also important considerations. Finally, the level of political or stakeholder interest or contentiousness of a particular issue may also affect the approach, available resources or complexity of the analysis.

7.2. International Guidelines for Considering Economic Impacts in Pest Risk Assessment

The SPS agreement deals with international trade and animal, plant and human health. The Agreement recognizes the need for WTO members to protect themselves from risks posed by the entry of pests and diseases, and seeks to minimize any negative effects of quarantine or food safety measures on trade.

The Agreement attempts to limit the use of SPS measures for protectionist purposes by requiring WTO members to ensure that phytosanitary measures are imposed only to the extent necessary to protect human, animal or plant life or health, are notified to trading partners and conform to international standards or are technically justified through risk assessment (see Chapter 4 for more detailed discussion of these points).

Economic considerations in phytosanitary risk assessment are briefly described in the SPS Agreement. In addition to briefly describing economic consequences, the Agreement (Article 5.1) also indicates that Members should take into account ISPMs, which are developed by the IPPC. IPPC standards provide more detailed guidance on economic analysis in risk assessment.

7.2.1. Economic analysis guidance in the SPS Agreement

The SPS Agreement explicitly endorses consideration of risk-related costs (e.g. potential production or sales losses or control and eradication costs) both in assessing risks and managing risks through the choice of an SPS measure to protect animal or plant health. The language in the Agreement suggests that consideration of producer impacts alone would be sufficient to comply with the letter of the SPS Agreement, and that choice of an SPS measure is not required to be justified by an analysis of the effects on producers, consumers, taxpayers and industries who use the regulated product as an input. The implications of considering only producer costs associated with imports of potentially risky commodities, and not the benefits of imports, are discussed in the next section.

Article 5.3 of the Agreement states:

> In assessing *the risk to animal or plant life or health and determining the measure to be applied for achieving the appropriate level of sanitary or phytosanitary protection from such risk*, Members shall take into account as relevant economic factors: the potential damage in terms of *loss of production or sales* in the event of the entry, establishment or spread of a pest or disease; *the costs of control or eradication* in the territory of the importing Member; and the relative *cost-effectiveness of alternative approaches to limiting risks*. (Italics added for emphasis.)

In addition, Article 5.6 states that Members must ensure that their measures are not more trade restrictive than necessary to achieve their appropriate level of protection (defined in note 1). What is meant by 'not more trade restrictive than necessary' and whether this term has implications for economic consequence analysis in PRA is a matter of interpretation. Roberts (in Anderson, 2001, p. 22) states that 'more elastic interpretations of the Agreement could view the incorporation of trade benefits as congruent with the objectives of "minimizing negative trade effects" and adopting "least trade restrictive" policies' articulated in Articles 5.4 and 5.6. Incorporation of trade benefits in economic consequence analysis tends to support adoption of less restrictive trade policies. However, decision criteria would have to be sufficiently explicit to support a finding that any potential variation in the ALOP is 'unarbitrary and justifiable'.

7.2.2. Economic impacts of SPS measures

Economic impacts of SPS measures depend on the baseline or initial situation. Is the initial situation one of free trade, where imports of a potentially risky commodity have historically been allowed with no SPS measures? If so, then imposition of an SPS measure will restrict existing trade. A situation like this could occur when a pest of quarantine significance to the importing country is found to occur in the exporting country, or when an importing country promulgates or revises import regulations requiring stricter measures for the commodity and/or pest of concern. Or is the initial situation one where imports of the potentially risky commodity are currently banned and an SPS measure is being contemplated that will permit imports, subject to satisfaction of certain safeguards or restrictions. If the latter is the case, then imposition of an SPS measure (or measures in the case with a systems approach[2]) can be trade liberalizing.

Situations like this can occur when technological advances make new phytosanitary technologies feasible (e.g. irradiation, modified atmospheres during sea transit or molecular diagnostics that permit identification of immature life stages). They can also occur when new production technologies with mitigative effects are adopted in an industry (e.g. high density baling of ginned cotton also crushes and kills arthropod pests

of concern, like boll weevil). Most often this situation occurs when an exporting country proposes a system of overlapping measures for one or several pests of concern that together can achieve an importing country's appropriate level of protection (e.g. a systems approach could include pre-harvest measures like trapping and chemical suppression in the field, post-harvest measures (like sorting, culling, fruit cutting, washing with a bactericide, cold treatment) and border measures including visual inspection.

Example 1: Adopting a phytosanitary measure

Assume an initial situation of free trade with respect to a certain commodity and pest/pathogen. The importing country (Country X) is proposing to require that lettuce from Country Y be washed with disinfectant before export and fumigated upon arrival with methyl bromide because of concerns regarding various pests of lettuce. To be consistent with the provisions of the SPS Agreement, before imposing these measures Country X should have conducted a risk assessment to demonstrate that these measures are not disguised barriers to trade. In evaluating risk, Country X should have considered not only the likelihood of harm from unmitigated lettuce imports from Country Y, but also the magnitude of potential damages that could occur to its domestic producers of lettuce and to domestic producers of other host commodities should their crops become infested with pests of concern, including loss of sales, and the costs of controlling or eradicating the target pests or diseases, and the relative cost-effectiveness of alternative approaches to limiting risks.

Assuming this has been done, Country X would be in compliance with the SPS Agreement and would be justified in imposing the measure. However, if the analysis stops here, decision-makers in Country X could be presented with an incomplete and skewed appraisal of the impacts of the measure on its citizens. Considering only the negative externality[3] associated with the trade (producer losses from possibly importing pests and pathogens) and not the benefits of the trade itself (generally expressed as higher quantities of lettuce available at lower prices) on consumers or related industries that use lettuce as an input (restaurant chains, for example), or negative effect of the particular measure on the environment (methyl bromide is an ozone depleter) may lead to suboptimal policy choices because the analysis does not consider the net effect of the measure to society. Criteria explicitly mentioned in the SPS Agreement for assessing economic consequences may be sufficient to justify that a measure is not a disguised barrier to trade, but are not sufficient to determine whether an SPS measure improves the welfare of all affected parties or is the best method for achieving a given level of risk reduction.

Example 2: Removing an import ban

A compelling example of the economic consequences of liberalizing phytosanitary trade restrictions is found in James and Anderson (1998). In their empirical analysis of importation of bananas into Australia the authors conclude:

> that even if disease importation were to be so severe as to destroy the profitability of the local industry, it is conceivable that such importation, through lowering prices might benefit consumers more than it would harm import-competing producers and environmental groups.

Their analysis underscores the importance of considering the positive impacts of trade on consumers as well as negative effects on producers. Empirical analysis of the consequences of lifting Australia's import ban on imported bananas indicated that the consumer gain from removing the ban would be likely to far outweigh the loss to banana growers even if diseases were to wipe out the industry. (Cost–benefit analysis and welfare analysis are discussed in more detail in a subsequent part of this chapter.)

The authors also develop the general case for four scenarios using standard supply and demand curve analysis. Their graph is reproduced in Fig. 7.1 with a tabular summary (for those who find graphical analysis difficult to follow) of societal impacts resulting from liberalized trade scenarios: (i) moving from an import ban to free trade; (ii) moving

from an import ban to a phytosanitary restriction; (iii) moving from an import ban to free trade with negative disease externalities (i.e. a pest or disease is imported along with the commodity); and (iv) moving from an import ban to a phytosanitary restriction with negative disease externalities (i.e. a pest or disease is imported). They demonstrate graphically that the net effects of liberalizing trade (either free trade or phytosanitary restrictions) are always positive when pests or pathogens are not imported. However, the direction of net welfare effects could be either positive or negative when there are disease effects to consider. This finding argues strongly for empirical analysis of economic consequences in cases where there is a higher probability of pest entry, spread and establishment and/or where the consequences of establishment are high. According to the James and Anderson (1998, p. 432) the net benefits of removing phytosanitary restrictions will tend to be positive when the situation in the importing country is characterized by the following combination of economic and pest/disease criteria:

- The higher the price in the importing country relative to the world price;
- The more responsive prices in the importing country are to changes in quantity that would result from increased imports (the more price elastic the demand and supply curves);
- The lower the probability of disease entry, establishment and spread in the absence of quarantine restriction;
- The smaller the losses associated with pest infestation (the smaller the inward shift in the supply curve).

Conversely, one would expect liberalized phytosanitary policies to result in lower or even negative net benefits the more internationally competitive domestic producers are (i.e. domestic price equal to or less than world price); the less price elastic the supply and demand curves, the larger the losses associated with pest importation; and the higher the probability of disease entry in the absence of quarantine restrictions.

Effects of removing a ban entirely or replacing it with a less stringent SPS

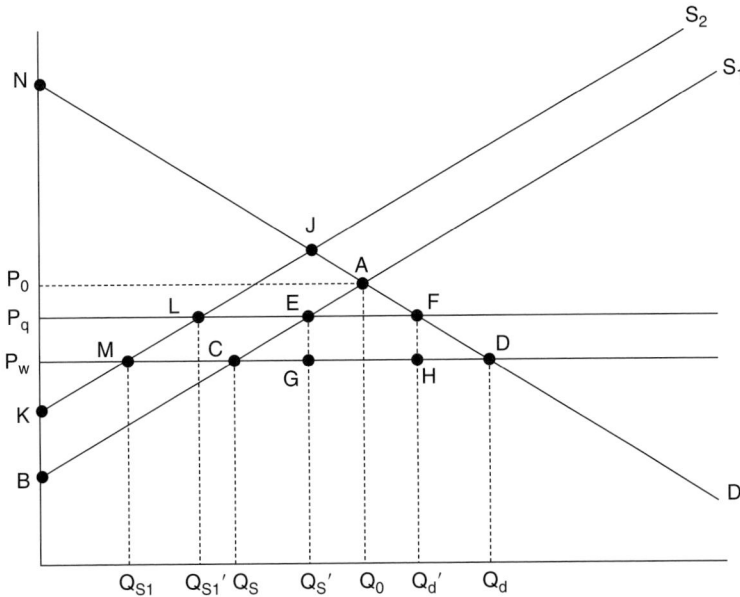

Fig. 7.1. Economic consequences of removing or liberalizing phytosanitary import restrictions. (Adapted from James and Anderson, 1999.)

measure are illustrated graphically in Fig. 7.1. S1 represents the supply curve in the absence of trade. The demand curve is represented by DN. Intersection of the supply, S1, and demand curves, D, at point A represents the initial market equilibrium. Consumers are demanding and producers supplying an amount, Q0, at a price, P0. At this initial market equilibrium consumer surplus is represented by the triangle beneath the demand curve and above the price line (NAP0). Producer surplus is represented by the triangle below the price line and above the supply curve (BAP0). Changes resulting from lifting the import ban under various assumptions are summarized in Table 7.1.

7.2.3. Why do benefits matter? What are the implications of including or omitting benefits?

It may be seen from James and Anderson's (1998) example that approaches to economic consequence analysis that do not take into account all potential benefits and costs may push countries to adopt overly conservative measures that limit the potential gains from

the SPS Agreement. That is, considering only the negative effects on producers could prompt decision makers to err on the side of imposing phytosanitary regulations that are technically justified, even if such measures do not always maximize social welfare.

The SPS Agreement omits any mention of benefits and this omission leads to many differences of opinion about the appropriate scope of economic consequence analysis in pest risk analysis. There may be several reasons for this omission. In considering whether benefits of imports can be considered as a legitimate factor in the SPS policy choice, Roberts (in Anderson, 2001, p. 22) refers to the overall purpose of the Agreement:

> The omission of explicit disciplines on benefits in the Agreement need not be interpreted as a prohibition on the inclusion of trade benefits as a factor in SPS regulatory decisions. It should be remembered that the Agreement is an international trade treaty whose purpose is to limit the use of putative scientific claims for protectionist purposes, and not to establish templates for risk management decisions.

In considering the omission of benefits from the Agreement she also poses an example of

Table 7.1. Changes resulting from lifting import ban under various assumptions (see also Fig. 7.1). (Adapted from James and Anderson, 1998, p. 430.)

Scenario	1. Baseline No Trade	2. Change from no trade to free trade	3. Change from no trade to restricted trade (i.e. SPS measure)	4. Change from no trade to free trade with disease	5. Change from no trade to restricted trade with disease
Price	Po	Decreases to Pw	Decreases to Pq (Pq is higher than Pw by compliance cost)	No change from scenario 2	No change from scenario 3
Quantity supplied Domestic	Supply curve = S1 Quantity supplied = Qo	Supply curve = S1 Quantity supplied decreases to Qs	Supply curve = S1 Quantity supplied decreases to Qs'	Supply curve = S2 Quantity supplied decreases to Qs1	Supply curve = S2 Quantity supplied decreases to Qs1'
Quantity supplied Imports	Not applicable	Increases to QsQd	Increases to Qs'Qd'	Increases to Qs1Qd	Increases to Qs1'Qd'
Quantity demanded Domestic	Qo	Increases to Qd	Increases to Qd'	No change from Scenario 2	No change from Scenario 3
Consumer surplus	PoAN	Increases to PwDN	Increases to NFPq	No change from Scenario 2	No change from Scenario 3
Producer surplus	PoAB	Decreases to PwCB	Decreases to BEPq	Decrease from Scenario 2 KMPw	Decrease from Scenario 3 KLPq
Gains from trade	Not applicable	Positive ACD	Positive EFA		
Net welfare effects		Free trade w/out disease externalities always superior to no trade	Restricted trade w/ out disease externalities always superior to no trade	Net welfare effects of free trade with disease externalities ambiguous	Net welfare effects of restricted trade with disease externalities ambiguous

(in Anderson, 2001, p. 22) potential problems associated with consideration of the gains from trade, i.e. violation of consistency principles of the Agreement, notably arbitrary and unjustifiable levels of protection (Article 5.5) and/or arbitrary and unjustifiable discrimination between Members (Article 2.3).

Consider the situation where the United States decides to allow imports of beef, but not poultry, because although the expected value of disease-related losses are the same for the two products, the benefits to consumers of importing beef outweigh the costs while the benefits of importing poultry do not (i.e. relative to foreign competitors, the US is a more efficient producer of poultry than of beef). Although the choice to allow only imports of beef might be efficient regulatory policy from a CBA (cost benefit analysis) perspective, some in the WTO community (and most certainly the country whose import request for poultry was turned down) could well view these choices as evidence of 'arbitrary and unjustifiable distinctions' in the level of protection that had resulted in discrimination, or a disguised restriction on trade.

A third perspective on the omission of more expansive guidance for economic assessment in the context of PRA may come from attitudes toward economic considerations in the broader risk assessment community. Hoffman (2011) argues that there may be concerns about fully integrating economic consequence analysis into risk assessment because concerns about the influence of economic interests on science may have affected the way people view the role of economic analysis (also an analytical discipline) in risk assessment. She references (p. 1346) National Research Council recommendations on risk analysis that state that 'the scientific findings and policy judgments embodied in risk assessment should be explicitly distinguished from the political, *economic*, and technical considerations that influence the design and choice of regulatory strategies.' (Italics added for emphasis.) She notes that when economic analysis is discussed in NRC reports it is always as part of risk management (See Fig. 7.2) and not risk assessment.

7.2.4. What does WTO case law say about the scope of economic consequence analysis in pest risk analysis?

According to Stanton (2001), almost no consideration was given to economic arguments or considerations during the first three disputes brought before the WTO dispute settlement body (the EC hormones ban dispute, Australian restrictions on salmon imports dispute, Japan varietal testing dispute) other than to calculate allowable retribution. The economic factors in the risk assessment and the way in which they were considered by Australia in the salmon dispute were implicitly accepted by the dispute panel. But according to Stanton: 'if in some future dispute the economic factors considered in a risk assessment were challenged, they could be subjected to intensive scrutiny by a panel' (2001, p. 69).

7.3. IPPC Guidance to Economic Consequence Analysis in Pest Risk Analysis

Several ISPMs, ISPM Nos 2, 5 and 11, either reference economic considerations or provide guidance that is applicable to economic analysis in a pest risk assessment. The overall importance of economic considerations in phytosanitary decision making is suggested by the number of key phytosanitary concepts that reference economic terms. In ISPM No. 5, *The Glossary*

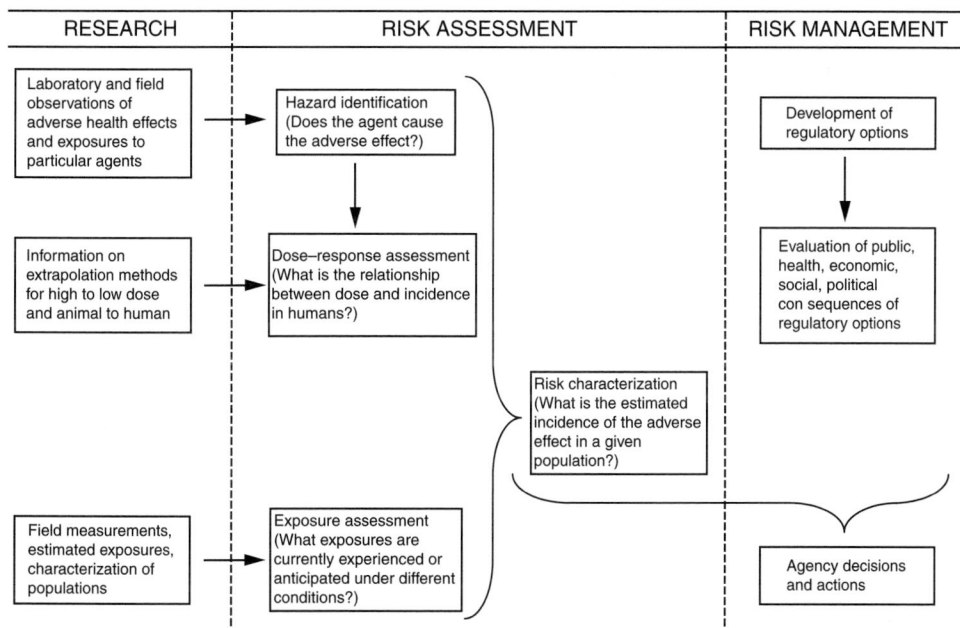

RESEARCH	RISK ASSESSMENT	RISK MANAGEMENT
Laboratory and field observations of adverse health effects and exposures to particular agents	Hazard identification (Does the agent cause the adverse effect?)	Development of regulatory options
Information on extrapolation methods for high to low dose and animal to human	Dose–response assessment (What is the relationship between dose and incidence in humans?)	Evaluation of public, health, economic, social, political con sequences of regulatory options
	Risk characterization (What is the estimated incidence of the adverse effect in a given population?)	
Field measurements, estimated exposures, characterization of populations	Exposure assessment (What exposures are currently experienced or anticipated under different conditions?)	Agency decisions and actions

Fig. 7.2. Overcoming barriers to integrating economic analysis into risk assessment. (Adapted from: *Risk Analysis*, 31, 9, pp. 1345–1355, 13 September 2011.)

of Phytosanitary Terms (hereafter Glossary), the phrases *economic impacts* or *economic consequences* are explicitly mentioned in definitions of several important phytosanitary terms: e.g. *pest risk, pest risk assessment, phytosanitary measure, phytosanitary regulation. Economic importance* and *economically important losses* are explicitly mentioned in the definitions of other terms, including the definition for the key phytosanitary concept of *quarantine pest.* The glossary does not include a definition for any of the terms related to economic impacts or economic importance but does contain a supplement that provides guidelines for understanding them (IPPC, 2010; ISPM No. 5, Supplement No. 2).

7.3.1. Supplement No. 2 to ISPM No. 5

The scope and purpose of this supplement is to provide clarification to ensure that economic terms are clearly understood and

References to economic terms in ISPM No.5, Glossary of phytosanitary terms (economic components underscored).

Endangered area	An *area* where ecological factors favour the *establishment* of a *pest* whose presence in the *area* will result in <u>economically important loss</u> (see Glossary Supplement No. 2)
Pest risk (for *quarantine pests*)	The probability of *introduction* and *spread* of a *pest* and the magnitude of the associated <u>potential economic consequences</u> (see Glossary Supplement No. 2)
Pest risk (for *regulated non-quarantine pests*)	The probability that a *pest* in *plants for planting* affects the *intended use* of those *plants* with an <u>economically unacceptable impact</u> (see Glossary Supplement No. 2)
Pest risk analysis (agreed interpretation)	The process of evaluating biological or other scientific and <u>economic evidence</u> to determine whether an *organism* is a *pest*, whether it should be regulated, and the strength of any *phytosanitary measures* to be taken against it
Pest risk assessment (for *quarantine pests*)	Evaluation of the probability of the *introduction* and *spread* of a *pest* and the magnitude of the associated <u>potential economic consequences</u> (see Glossary Supplement No. 2)
Pest risk assessment (for regulated non-quarantine pests)	Evaluation of the probability that a *pest* in *plants for planting* affects the *intended use* of those *plants* with an <u>economically unacceptable impact</u> (see Glossary Supplement No. 2)
Phytosanitary measure (agreed interpretation)	Any *legislation, regulation* or *official* procedure having the purpose to prevent the *introduction* and/or *spread* of *quarantine pests*, or to <u>limit the economic impact of</u> *regulated non-quarantine pests*
Phytosanitary regulation	*Official* rule to prevent the *introduction* and/or *spread* of *quarantine pests*, or to <u>limit the economic impact</u> of *regulated non-quarantine pests*, including establishment of *procedures* for *phytosanitary certification* (see Glossary Supplement No. 2)
Quarantine pest	A *pest* of <u>potential economic importance</u> to the *area endangered* thereby and not yet present there, or present but not widely distributed and being *officially controlled*
Regulated area	An *area* into which, within which and/or from which *plants, plant products* and other *regulated articles* are subjected to *phytosanitary regulations* or *procedures* in order to prevent the *introduction* and/or *spread* of *quarantine pests* or to <u>limit the economic impact</u> of *regulated non-quarantine pests* (see Glossary Supplement No. 2)
Regulated non-quarantine pest	A *non-quarantine pest* whose presence in *plants for planting* affects the *intended use* of those *plants* with an <u>economically unacceptable impact</u> and which is therefore regulated within the territory of the importing contracting party (see Glossary Supplement No. 2)

consistently applied, and to illustrate certain economic principles as they relate to the IPPC's objectives, in particular but not limited to, environmental considerations. The supplement clearly states that the IPPC can account for environmental concerns in economic terms using monetary or non-monetary estimates and that market impacts are not the sole indicator of pest consequences.

Section 4 of the supplement to the glossary, *Economic Considerations in PRA*, discusses types of economic effects and costs and benefits. It describes a relatively inclusive approach to economic considerations in pest risk analysis, indicating that all economic effects (not just market related), both costs and benefits, and both direct and indirect effects, should be considered in a pest risk analysis. It affirms the cost–benefit criteria for decision making, whereby policies should be pursued if benefits are as least as large as costs, and indicates that judgements about the preferred distribution of costs and benefits are a policy choice to be made outside the context of the economic analysis.

7.3.2. ISPM No. 2 *Framework for Pest Risk Analysis* (2007)

ISPM No. 2 does not give specific guidance on how economic impacts should be conceptualized or measured, but describes the stages in a pest risk assessment and indicates where it is appropriate to consider economic factors.

ISPM No. 2 describes three stages of pest risk analysis: initiation, pest risk assessment and pest risk management. It focuses primarily on the initiation stage and general issues affecting the conduct of a pest risk analysis, and also describes the progression from one stage of the analysis to the next and the steps in each.

The analysis may stop after Stage 1, if it is determined that an organism is not a pest or that pathways do not carry pests, in which case an economic analysis would not be required. However, when an organism is determined to be a pest the analysis proceeds to Stage 2, pest categorization,

which includes an assessment of the potential for introduction and spread *and an assessment of economic impacts* [emphasis supplied]. If this analysis determines an unacceptable level of risk, then the analysis proceeds to Stage 3, pest risk management, which determines whether or not appropriate phytosanitary measures to reduce pest risk to an acceptable level are available, cost-effective and feasible. The term cost-effective implies that the cost and effectiveness of mitigation options have been considered and compared. In addition, ISPM No. 2 indicates that PRA documentation should include *evidence of economic impact* [emphasis supplied], and evaluation of risk management options.

7.3.3. ISPM No. 11 Pest Risk Analysis for Quarantine Pests, Including Analysis of Environmental Risks and Living Modified Organisms (2004)

Section 2.3, of ISPM No. 11, titled 'Assessment of potential economic consequences', contains the most fully elaborated description of the process for assessing economic consequences in the pest categorization stage of a risk assessment, and offers practical and specific guidance for a risk assessment practitioner, but contains such broad guidance that it may leave many questions unanswered about what should be measured and how it should be measured.

Consistent with the IPPC's emphasis on plant health, the guidance in ISPM No. 11 states that plant pest effects on human or animal health (i.e. allergenicity or toxicity) should be considered, as appropriate, by other agencies.

The guidelines discuss situations in which a detailed analysis of economic consequences may or may not be necessary. If it is widely agreed that pest introduction will have unacceptable consequences, detailed analysis may not be necessary. On the other hand, it may be necessary to examine economic factors in greater detail when the level of consequences is in question, or when consequences are needed to evaluate the strength of measures, or to assess the relative benefits of exclusion versus control.

Qualitative assessments are permissible, but wherever appropriate, quantitative data that provide monetary values should be obtained. The guidelines describe three commonly used quantitative techniques for estimating economic consequences that may be appropriate depending on the extent of the effects on the economy. Partial budgeting, which measures adjustments by producers to a pest incursion, is appropriate to use if economic effects induced by the action of the pest are generally limited to producers and are considered to be minor.

A partial equilibrium approach, which measures the net effects from pest impacts on producers and consumers in affected sectors, is appropriate to use if there is a significant change in producer profits or consumer demand and impacts are not experienced throughout the entire economy. General equilibrium analysis, which measures changes to interrelated markets throughout the entire economy, is appropriate to use when pest infestation could have significant impacts on economy-wide factors such as wages, interest rates or exchange rates. Each of these approaches has benefits and drawbacks that are discussed later in this chapter.

Environmental effects

The importance of considering environmental impacts, even when they may be difficult or impossible to quantify or monetize, is emphasized in ISPM No. 11. Non-market environmental impacts such as ecosystem stability, biodiversity and tourism and some commonly accepted approaches to conceptualizing their value (e.g. use values, which include tourism, recreation and hunting/fishing, and non-use values, which include existence, bequest, and option values) are described. Valuing environmental impacts using market-based approaches, surrogate markets, simulated markets and benefit transfer are mentioned.[4]

Direct and indirect effects

Direct effects, which include effects of the pest on the potential host or the environment, should be considered and total crop area and/or potentially endangered area should be identified. Examples of direct effects on cultivated hosts could include crop losses, control measures and effects on production practices. Direct effects of the pest on the environment could include reduction of keystone species or endangered native plants. Examples of indirect effects of the pest in the pest risk analysis area include effects that are not host-specific, such as effects on domestic and export markets (i.e. loss of export markets), changes to demand because of quality changes in the commodity and social or other effects.

Market and commercial effects, especially consumer impacts

In the discussion of indirect effects, changes to domestic or foreign consumer demand, especially market access effects, and changes to consumer demand for a product stemming from quality changes are mentioned. *ISPM No. 11 indicates that commercial effects, both positive and negative, should be identified and quantified, including effects of pest induced changes on producer profits on in quantities demanded by domestic or international consumers* [emphasis supplied]. The guidance indicates that partial equilibrium analysis is necessary to measure welfare changes, or the net change arising from pest impacts on consumers and producers, but it does not explicitly mention consideration of gains from trade that could occur from importing a potentially risky commodity.

7.4. Summary of SPS Agreement and ISPM Guidance on Economic Consequences

The SPS Agreement describes a more limited set of factors to be considered in economic assessments than do the ISPMs. This distinction is important because different approaches described in the SPS Agreement and the ISPMs (i.e. estimating negative impacts to producers as opposed to estimating both costs and benefits) will affect what is measured, how results are interpreted, and could support different conclusions regarding risk management by decision makers.

The SPS agreement explicitly endorses consideration of risk-related impacts to producers in the importing country. The ISPMs describe a very broad range of approaches to economic consequence analysis in a pest risk analysis, with ISPM No. 11 emphasizing measurement of environmental impacts and endorsing a continuum of approaches that range from 'no detailed analysis' if consequences are widely accepted to be unacceptable, to qualitative analysis, to various approaches for quantitative analysis, which include consideration of relevant impacts on consumers, producers, domestic and foreign markets.

Based on available guidance in the SPS Agreement and the ISPMs, and in the absence of any clarifying WTO jurisprudence or case law, it may be concluded that risk assessment practitioners have considerable latitude in determining how to approach economic consequence analysis in pest risk analysis. This latitude would be subject to the conditions that any phytosanitary measures based on a risk assessment and economic consequence analysis should not violate the consistency provisions of the SPS Agreement by arbitrarily or unjustifiably discriminating between Members and should not be applied in such a way as to constitute a disguised restriction to trade.

7.5. Country Approaches to Economic Assessment in Pest Risk Analysis

In practice how have countries approached economic consequence analysis in risk assessment? Some countries have published policy directives that define the scope of economic factors that may be considered in their risk assessments. For example, Biosecurity Australia (2003) focuses on whether an imported commodity exposes affected industries and the environment to an unacceptable level of risk:

> In keeping with ... Australia's obligations as a member of the WTO, economic considerations are taken into account only in relation to matters arising from the potential direct and indirect impact of pests and diseases that could enter, establish or spread in Australia as a result of importation.

> The potential competitive economic impact of prospective imports on domestic industries is not within the scope of IRAs.

Many countries, including the USA, those of the European Union, and Australia follow a qualitative approach to risk assessment, in which the economic factors to be considered and the methods for considering them are clearly specified and consistently applied. The US approach to considering consequences in a risk assessment is considered in detail in other chapters of this book. In brief, evaluation of potential consequences of pest introduction comprises five risk elements (climate–host interaction, host range, dispersal potential, economic impact and environmental impact) that are each given a score of high, medium or low. A cumulative score for the five risk elements is calculated. There is a similar process for ranking likelihood of introduction, which evaluates the quantity of the commodity imported and the potential for survival and suitable hosts and habitat. An overall rating of pest risk potential, rated high, medium or low, is based on the component scores for consequence of pest introduction and likelihood of introduction.

In practice, because of time, resource (skilled analysts) and data limitations, exceptions to qualitative economic analysis in pest risk analysis tend to result where imposition or liberalization of a phytosanitary measure is particularly complex or contentious due to potential effects on domestic producers or trading partners, or both, or where there are many potential mitigation options to be evaluated. In these cases a more detailed quantitative assessment of costs and benefits is likely to be undertaken.

7.6. Qualitative versus Quantitative Analyses

Some question whether advanced quantitative techniques for assessing economic benefits bring added value to a pest risk analysis, and whether the added costs in terms of data and resources, is justified. A qualitative analysis is certainly consistent with international guidelines in ISPMs and may be all that can be supported by available data or modelling expertise and time constraints. However, it is important that qualitative analyses be based on a clear and well-reasoned framework – otherwise results can be inconsistent across analyses, may not be transparent to either domestic decision makers or trading partners, or the results could be '(ab)used for political or protectionist goals' (Soliman *et al.*, 2010, p. 519).

7.7. Economic Analysis in Choosing Quarantine Policy Options

Examples of *ex ante* quantitative analyses of economic consequences in pest risk assessment that have been used or could be used to inform quarantine policy decisions are found less frequently than *ex post* analyses that estimate impacts of quarantine policy decisions that have already been taken or that illustrate desirable approaches or methodologies are more common. Orden *et al.* (2001) highlight the potential for complementarity between science-based risk assessment and economic-based cost–benefit analysis in regulatory decision making and argue for fuller integration of these approaches, especially in choosing least trade restrictive quarantine policy options.

A frequently cited example of an *ex ante* quantitative analysis is that examining the importation of Mexican avocadoes to the USA (USDA, US Federal Register, 2004). This assessed the potential economic impacts of removing a partial phytosanitary restriction on avocadoes using a static, partial equilibrium model. The economic analysis was published as part of the US federal rule to justify a change in quarantine policy. Given

model assumptions, removal of an import ban was estimated to lead to a 267% increase in imports of Mexican avocadoes over five years, resulting in losses for US and Chilean avocado producers and gains for US consumers, who benefitted from a greater availability of avocadoes at a lower price. Estimates yielded a net welfare gain to affected sectors in the US ranging from roughly US$31–$33 million depending on the model scenario.

Another example of an *ex ante* analysis deals with the impacts of modifying a US quarantine ban on Argentine lemons. Cororaton *et al.* (2011) analysed the potential economic effects that modifying a quarantine ban on Argentine lemons would have on US production and consumption of lemons and on other trading partners who supply the US market. They use a partial equilibrium simulation approach to model the entry of Argentine lemons into the US under three scenarios based on differing quarantine policy regarding geographic and temporal access to the US market and different assumptions regarding supply from other citrus exporting countries (Mexico, Chile and Spain).

There are many other examples of quantitative analyses that illustrate the approaches described in the SPS agreement and the ISPMs but which have not, to our knowledge, been used as part of a formal risk assessment to inform quarantine policy decisions. Some of these studies are referenced below.

7.8. Valuing Environmental Impacts in Pest Risk Assessment

Despite the emphasis in ISPM No. 11 on estimating environmental effects in risk assessments, there are few examples of *ex ante* qualitative or quantitative analyses that have valued environmental impacts related to plant pest establishment or imposition of control measures using either market or non-market techniques. Paine *et al.* (2003) in *Ash Whitefly and Biological Control in the Urban Environment* quantify the benefits of using biological control to preserve the aesthetic value of ash and ornamental pear trees. The ash whitefly (*Siphoninus phillyreas* (Haliday)) attacks agricultural crops including pomegranate, pear and apple, but the focus of this analysis is the ash whitefly's potential as a pest in the urban landscape. The authors estimate the costs and benefits of biological control, using a widely recognized landscape appraisal technique to value trees. Benefits are conceptualized as the difference between appraised values of primary host trees (ash and ornamental pear) before and after defoliation. Total benefits, calculated as change in appraised value per host tree times the number of affected host trees, ranged from $312–412 million, while costs (which included salary, and costs associated with collection, importation, rearing and monitoring of parasites) totalled $1.22 million.

Klotz *et al.* (2003) in *An Insect Pest of Agricultural, Urban, and Wildlife Areas: The Red Imported Fire Ant* is an example of how a qualitative assessment of environmental impacts can be combined with a quantitative assessment of market impacts. The authors consider benefits and costs of eradicating the red imported fire ant (RIFA). Benefits are conceptualized as losses averted from establishment of the RIFA and costs are conceptualized as the cost of eradication. The authors estimate the losses that would accrue from RIFA establishment on urban households (mounds, damage to electrical equipment, medical and veterinary expenses) and on nursery stock and agricultural producers under varying assumptions regarding acreage affected and severity of impacts. Also incorporated are a quantitative measure of rangeland effects ($ losses per acre) and a qualitative assessment of potential effects on endangered species, which would occur primarily through displacement of native species and predation on bird and reptile hatchings.

7.9. Quantitative Techniques for Estimating Economic Consequences

Roberts *et al.* (1999) describe an economic modelling framework for quantifying the trade effects of technical trade barriers. The authors derive the trade and welfare effects of SPS measures under various assumptions, examining cases where measures are imposed as regulatory protection (no actual phytosanitary risks associated with importation); are applied to all exporters versus only one exporter; affect the supply curve (importation of pests and pathogens along with commodity); where measures affect the demand curve (consumers benefit from food safety regulation), from the perspective of importers and exporters, large countries and small countries. In addition, they note that pest and disease control involve probabilities rather than certainties so that costs and benefits may need to be expressed as means and variances and that much of the skill in modelling SPS barriers *will be in the translation of scientific knowledge of animal and plant health effects to probabilities of loss and valuations of that loss.*

Soliman *et al.* (2010), in their thorough and useful review of the scope, data requirements and appropriateness of quantitative methods referenced in ISPM No. 11 for assessing economic impacts in a risk assessment discuss important considerations regarding qualitative and quantitative approaches.

7.9.1. Partial budgeting

Quantitative economic analysis techniques like partial budgeting, which emphasize economic impacts of pest entry, establishment and spread on a producer's budget or on aggregated producers' budgets are relatively simple to design, implement and understand but cannot capture effects in interrelated markets and are most appropriate when impacts in other sectors are not expected to be significant. While this approach could address effects on affected producers, the impacts on the environment, on consumers or lost exports due to trade bans would need to be calculated separately, outside the partial budgeting framework.

Other important limitations to this approach are that prices, both output and input, are exogenous to the model (derived from external sources, not calculated by the model) and that this approach does not distinguish between effects on different categories of producers, i.e. infested versus non-infested producers. Ferris *et al.* (2003) in *Risk Assessment of Plant-Parasitic Nematodes* analyse the risk posed by five representative nematodes of economic importance (rice foliar, sting, reniform, burrowing and golden nematodes) that are exotic to the state of California. This analysis illustrates one of the inherent limitations of a partial budgeting approach, i.e. the inability to consider impacts on non-infested producers, when selecting an optimal response to nematode infestation. The authors provide a qualitative assessment of risk focusing on factors affecting introduction and spread, available intervention and management strategies, affected commodities and sectors.

The associated economic analysis compares grower costs associated with eradication of newly introduced nematodes versus management of established nematode infestations for each of the five nematodes and their associated annual and perennial hosts. For all nematodes and crops assessed, the costs of eradication for individual producers were greater that the costs of controlling an established infestation. Impacts on non-infested producers of choosing not to eradicate a nematode infestation are not considered in the analysis. However, the authors note that while eradication might not be the optimal choice for an individual producer, it might make sense for a public entity when costs associated with spread of nematode infestations to other locations are considered.

7.9.2. Partial equilibrium analysis

Partial equilibrium analysis is a quantitative approach that considers impacts, both positive and negative, to affected consumers and producers. It can be used to model the impacts of introduction of a pest or disease as in the graphic example shown by Soliman *et al.* (2010), and can also be used to evaluate the economic consequences of risk management options, for example unrestricted access for a potentially risky commodity versus geographic or temporal access limitations.

Partial equilibrium analysis accounts for the effect of a 'shock' or 'intervention' on markets as well as the responses by consumers and producers to changing quantities and prices. These effects can be both positive and negative. Changes resulting from increased supply that puts downward pressure on prices typically benefit consumers (increased consumer surplus) and have negative effects on producers (reduced profit or producer surplus). Estimating the slopes of supply and demand curves for affected commodities (in order to calculate how responsive quantity demanded and supplied are to changes in price) can require econometric estimation of price elasticities of supply and demand if the analyst is not fortunate enough to find recently published estimates of these important parameters.

Both partial budgeting and partial equilibrium analyses require a thorough understanding of effects of a pest introduction on producers; and, in addition, partial equilibrium analysis requires an understanding of consumer demand for the commodity, understanding of economic theory, and economic and possibly econometric modelling expertise, and incorporates data and parameters that may be difficult to locate or estimate.

Partial equilibrium analysis is often used to evaluate societal impacts resulting from regulatory interventions that affect supply or demand. The decision rule for adopting a proposed policy is net benefits (benefits – costs) greater than zero, or gainers able in theory at least to compensate losers. In reality, mechanisms for redistributing benefits from gainers to losers are rarely employed, so that while everyone may be made better off in theory, there could still be losers in reality.

Analyses discussed in previous sections (James and Anderson, 1998; USDA, 2004; Cororaton *et al.*, 2011) all employ a partial equilibrium approach.

An interesting recent article (Peterson and Orden, 2008) uses a partial equilibrium approach to model economic impacts of avocado imports from Mexico to the USA. It extends the *ex ante* analysis carried out prior to the market opening by considering compliance costs to Mexico, pest risks and US producer losses from trade-related pest infestations. The model simulates pest outbreak frequencies and control costs under three scenarios: continuation of the 2004 systems approach measures; elimination of 2004 systems approach measures; and elimination of fruit fly compliance measures. The authors found that when all compliance measures were removed, control costs in California orchards became significant especially for frequent outbreaks of stem weevil, suggesting that abandoning the systems approach would be questionable based on economic and pest risk criteria. Estimated welfare gains from removing geographic and seasonal restrictions in 2004 were confirmed.

7.9.3. General equilibrium

General equilibrium models can be used to assess economy-wide impacts in all interrelated markets that would result from an event, for example a supply shock in the case of pest infestation, in one sector of the economy. Their use is appropriate in cases where there are significant interrelationships between the affected sector and other sectors in the economy. The relationships (supply, demand, substitution for example) between sectors (e.g. multiple livestock or grain sectors) would be described by functional relationships (equations) that represent the effects that changes in one sector would have on another sector.

Development and maintenance of general equilibrium models are extremely time- and resource intensive as they require accurately describing and validating complicated market relationships within and between numerous sectors. In addition, the few existing general equilibrium models have tended to include agricultural sectors such as livestock, grain, cotton, and have not typically included model components that would address the effects of pest infestation on other host commodities that are often of interest to phytosanitary risk practitioners, e.g. citrus, tree fruits and nuts, row crops, summer vegetables. Soliman *et al.* (2010) reference examples of general equilibrium models.

7.10. Summary

Guidance found in the SPS Agreement and the ISPMs describe a broad range of approaches and techniques for assessing economic impacts in pest risk assessment. Most countries use a qualitative approach to estimating economic impacts in risk assessment. However a number of quantitative approaches can be taken. Time, resources for and complexity of these approaches vary, as do the implications of including or excluding gains from trade from the analysis. There is no clarifying WTO case law or jurisprudence relating to assessment of economic impacts. Soliman *et al.* (2010) argue that despite its limitations, partial budgeting would be the preferred quantitative approach for basic economic analysis because it is not data- or resource intensive and is relatively easy to understand and explain to decision makers. However, this approach is not without shortcomings. Other approaches, like partial equilibrium analysis, which expand the frame of analysis beyond impacts on producers, can give a more complete perspective on the economic effects of adopting a quarantine policy, but may not be feasible in all cases.

Given the time and resource costs of developing even the simplest quantitative models and the time and resource constraints facing most risk assessment practitioners, it would appear that collaboration(s) between economic modellers and risk assessment practitioners to develop quantitative models with general applicability to host commodities of phytosanitary priority could be of value.

Notes

[1] According to Goh and Ziegler, 'A member's appropriate level of protection reflects its individually acceptable level of risk. It is a political determination which seeks to balance the economic benefits of trade to the particular Member against the potential biological, economic and environmental consequences of pest and disease establishment' (Goh and Ziegler, 6 in Anderson, 2001, p. 7).

[2] The IPPC (ISPM No. 5, IPPC, 2010) defines a systems approach as 'The integration of different risk management measures, at least two of which act independently, and which cumulatively achieve the appropriate level of protection against regulated pests.'

[3] An externality is a cost or benefit, not transmitted through prices, incurred by a party who did not agree to the action causing the cost or benefit. An example of a positive externality is a farmer's orchard being pollinated by his neighbour's bees. An example of a negative externality is disease being transmitted from one producer to another.

[4] While not explicitly described in ISPM No. 11, examples of direct market effects on consumers and producers from environmental impacts would include productivity changes that could arise from negative environmental impacts. Productivity changes could be measured via cost functions or damage functions. Indirect or surrogate market effects on consumers resulting from an environmental change (in water quality for example) could be measured via hedonic pricing (estimating the implicit values of environmental quality characteristics embedded in property values, such as prices of lakeside properties adjoining polluted and unpolluted waters), or by the travel cost method (valuing environmental quality at public recreational sites by observing the relationship between visits to recreational sites with clean water and the cost of these visits). In cases where no direct or indirect market data are available, environmental quality changes can be measured using bidding or survey formats such as contingent valuation (CV) or contingent ranking (CR) to elicit statements of monetary valuation from users and non-users of environmental goods. The basic objective of CV and CR is to construct an individual's demand curve for a possible change in environmental quality from survey questions or bidding experiments. A key assumption relating to CV and CR is that individuals can and will honestly reveal personal preferences in an experimental setting.

References

Anderson, K (ed.) (2001) *The Economics of Quarantine and the SPS Agreement*, Center for International Economic Studies, Adelaide.

Biosecurity Australia (2001) *Guidelines for Import Risk Analysis*. Agriculture, Fisheries and Forestry Australia/ Biosecurity Australia, Canberra.

Cororaton, C.B., David, O. and Peterson, E. (2011) Economic Impact of Potential U.S. Regulatory Decisions Concerning Imports of Argentine Lemons, GII Working Paper No. 2011-1, Virginia Polytechnic Institute and State University.

Ferris, H., Jetter, K., Zasada, I., Chitambar, J., Venette, R., Klonsky, K. and Becker, J.O. (2003) Risk assessment of plant-parasitic nematodes. In: Sumner, D.A. (ed.) *Exotic Pests and Diseases: Biology and Economics for Biosecurity*, Iowa State University Press, Ames, Iowa, pp. 99–119.

Goh, G. and Ziegler, A. (2001) Implications of recent SPS dispute settlement cases. In: Anderson, K (ed.) *The Economics of Quarantine and the SPS Agreement*. Center for International Economic Studies, Adelaide, pp. 174–182.

Hoffman, S. (2011) Overcoming barriers to integrating economic analysis into risk assessment. *Risk Analysis* 31, 9, 1345–1355.

IPPC (2004) International Standards for Phytosanitary Measures, Publication No. 11: *Pest Risk Analysis for Quarantine Pests Including Analysis of Environmental Risks and Living Modified Organisms*. Secretariat of the International Plant Protection Convention (IPPC), Food and Agriculture Organization of the United Nations, Rome.

IPPC (2007) International Standards for Phytosanitary Measures, Publication No. 2: *Framework for Pest Risk Analysis*. Secretariat of the International Plant Protection Convention (IPPC), Food and Agriculture Organization of the United Nations, Rome.

IPPC (2010) International Standards for Phytosanitary Measures, Publication No. 5: *Glossary of Phytosanitary Terms*. Secretariat of the International Plant Protection Convention (IPPC), Food and Agriculture Organization of the United Nations, Rome.

James, S. and Anderson, K. (1998) On the need for more economic assessment of quarantine policies. *Australian Journal of Agricultural and Resource Economics* 42, 4, 425–444.

Klotz, J.H., Jetter, K., Greenberg, L., Hamilton, J., Kabashima, J. and Williams, D. (2003) An insect pest of agricultural, urban, and wildlife areas: The red imported fire ant. In: Sumner, D.A. (ed.) *Exotic Pests and Diseases: Biology and Economics for Biosecurity*. Iowa State University Press, Ames, Iowa.

Olson, L.J. (2006) The economics of terrestrial invasive species: a review of the literature. *Agricultural and Resource Economics Review* 35, 178–194.

Orden, D., Narrod, C. and Glauber, J.W. (2001) Least trade-restrictive SPS policies: an analytic framework is there but questions remain. In: Anderson, K. (ed.) *The Economics of Quarantine and the SPS Agreement*, Center for International Economic Studies, Adelaide, pp. 178–215.

Paine, T.D., Jetter, K., Klonsky, K., Bezark, L. and Bellows, T. (2003) Ash whitefly and biological control in the urban environment. In: Sumner, D.A. (ed.) *Exotic Pests and Diseases: Biology and Economics for Biosecurity*. Iowa State University Press, Ames, Iowa.

Peterson, E.B. and Orden, D. (2008) Avocado pests and avocado trade. *American Journal of Agricultural Economics* 90, 2, 321–335.

Pimentel, D., Lach, L., Zuniga, R. and Morrison, D. (2000) Environmental and economic costs of nonindigenous species in the United States. *BioScience* 50, 53–65.

Pimentel, D., McNair, S., Janecka, J., Wightma, J., Simmonds, C., O'Connell, C., Wong, E., Russel, L., Zern, J., Aquino, T. and Tsomondo, T. (2001) Economic and environmental threats of alien plant, animal, and microbe invasions. *Agriculture, Ecosystems and Environment* 84, 1–20.

Roberts, D. (1998) Implementation of the WTO Agreement on the application of sanitary and phytosanitary measures, Agriculture in the WTO/WRS-98-44/December 1998, USDA/ERS, pp. 27–33.

Roberts, D. (2001) The integration of economics into SPS risk management policies: issues and challenges. In: Anderson, K. (ed.) *The Economics of Quarantine and the SPS Agreement*. Center for International Economic Studies, Adelaide, pp. 9–28.

Roberts, D., Josling, T.E. and Orden, D. (1999) A framework for analyzing technical trade barriers in agricultural markets. Market and Trade Economics Division, Economic Research Service, US Department of Agriculture. Technical Bulletin No. 1876.

Soliman, T., Mourits, M.C.M., Lansink, O. and van der Werf, A.G.J.M. (2010) Economic impact assessment in pest risk analysis. *Crop Protection* 29, 517–524.

Stanton, G. (2001) The WTO dispute settlement framework and operation. In: Anderson, K. (ed.) *The Economics of Quarantine and the SPS Agreement*. Center for International Economic Studies, Adelaide, pp. 53–74.

Stein, B.A. and Flack, S.R. (1996) *America's Least Wanted: Alien Species Invasions of U.S. Ecosystems*. The Nature Conservancy, Arlington, Virginia.

US BEA (2011) GPD by industry accounts: gross output by industry. United States Bureau of Economic Analysis. http://www.bea.gov/iTable/iTable.cfm?ReqID=5&step=1 (30 October 2011).

USDA (2004) Economic analysis final rule: allow fresh haas avocados grown in approved orchards in approved municipalities in Michoacan, Mexico, to be imported into all states year-round (APHIs Docket No. 03-022-3). USDA, Washington, DC, 30 November.

WTO (1994) *The WTO Agreement on the Application of Sanitary and Phytosanitary Measures*. World Trade Organization, Geneva.

8 Types and Applications of Pest Risk Analysis

Christina Devorshak

8.1. Introduction

The SPS Agreement is a trade agreement that makes provision for plant protection and the IPPC is a plant protection agreement that makes provision for trade. Due to both the SPS Agreement and the IPPC, pest risk analysis is widely used to justify phytosanitary measures applied to the international movement of commodities, in particular agricultural commodities. Therefore, pest risk analysis in relation to international trade is the most common application of pest risk analysis. However, there are numerous other applications for pest risk analysis, including for the purpose of supporting various types of domestic (national) or sub-national programmes. This includes everything from surveillance programmes to eradication programmes to supporting exports of agricultural commodities to other countries.

The decision to carry out a pest risk analysis can arise for different reasons. One of the most common reasons is a request by an exporting country to export a commodity to an importing country. In this case, the importing country may evaluate the risk of pests moving on a given commodity from the exporting country. Another reason for doing a pest risk analysis is to evaluate a new pest that may be potentially introduced, or already introduced, into a country. In this

case, the country may wish to determine whether the pest requires specific action to either prevent its introduction or, if already introduced, whether the pest should be managed or eradicated. A third reason for starting a pest risk analysis may arise from changes in policy. For instance, an NPPO may decide to revise its requirements on a particular type of commodity (e.g. wood packing material or nursery stock), thereby necessitating a pest risk analysis to determine the most appropriate measures.

8.2. Types of Pest Risk Analysis

Pest risk analysis is most commonly applied for four different purposes (IPPC, 2004, 2007), namely:

- Analysing risks for organisms;
- Analysing risks for pathways;
- Analysing risks for commodities;
- Supporting new policies or changes to existing policies;
- Prioritizing resources.

At the most basic level, a single pest may be analysed in a pest risk analysis, and this is often referred to as an 'organism-specific pest risk analysis'. For imports and exports of agricultural products, pest risk analyses are conducted for the commodities in question. In this case, the 'commodity pest risk analysis' typically considers many organisms that may move with a particular commodity in trade. Alternatively, a pest risk analysis may be very broadly applied to consider pathways. A 'pathway pest risk analysis' typically considers different types of pathways, or means by which a pest can spread from one place to another. A pathway pest risk analysis can examine a single pathway, or multiple pathways, and may be concerned with a single pest or multiple pests. A commodity is one type of pathway (and a commodity pest risk analysis is one type of pathway pest risk analysis), however many other pathways may be the subject of an analysis, including natural spread. A pest risk analysis may also be done to support the revision of existing policies or implementation of new policies. For instance, a country may conduct a pest risk analysis on the level of trade in various ports to determine whether a change in port policies is needed.

Furthermore, any of these different pest risk analyses may incorporate a variety of methods and processes to analyse risk. The different methods, discussed in detail in different chapters in this text, include the use of mapping and geographic information systems, economic analysis, quantitative analysis and qualitative analysis. The methods that are used for the analysis depend largely on the type and scope of the problem being analysed, the data available to conduct the analysis and the overall need for the analysis in terms of detail, time and audience.

The different types of pest risk analyses vary in complexity, depending on the nature of the problem, the audience or application of the pest risk analysis and the availability of scientific and other information to support the analysis. This chapter provides an overview of the different applications of pest risk analyses. The general processes for conducting the different types of pest risk analyses are described in Part 3: Pest Risk Assessment Methods (Chapters 9–12). Chapter 20 also describes special considerations including environmental concerns in pest risk analyses.

8.3. Organism Pest Risk Analysis

Organism pest risk analyses are analyses that are conducted on specific species. This is usually an individual species (e.g. a fruit fly species, a weed, a beneficial organism, etc.) but may sometimes be applied to a group of organisms as whole (e.g. bark beetles, *Phytophthora* spp., etc.). An organism pest risk analysis may be done for organisms that are not yet present in a country to determine whether the pest is likely to be introduced to a country and the likely consequences if the pest were to be introduced. Organism pest risk analyses may also be carried out on organisms that have already been introduced, either intentionally (e.g. plants for planting or beneficial organisms), or unintentionally (e.g. a new pest species).

In the case of organisms that are not yet present in an area, the purpose of an organism pest risk analysis is usually to determine whether (and to what extent) measures should be taken to prevent its introduction to a new area. For organisms that have already been introduced, the purpose of the pest risk analysis is to determine whether measures should be taken, and how likely they are to be successful, to control or eradicate that pest.

The scope of organism pest risk analyses, as with all pest risk analyses, is determined primarily by the purpose of the pest risk analysis. As mentioned above, organism pest risk analyses may be done on individual species, and in those cases, the scope is usually well defined. Questions that may be asked in the analysis include:

- How likely is the pest to enter into a new area.
- What pathway(s) is the pest likely to follow leading to its introduction to a new area.
- How likely is the pest to establish in the new area.
- What pathways(s) is the pest likely to follow after its introduction to a new area (i.e. can the pest spread after it is introduced).
- What type of impacts can be expected (economic and/or environmental).

- What is the potential magnitude of the consequences.
- What measures can be applied to prevent introduction or to control a pest if it is introduced.
- How feasible are the measures, and are they cost-effective.

One of the best known pest risk analysis models for organism-level pest risk analyses has been developed by the European and Mediterranean Plant Protection Organization as the EPPO pest risk analysis Decision Support Scheme for Quarantine Pests (EPPO, 2007).

8.3.1. Types of organisms

Different types of organisms are analysed in organism pest risk analyses. Most commonly, organisms that are not yet present in an area or country, known as quarantine pests, are subject to analysis. This is because NPPOs want to know the risk associated with an organism before it has been introduced so that they can take appropriate actions to prevent its entry, if necessary. In some cases, pests may be considered to be 'imminent' threats – for example a new pest that becomes regularly intercepted in trade or a new pest in a neighbouring country are considered to be imminent. In these situations, the analysis is considered to be urgent since the likelihood of introduction (and therefore the risk) is perceived to be much higher.

A pest that has been introduced to a country may still be a quarantine pest, if it has only a limited distribution in the new area. Once the pest is present and established, a pest risk analysis may still be helpful in determining whether specific measures should be taken to control or eradicate that pest. Other types of organism pest risk analyses include beneficial organisms and weeds. These organisms are the subject of their own chapters.[1]

8.4. Pathways

The term 'pathway' is a very broad term – according to the IPPC it is 'any means that allows entry and spread of a pest' (IPPC, 2010). Pest risk analyses for pathways vary from relatively simple, short analyses to long and complex analyses, depending on the problem or question being addressed. As mentioned above, commodities are one type of pathway that allows pests to enter and spread in new areas, and the most common type of pathway that is analysed. However, other types of pathways exist, and can present significant risks that need to be analysed and potentially managed. As with organism and commodity pest risk analyses, the purpose of a pathway pest risk analysis is to estimate the level of risk posed by a particular situation or activity, and then determine whether measures would be feasible, effective and should be applied to manage the risk.

The scope of pathway pest risk analyses is as variable as the pathways themselves, although some generalities can be made. If a pathway is 'any means that allows entry and spread of a pest', we can see that in a pathway pest risk analysis, as with other analyses, the most basic unit for analysis would be a single pest, on a single pathway. Other possibilities include analysing multiple pests on a single pathway, a single pest on multiple pathways or multiple pests on multiple pathways.

Some pathway pest risk analyses analyse only specific times or events in a pathway – for instance, a pathway pest risk analysis may only examine likelihood of entry (e.g. examining how likely pests are to cross a particular border). In such cases, specific pests may not be identified in the analysis; instead, the analysis may focus on analysing the aspects of a border that make it more or less likely that pests may enter via that pathway. Alternatively, a pathway pest risk analysis may examine how much a pest could spread once it has entered and established in an area – in this case, components of most pest risk analyses such as likelihood of entry and establishment may be either assumed or already have occurred and are not included in the analysis.

Figure 8.1 shows a graphic representation of a pathway for pest introduction and spread, and provides us with a framework to consider what kind of events must occur

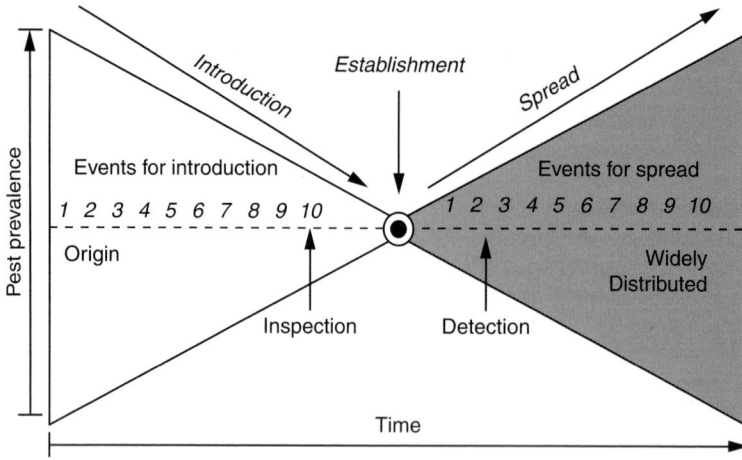

Fig. 8.1. Schematic representation of a pathway showing hypothetical events leading to pest introduction and spread.

for a pest to be introduced and spread. One of the first things we see in Fig. 8.1 is that pest prevalence may be very high at its origin (the far left side of the figure). Over time, and through a series of events (events for introduction 1–10), the prevalence may go down prior to entry (sometimes referred to as 'attrition'). Attrition, or reduction in the numbers of individual pests in the pathway, may be due to natural mortality, predation, lack of fitness, unfavourable conditions for the pest or human-mediated mortality (e.g. from treatments, culling, etc.). Even without any human intervention to manage the pest, we will often see a reduction in pest numbers simply because we know that pest populations do not show 100% survival over time due to natural mortality.

If we consider Fig. 8.1 to represent a pathway, for instance the importation of a fresh fruit commodity like mangos or apples, we can begin to define the specific events prior to export of the commodity and immediately after the importation of the commodity. These events may include infestation of the commodity in the field, harvesting infested fruit, culling infested fruit in a packing house, loading infested fruit into boxes, storing and then shipping infested fruit, etc.

Not only may we be able to define specific events along the pathway, we can measure those events – the prevalence of pest in the field for instance, or the effect of particular pest management measure (e.g. mortality of a treatment) – this becomes important if we apply quantitative methods in our analysis (addressed in Chapter 10). Furthermore, this diagram demonstrates that, particularly for pathway analyses, we may focus on only certain parts of the pathway. For instance, if we are concerned *only* with pest spread, after introduction, our pathway analysis would focus on events we have identified on the right side of Fig. 8.1 (e.g. 'events for spread').

8.4.1. Types of pathways

Movement of goods and conveyances

The movement of goods and conveyances results in creating a myriad of pathways for the human-mediated introduction of new pests. The most common pathways subject to pest risk analysis are imported and exported commodities (such as fruits, vegetables, nursery stock and other plant products). These are covered in detail in the section 8.5 on 'Commodity pest risk analyses' in this chapter. In addition to these kinds of commodities, there are numerous other pathways that result from the movement of goods and conveyances, and examples of these are provided below.

Commodities are shipped in a variety of ways – over land (rail and truck), by air (air cargo) or by sea (maritime cargo) in containers. When products are packed into containers, they are usually held in place on wooden pallets, and braced with pieces of wood (called dunnage). This wood, referred to collectively as 'wood packing material', is a significant pathway for the movement of wood pests. Both the Asian long-horned beetle and emerald ash borer were introduced into the USA via the wood packing material pathway. In order to address the risks associated with the pathway of wood packing material, the IPPC produced an ISPM (ISPM No. 15 *Guidelines for Wood Packing Material*; IPPC, 2009) that outlines measures that countries should use to mitigate risks associated with wood packing material. Although these measures (kiln-drying, debarking and fumigating) do not mitigate the risk of *all* wood pests, the standard establishes a baseline risk and the measures are aimed at managing that risk to an agreed upon acceptable level (see Chapter 4 for more on *harmonization* of measures).

In addition to the wood packing material serving as a pathway, the containers and methods of shipping can also become pathways for pests moving large distances. Asian gypsy moths are documented to lay eggs on ships, and shipping containers and egg masses can be moved from one place to another, allowing the gypsy moth to spread (USDA-APHIS, 2003). Similarly, snails have been known to adhere to railcars, and spread along rail lines and across continents.

Other types of goods or products not traditionally viewed as posing pest risks may serve as important pathways for pests. New Zealand examined the risks associated with the importation of vehicles from Asia. In its analysis, New Zealand determined that several hundred types of organisms could be carried into New Zealand via this pathway – these organisms included reptiles, amphibians,

arthropods, molluscs, nematodes and pathogens. As a result of the analysis, New Zealand determined that measures to mitigate risks associated with importation of vehicles should be required (MAF, 2007). Another example is the risks associated with snails being introduced on imported ceramic and marble tiles (used, for instance, in home decor). Snails are attracted to the calcium in ceramic and marble, and can adhere very tightly to tiles that are moved in trade. Although not traditional agricultural commodities, both vehicles and tile are effective pathways for moving important agricultural pests.

8.4.2 Natural pathways

There are many different types of pathways through which a pest may move from one place to another, sometimes over great distances. One of the most important, but often neglected pathways, are natural pathways. Natural pathways can be any naturally occurring event or characteristic, including the pest's innate ability to move on its own (e.g. flying insects), a pest's ability to move via wind or water, or any combination of those factors. Different types of rusts (plant pathogens) have been documented to move enormous distances by wind, even blowing across the ocean to spread from one continent to another. In other cases, small insects may fly or be blown by winds (including hurricanes) over relatively large distances.

There is a perception that natural pathways do not need to be analysed because natural pathways typically cannot be regulated or managed like other (human-mediated) pathways. However, analysing the natural pathways by which a pest can move establishes the baseline risk for a given situation – and is therefore an essential component in many pathway pest risk analyses. As an example, analysis of the natural pathways for soybean rust demonstrated that the disease would likely spread on air currents from South America to North America. No measures were available that would prevent the entry and establishment of soybean rust via natural pathways; therefore it was considered to be technically unjustified to require measures to prevent human-mediated spread because the baseline risk (from natural pathways) was not preventable.

Although natural pathways may be important for many types of pests, pathways that are human mediated are usually the ones that are of most concern to NPPOs. 'Human-mediated' means that the actions of humans, including travel, trade, smuggling, mail and various other activities create pathways by which pests can move. While trade related pathways (e.g. commodities) are important means for pests to spread, other types of human-mediated pathways can pose risks, including smuggling and travel.

Smuggling

One of the most difficult pathways to measure and analyse is smuggling. Some products are prohibited from being imported into a country because of the risk that product poses (either from the product itself in the case of weeds, or because of the pests it may carry). Prohibition is the most extreme option for managing risk, but may create hidden risks as a result of the prohibition. If a product cannot legally be imported, then people may try to illegally import, or smuggle the product into the country. This presents a greater risk because smuggled products are not subject to any phytosanitary measures. In addition, smuggled products may not be 'commercial grade' – that is, they may be of lower quality and carrier more pests than products produced specifically for commercial purposes. Although difficult to definitively prove, there are several important pests that were probably introduced into the United States through smuggled products. This includes the citrus disease, such as citrus huanglobing (or citrus greening), that was probably introduced into Florida on smuggled nursery stock (USDA-ARS, 2006).

Smuggling is a difficult pathway to analyse because it is difficult to gather data on the types of products that are smuggled, their origin and the types of pests that may spread with the smuggled product. In the USA, the Customs and Border Protection organization monitors our ports of entry for products being brought in illegally (e.g. smuggled or without a permit). They collect data on these products and look for significant trends between smuggled products, origins, destinations, etc. These types of data are the most useful in predicting future smuggling patterns, but smuggling as a whole remains a very complex, but poorly understood pathway.

Travel

More and more people are travelling internationally, and it is common for travellers to want to bring home souvenirs from their holidays. However, many of these souvenirs may present hidden risks – fresh fruits or flowers may be carrying important pests, as can other types of souvenirs such as wooden handicrafts. Like smuggled products, these souvenirs are not subject to phytosanitary measures, so any pests that are present have a high likelihood of surviving movement from one place to another. This is particularly the case when that method of transportation is via air travel because a person can travel literally thousands of miles in a matter of hours.

When passengers arrive in a foreign country, they are often required to declare whether they are carrying certain types of products (fresh fruit or flowers, meat, wooden handicrafts, etc.). In addition, suitcases are often subject to various types of inspection including hand searches, X-ray scans and even sniffer dogs that are specially trained to detect illegal products. If a person is caught with illegal products that they have not declared, they may be subject to significant legal penalties. Chapter 16 (Risk Communication) discusses how countries communicate with travellers regarding reducing risk via the particular pathway.

8.5. Commodity Pest Risk Analysis

Pest risk analyses for commodities are a special group of pathway pest risk analyses that are focused on analysing the risk of pests entering an area or a country by moving with a specific commodity. Pest risk analyses conducted for imported commodities are often referred to as 'commodity pest risk analyses' (or CRAs), 'import risk analyses' (or IRAs), or 'pathway initiated pest risk analyses'. Chapter 4 highlighted international obligations under both the WTO and IPPC, in particular that countries are obligated to provide a technical justification for measures. Most often, the technical justification is provided in the form of a pest risk analysis. When an exporting country makes a request to an importing country for market access for a commodity, the importing country should conduct a pest risk analysis to determine whether and to what extent phytosanitary measures should be applied to that commodity.

The purpose of a commodity pest risk analysis is to analyse the risk of pests associated with a specific commodity moving from one area to another. Commodities may be imported into a country from many places and the types of pests associated with a given commodity are dependent on the origin of that commodity. If it is determined that the commodity poses risks for moving pests, then appropriate measures can be put in place to manage those risks.

A commodity can be anything moving from one place to another, particularly in the context of international trade. Common commodities that are analysed for pest risk include fresh fruits and vegetables for consumption, nursery stock and plants for planting, seeds and germplasm, various types of wood products (e.g. wood packing material, raw lumber, logs, wood chips, etc.) and handicrafts made from plant materials (e.g. potpourri, baskets, decorative items). These types of commodities are well documented as providing pathways for pests to move great distances.

However, other types of commodities may also serve to transport pests even if they are not of agricultural origin. Snails are documented to adhere tightly to ceramic and marble tiles (as described in Section 8.4.1), attracted to the calcium in the tiles, and travelling great distances wherever they are shipped. Similarly, cars, vehicles and other types of machinery (e.g. farm equipment, military equipment) may move pests – pests may crawl into hidden spaces and/or lay eggs on the vehicles, or soil containing pathogens may be carried with vehicles from one place to another.

8.5.1. Types of commodities

In plant protection, it is well understood that some types of commodities typically pose greater risks than other types of commodities. The IPPC has produced a standard (ISPM No. 32 *Categorization of Commodities According to their Pest Risk*) that provides guidance on the level of risk posed by various types of commodities. The level of risk that a commodity presents is dependent on the:

- level of processing that the commodity is subjected to;
- the intended use of the commodity (e.g. consumption, planting, etc.).

In general, commodities for which the intended use is consumption (e.g. fresh fruits and vegetables, many types of grains, etc.) are considered to be inherently lower risk than commodities that are intended to be planted in the environment (e.g. plants for planting, seeds, etc.). This is because the pathway for consumption reduces (but does not eliminate) the likelihood that any pests, if present, would be able to escape into the environment and find suitable host material.

Products like seeds, germplasm and plants intended for planting are often regarded as higher risk than other categories. Pests that are present on seeds, germplasm or plants for planting have a higher likelihood of successfully entering, establishing and eventually spreading because they are intentionally introduced into the environment – therefore increasing the potential impacts associated with those pests. Products that are processed – oils, wood veneer, milled grain, cotton fibre, canned or processed vegetables, etc. – are considered to be lower risk because the processing often eliminates or kills any pests that are present.

8.5.2. Pest risk analysis for imported commodities

Usually, the decision by an importing country to conduct a pest risk analysis on a specific commodity is made in response to a specific request by an exporting country for permission to export a specific product. However, commodity import pest risk analyses may also be carried out in response to other types of activities, such as the movement of military equipment – particularly when military equipment is being returned to its country of origin after being deployed to another country. This is done to ensure that the equipment does not carry exotic pests back to the country of its origin.

The scope of a commodity pest risk analysis is most often defined as a commodity coming from a specific origin. For instance, an importing country may look at a specific type of fruit (e.g. papaya, oranges, mangos, apples, etc.) coming from a specific area or country (e.g. Australia, Brazil, India, USA, etc.). This is the most common and simplest approach that countries use. Less commonly, countries may decide to look at a commodity in a broader sense. This could be looking at an entire class of a commodity (e.g. *Citrus* spp. instead of just oranges, all 'stone fruit' instead of just peaches, etc.) from a specific origin, or it may be looking at a specific commodity, coming from many origins (e.g. mangos from any origin). The broad approach is usually more time-consuming and complicated, but once the pest risk analysis is done, it may be more comprehensive and reduce work in the future.

Different countries use various approaches to analysing commodities. However, the processes that are used should be consistent with the requirement of the IPPC pest risk analysis standards. Countries like Canada, Australia, New Zealand and the USA (USDA, 2000; Biosecurity Australia, 2001; CFIA, 2001; MAF, 2006; Burgman et al., 2010) each have their own commodity pest risk analysis guidelines that analysts use to analyse risks associated with commodities. Because commodity pest risk analyses are often shared between the importing country (that prepares the pest risk analysis) and the exporting country (that is exporting the commodity), it is important for the commodity pest risk analysis process to be transparent and justifiable. And like the EPPO pest risk analysis scheme discussed above, following guidelines for commodity pest risk analyses creates greater consistency between different pest risk analyses. Details on the methods used to conduct such analyses are provided in Chapters 9 and 10.

8.5.3. Pest risk analysis support for exported commodities

Under both the SPS Agreement and the IPPC, countries are obligated to base measures on international standards, or justify measures through risk analysis. Pest risk analyses are usually prepared by importing countries in response to a request from another country to export a particular commodity. The pest risk analysis becomes the technical justification for any requirements associated with the importation of a commodity, subject to specific risk mitigations.

Both agreements encourage cooperation and sharing information for the purpose of promoting safe and fair trade (see Chapter 4). The importing country is obligated to demonstrate risk and identify options for risk management. The exporting country should share information to facilitate the risk analysis process, including information on pests and risk mitigation procedures that occur in the exporting country. The exporting country is often in the best position to provide accurate information on issues such as pest distribution and commodity handling. An exporting country may provide information and analysis to an importing country to facilitate the risk analysis process and promote a technical dialogue.

The purpose of an export pest risk analysis is to provide scientific and technical support for a specific export. Export pest risk analyses are used to negotiate trade issues with other countries based on technical and scientific information and analysis.

As with any pest risk analysis, the information and analysis should be technically sound, comprehensive and defensible. Depending on the issue, the analysis may provide a general review of pests associated with a given commodity, potential risk mitigation options for a commodity and its pests, or an in-depth analysis of a specific pest problem that is hindering the export of a commodity to a foreign country.

It is important to understand, however, that export pest risk analyses usually do not include an assessment of consequences since the judgements that must be made to determine consequences are most appropriately done by the importing country. And because the consequences should be analysed by the importing (rather than exporting) country, the export pest risk analysis cannot make a determination as to the acceptability of the risk for the importing country. Therefore work on pest risk analysis for exports usually only results in portions of a pest risk analysis being provided (e.g. an analysis of likelihood of entry, or preparation of a pest list for a commodity).

Types of risk analyses support for exports

The majority of export risk analyses are one of three types: pest lists, review and response to other countries' pest risk analyses, and pest specific 'in-depth' pest risk analyses.

PEST LISTS. Importing countries often request an exporting country to provide a list of pests associated with a commodity before beginning their own pest risk analysis process. This list should include the most important common pests associated with a commodity. Pest lists may be the first step in a long process and can lead to additional specific analyses at a later date. Pest lists may also include, in addition to a basic list of pests, information on risk mitigation, pest distribution, commodity handling or other relevant information that the exporting country thinks will be useful to provide for the importing country's consideration. In some cases, the importing country may have requested the information; in other cases, the exporting country may volunteer the information.

REVIEW AND RESPONSE TO OTHER COUNTRIES' PEST RISK ANALYSES. An importing country should provide a risk analysis as the technical justification for any requirements on imported commodities. Before any final decisions are made, the exporting country typically has an opportunity to review and respond to a pest risk analysis. At this time, the exporting country may question judgements made by the importing country, request clarification on the analysis, provide additional information to clarify certain issues or provide an alternate analysis (e.g. a pest specific pest risk analysis).

PEST SPECIFIC 'IN-DEPTH' PEST RISK ANALYSES. In some cases, exports of a particular commodity may be prevented due to one or more specific pests that may follow the pathway of that commodity. The exporting country may prepare an in-depth analysis on these pests to support the negotiation process in order to facilitate market access in the importing country.

8.6. Supporting New Policies or Changes to Existing Policies

Pest risk analysis is also applied for a variety of purposes that are not directly related to international trade. One of the most important non-trade applications of pest risk analysis is to provide technical and scientific support to new or revised national policies. This includes:

- National reviews of phytosanitary regulations, requirements or operations;
- Evaluating the need for an official control programme;
- Evaluation of phytosanitary activities such as implementing pest free areas, areas of low pest prevalence or eradication programmes;
- Evaluation of a regulatory proposal of another country or international organization;

- Evaluation of a new system, process or procedure (or review of existing programmes);
- Technical and scientific inputs for international disputes on phytosanitary measures;
- The phytosanitary situation in a country changes or political boundaries change;
- Identification of research needs.

NPPOs carry out a range of activities to prevent the introduction and spread of pests – from excluding pests from entering to managing pests once they have been introduced, and all the stages in between. These activities are driven by policies, laws and regulations. Pest risk analysis is a useful means for evaluating and informing various types of policies – to predict likely outcomes, evaluate efficacy or provide other insights into policies and programmes administered by NPPOs.

For example, an NPPO may want to evaluate its inspection programme at various ports of entry – to determine whether different ports are exposed to different types or levels of risk, and whether the inspection resources allocated to given ports are commensurate with the level of risk those ports experience. Another example of a policy-level analysis would be if a country were to re-evaluate a given set of regulations for a particular commodity class (e.g. wood packing material, nursery stock or seeds as commodity classes). This is currently the case with the USDA – its regulations for 'Q-37' material (plants for planting) have undergone a revision, largely to make decisions for importation on plants for planting more risk-based and technically justified (see also Chapter 3 for information on the original 'Q-37' laws).

Like pathway pest risk analyses, the types of pest risk analyses that are done to support new and revised policies will vary, depending on the problem to be addressed. In some cases, only parts of an analysis may be done (e.g. only consequences may be evaluated for a pest that is established and under consideration for eradication). In other cases, a broad 'pathway' approach to pest risk analysis may be used (as might be

the case in evaluating inspection levels at a group of ports). The methods used to conduct pest risk analyses for policies are therefore highly variable, reflecting the range of applications for pest risk analysis support to decision making.

8.7. Prioritization of Resources

As mentioned above, NPPOs must carry out a range of activities to prevent the entry and spread of pests. Some of these activities are:

- Preventing entry of pests into an area (through quarantines, inspections, treatments, etc.);
- Preventing or managing spread of a pest once it is introduced;
- Surveying for pests that have not been introduced (e.g. general surveillance and intelligence gathering);
- Surveying for new pests within their territories (e.g. detection surveys);
- Monitoring existing programmes (including monitoring or delimiting surveys);
- Establishing and maintaining pest free areas or areas of low pest prevalence;
- Conducting official control programmes.

Ideally, NPPOs would have limitless resources to address all of these activities all of the time. However, realistically, NPPOs work with limited resources and must prioritize the types of activities that can be accomplished given time, staff and financial constraints. Pest risk analysis provides an objective, science-based and technically defensible approach to support prioritization of resources.

Pest risk analyses that are done to prioritize resources – like some pathway pest risk analyses or pest risk analyses for policies – are often partial. They may only look at certain aspects of risk (e.g. only analyse consequences, or only analyse likelihood of entry), or look at certain specific characteristics of a group of pests. This approach is used by the US Department of Agriculture to prioritize pests for domestic surveillance in the Cooperative Agricultural

Pest Surveillance (CAPS) program. The CAPs program organizes domestic surveillance activities for pests that are considered by experts to be threats to US agriculture and the environment.

The goal of such surveillance is early detection of serious pests if they are introduced, but before they become widely distributed. Because of resource constraints, it is not possible to conduct surveillance for *all* pests that may be harmful if introduced into the USA. Thus, a mechanism is needed to prioritize which pests are the most important and for which surveillance programmes will be conducted. Prioritization processes are also used for other purposes. The USDA has developed a weed risk analysis tool that ranks weeds (see Chapter 18), and the EPPO and the State of Victoria Department of Primary Industries in Australia also have applied prioritization schemes for ranking weeds (Weiss and McLaren, 2002; Brunel *et al.*, 2010).

8.8. Specialized Analyses

Other organisms that do not fall under the traditional purview of pest risk analysis (particularly commodity-based pest risk analyses) include pest risk analysis for beneficial organisms and pest risk analysis

for weeds. Both of these types of analyses present unique challenges, and modified approaches to pest risk analysis are used in each case. Beneficial organisms are covered in detail in Chapter 17 and pest risk analysis for weeds is covered in Chapter 18. Conducting pest risk analysis for invasive alien species and LMOs, including assessment of environmental impacts, is addressed in Chapter 20.

8.9. Summary

In this chapter we learned about the different applications of pest risk analysis. NPPOs use pest risk analysis for a variety of reasons – analysing risks associated with pathways (including commodities); analysing risks associated with specific pests; and for providing analytical support for new or revised policies. In addition, there are analytical methods that have been specially adapted for prioritizing resources. In short, almost every aspect of work that an NPPO must carry out can be supported by risk analysis – from making decisions on how to spend resources, to deciding on measures for imported commodities. The applications of pest risk analysis in plant protection continue to expand as NPPOs learn to apply pest risk analysis beyond the traditional applications for trade.

Note

[1] Chapter 17 discusses pest risk analysis for beneficial organisms; Chapter 18 discusses pest risk analysis for weeds, and Chapter 20 provides more information on pest risk analysis for 'living modified organisms' (LMOs) and invasive alien species.

References

Biosecurity Australia (2001) *Guidelines for Import Risk Analysis*. Agriculture, Fisheries and Forestry Australia/ Biosecurity Australia, Canberra.

Brunel, S., Branquart, E., Fried, G., van Valkenburg, J., Brundu, G., Starfinger, U., Buholzer, S., Uludag, A., Joseffson, M. and Baker, R. (2010) The EPPO prioritization process for invasive alien plants. *EPPO Bulletin*, 40, 407–422.

Burgman, M., Mittinty, M., Whittle, P. and Mengersen, K. (2010) *Comparing Biosecurity Risk Assessment Systems*. Final Report. ACERA Project 0709. University of Melbourne, Melbourne.

CFIA (2001) *Canadian PHRA Rating Guidelines*. Plant Health Risk Assessment Unit Science Advice Division, Canadian Food Inspection Agency, Ottawa.

EPPO (2007) *Guidelines on Pest Risk Analysis: Decision-support Scheme for Quarantine Pests*. EPPO Standard PM 5 / 3(3) EPPO, Paris. Available online at: http://archives.eppo.org/EPPOStandards/PM5_PRA/PRA_scheme_2007.doc, accessed 15 December 2011.

IPPC (2004) International Standards for Phytosanitary Measures, Publication No. 11: *Pest Risk Analysis for Quarantine Pests Including Analysis of Environmental Risks and Living Modified Organisms*. Secretariat of the International Plant Protection Convention (IPPC), Food and Agriculture Organization of the United Nations, Rome.

IPPC (2007) International Standards for Phytosanitary Measures, Publication No. 2: *Framework for Pest Risk Analysis*. Secretariat of the International Plant Protection Convention (IPPC), Food and Agriculture Organization of the United Nations, Rome.

IPPC (2009) International Standards for Phytosanitary Measures, Publication No. 15: *Regulation of Wood Packaging Material In International Trade*. Secretariat of the International Plant Protection Convention (IPPC), Food and Agriculture Organization of the United Nations, Rome.

IPPC (2010) International Standards for Phytosanitary Measures, Publication No. 5: *Glossary of Phytosanitary Terms*. Secretariat of the International Plant Protection Convention (IPPC), Food and Agriculture Organization of the United Nations, Rome.

MAF (2006) *Biosecurity New Zealand Risk Analysis Procedures*. Ministry of Agriculture and Forestry, Wellington.

MAF (2007) *Import Risk Analysis: Vehicles and Machinery*. Ministry of Agriculture and Forestry, Wellington.

USDA (2000) *Guidelines for Pathway-initiated Pest Risk Assessments*, Version 5.02. Plant Protection and Quarantine, Animal and Plant Health Inspection Service, United States Department of Agriculture, Washington, DC.

USDA – APHIS (2003) Asian Gypsy Moth factsheet. Available online at: http://www.aphis.usda.gov/publications/plant_health/content/printable_version/fs_phasiangm.pdf, accessed 15 December 2011.

USDA – ARS (2006) *Recovery Plan for Huanglongbing (HLB) or Citrus Greening caused by 'Candidatus' Liberibacter africanus, L. asiaticus, and L. americanus*. Available online at: http://www.ars.usda.gov/SP2UserFiles/Place/00000000/opmp/CitrusGreening61017.pdf, accessed 15 December 2011.

Weiss, J. and McLaren, D. (2002) Victoria's pest plant prioritisation process. In: Jacob H.S., Dodd, J. and Moore, J.H. (eds) *Proceedings of the 13th Australian Weeds Conference, Plant Protection Society of Western Australia*, Perth, pp. 509–512.

Part III

Pest Risk Assessment Methods

9 Qualitative Methods

Christina Devorshak

_____

9.1. Introduction

Now that we understand the background for pest risk analysis, we can start talking about *how* to do pest risk analysis. There are many different methods and approaches – varying from simple narrative reports to in-depth quantitative analysis using complex models. The choice of method is driven by many factors:

- Nature of the issue to be analysed;
- Purpose of the pest risk analysis;
- Intended audience;
- Availability, type and quality of information, evidence and data;
- Time available.

In a perfect world, we would have ample information, plenty of time and resources and a clearly stated purpose and audience for every pest risk analysis. However, in reality, we are faced with time, information and resource constraints, and sometimes the audience for and purpose of the analysis are not so clearly defined. All these factors will combine to determine (or limit) what sort of analysis should be done, and influence the method(s) we select.

Although there are several international and regional standards dealing with the topic of pest risk analysis, the standards are not prescriptive over how we analyse risk. The standards instead provide broad guidelines for the specific elements that should be considered in the risk analysis, and outline the minimum requirements that should be met in preparing and documenting a pest risk analysis. This means that it is up to the analyst, or the organization, to use their best judgement in selecting the most appropriate method(s).

General methods for pest risk assessment (qualitative and quantitative) are described in the current chapter and Chapter 10, respectively, and the pest risk analysis process is covered in detail in Chapter 11. This chapter includes a discussion of qualitative and semi-quantitative methods. The first part of this chapter focuses on methods to describe and rate risk elements qualitatively and semi-quantitatively; the second half of this chapter provides an overview of different approaches to pest risk analysis that use qualitative and semi-quantitative methods.

9.2. Qualitative and Quantitative Methods

In the broadest sense, we can think of methods as being qualitative or quantitative.

Qualitative risk analysis methods use subjective information and judgements, based on evidence, experience and expert judgement, to provide a description of the risk. Quantitative methods use objective information (usually numerical) and provide a numerical estimate of the risk.

Depending on the problem and availability of data, we may decide to use qualitative, quantitative or a combination of qualitative and quantitative methods. Analyses that incorporate detailed economic information, for instance, are usually quantitative, at least in part. To complicate matters, many qualitative methods convert ratings into numerical values, or use ratings to represent numerical ranges. Ratings are then 'quantified' and the analysis becomes, in effect, 'semi-quantitative'. Thus we often see that even if a process is predominantly qualitative or quantitative, most analyses include elements of both.

Qualitative methods are often effective, easy to understand, can be done with a minimum of information and resources, and may be useful for identifying subjects that may need more in-depth analysis. Qualitative methods can be applied to a variety of approaches for analysis, including:

- Generic rating systems (e.g. Low – Medium – High);
- Keys or schemes;
- Weighted evaluations;
- Narrative descriptions and reports;
- Decision sheets.

Because quantitative methods use objective information (e.g. numerical data), the information requirements and other necessary resources for conducting a quantitative pest risk analysis are usually much higher compared with a qualitative analysis. However, if we have the necessary information and data available, quantitative methods are useful to:

- Support qualitative work;
- Compare risks 'before and after';
- Build expert consensus;
- Express specific uncertainty;
- Evaluate incremental changes;
- Evaluate complex systems.

Some common approaches to quantitative analysis include probabilistic scenario analysis (including event and fault tree analysis), Monte Carlo methods and other models. Bayesian approaches are useful for working with limited data, as well as for showing inter-dependence of events. Quantitative approaches can be extremely powerful tools, but are not always the most appropriate methods to use. Quantitative approaches may not be well suited when:

- Most of the information is subjective;
- The audience does not understand the analysis;
- We lack resources;
- The techniques are poorly understood;
- There are not clear 'cause and effect' linkages.

Qualitative and quantitative approaches each have advantages and disadvantages. Qualitative methods are often easier to understand and apply, but may be less transparent and the results may be less precise. Quantitative methods may be more explicit in presenting the risk (including uncertainty) but still incorporate assumptions and judgements. Quantitative approaches may present a precise result, but such results may be easily misinterpreted if the underlying assumptions are not clearly presented and understood. The relative merits of using qualitative versus quantitative methods are the subject of considerable debate in the risk analysis community. Neither method is more 'right' or 'wrong' than the other; or 'better' or 'worse', and it is beyond the scope of this text to attempt to solve the debate here. However, as risk analysts, we should understand the applications and limitations of each in making the choice to use a qualitative or quantitative method.

This chapter will describe various types of qualitative approaches, and some semi-quantitative methods; Chapter 10 will describe quantitative approaches. However, readers should be mindful there is a continuous spectrum between quantitative and qualitative methods, and designating a method as qualitative, semi-quantitative or quantitative is somewhat artificial. Frequently models make use of combinations of methods, and we often see 'semi-quantitative' methods where qualitative information is quantified.

Before we go on to describe some of the specific approaches, we need to understand the use of descriptive terms and ratings, and how they are used in analyses. Table 9.1 provides examples of terms used in rating risk (including probabilities, consequences and uncertainty). This list includes some of the most common terms, from lowest to highest – but it is possible that other terms are used in rating risk.

9.3. Descriptors, Scoring and Ratings

We already have discussed phytosanitary terminology in previous chapters. The language of risk analysis, as a broad field, is highly variable and inconsistent. Particularly in qualitative analyses, there is no shortage of terms to refer to levels of probability and consequences, as we can see in Table 9.1. In addition to a diverse set of terms used to describe risk, there is also a lot of variability in how such terms are interpreted. In some cases, ratings are assigned, with no numerical probability ranges associated with the ratings (e.g. a purely qualitative description). In other cases, ratings are assigned specific probability ranges (e.g. using a 'semi-quantitative' approach). Organizations and analysts in many fields have attempted to assign specific probabilities to terms used to describe risk – and just as the terms are variable, the numerical probabilities assigned to the terms show a lot of variability as well.

Sherman Kent was an analyst for the US Central Intelligence Agency during the 1960s,

Table 9.1. Examples of terms used for rating probability, consequences and uncertainty. (Adapted from Kesselmann, 2008 and Hamm, 1991.)

	Descriptive terms		
	Probability	Consequences	Uncertainty
	Impossible		Uncertain
	Improbable	None	
	Uncertain	Negligible	Very low confidence
	Negligible		
	Rare		Moderately uncertain
	Very unlikely	Inconsequential	
	Highly unlikely	Insignificant	Low confidence
	Not very probable		
	Fairly unlikely	Minimal	Low
	Somewhat unlikely		
	Seldom	Minor	Medium
	Unlikely		
	Low	Low	
	Possible		Medium confidence
	Slightly less than half the time	Medium	
	Toss-up		Moderately certain
	Medium	Moderate	
	Slightly more than half the time		
	Better than even	Significant	High confidence
	Probable		
	Quite likely	Major	High
	Likely		
	A good chance	High	Very high confidence
	Highly probable		
	High	Massive	
	Almost certain		
	Definite		Certain
	Certain	Catastrophic	

Approximate level of probability, consequences or uncertainty — Lowest→ ←Highest at the height of the Cold War, and was responsible for analysing military intelligence. Many of the analyses relied on what are termed 'words of estimative probability', and Kent attempted to designate specific meaning and values to the vocabulary used in the intelligence reports (Kent, 1964). This represented one of the first attempts to designate specific probabilities to verbal expressions. We can see in Table 9.2 that Kent matched terms to numerical probability ranges, but in many qualitative risk assessments terms are described without any numerical probabilities.

Different academic fields use different ranges for what might be considered 'high', 'low' or some other descriptor. In discussing probabilities associated with climate change, the Inter-Governmental Panel on Climate Change (IPCC) established the following criteria (Patt and Schrag, 2003; IPCC, 2007):

- Virtually certain >99% probability of occurrence;
- Extremely likely >95%;
- Very likely >90%;
- Likely >66%;
- More likely than not >50%;
- Very unlikely <10%;
- Extremely unlikely <5%.

The IPCC goes a step further to provide probability range estimates for uncertainty (or confidence as it is referred in IPCC documents):

- Very high confidence: at least a 9 out of 10 chance of being correct;
- High confidence: about an 8 out of 10 chance;
- Medium confidence: about a 5 out of 10 chance;
- Low confidence: about a 2 out of 10 chance;
- Very low confidence: less than a 1 out of 10 chance.

Note that just in comparing these three approaches, we see a lot of variability in the terms that are used, and in what the terms mean. In some cases, identical terms are used, but have very different meanings. 'Likely' for medical consent means greater than 50% probability, while 'likely' for the

Table 9.2. Kent's words of estimative probability. (From Kent, 1964.)

100% certainty		
The general area of possibility		
93%	give or take about 6%	Almost certain
75%	give or take about 12%	Probable
50%	give or take about 10%	Chances about even
30%	give or take about 10%	Probably not
7%	give or take about 5%	Almost certainly not
0% Impossible		

IPCC means greater than 66%. If we look at how different plant protection organizations apply terms and criteria, we would probably see just as much variability. So much variability means that different individuals and organizations will rate risk and interpret results very differently. It is also absolutely essential to understand that the numerical probabilities in qualitative assessments are only rough estimates, and that the rating is the result of the judgement of the analyst.

What does this mean in practical terms though? Although it would be convenient if all the NPPOs could agree on a single set of ratings and explanations for those ratings, this is unrealistic and beyond the scope of harmonizing pest risk analysis methods. However, it is important that any ratings that are used are transparent (i.e. based on science and technically justified) and applied consistently. This sounds simple enough, but in practice can be challenging since even individual analysts may vary in their attitudes towards certain types of risks.

One mechanism to improve consistency is to provide very specific guidance to analysts on the criteria for different ratings. Considerable effort has been put into developing such guidance – for example, a European project on improving pest risk analysis procedures (PRATIQUE) included a work package on improving consistency in ratings. The solution was viewed to be the development of highly specific and detailed guidance on what constitutes different scores, throughout the entire pest risk analysis 'scheme'.

Rating and scoring systems are ingrained in the methods for risk analysis, and particularly for pest risk analysis. Although there are some clear disadvantages, and potential problems, associated with using such systems, they remain one of our most useful tools for analysing and communicating risk. Audiences – decision makers, stakeholders and trading partners – are comfortable with rating systems and often have a high level of confidence in the results, even if the results may be vague or imprecise (Hamm, 1991). And because we often lack detailed information on pests, the more generalized approach of rating systems is often the most appropriate.

The question remains however – what is the best rating system to use? There is no single answer to that question. Some NPPOs use three levels (e.g. high, medium, low) while others use six or more levels (e.g. negligible, very low, low, medium, high, very high). The choice of the number of ratings to use is relatively arbitrary – the rating itself may be overly refined when refined data are lacking; but using few ratings may mean it is harder to distinguish between levels of risk. Whatever descriptors and number of descriptors that are used, ratings should be:

- Technically justified;
- Transparent;
- Consistently applied.

9.4. Criteria for Ratings

The pest risk analysis standards provide guidance on the elements that must be considered in an analysis (see Chapter 11). The simplest analyses typically use each one these elements as broad questions, each of which can have an associated rating. The questions address:

- Likelihood of entry (Is a quarantine pest associated with a pathway? Can it survive the pathway? Will it be detected or eliminated before entry?);
- Likelihood of establishment (Will the pest be able to transfer from the pathway to a suitable environment and be able to find host plants?);

- Likelihood of spread (Is the pest capable of moving over large areas? Does it reproduce quickly and in large numbers?);
- If and to what extent the pest has consequences (economic, environmental and/or social consequences).

Each one of those questions and subquestions can be rated. We could simply choose to leave it up to individual analysts to decide for themselves what constitutes a 'high', a 'negligible' or a 'medium', etc. based on that analyst's experience. However, because analysts bring different backgrounds and experiences to their work, we may end up with very different viewpoints resulting in considerable variability between individuals – and our analyses might not be transparent, objective or consistent.

A better approach is to develop and provide specific criteria for rating each one of these questions and apply these criteria consistently between pest risk analyses to the extent possible. For each question, our model can describe a set of parameters for what constitutes specific ratings. For instance, in looking at the likelihood that a pest can find suitable host material, we may designate scores based on the presence of the number of species the pest can feed on, availability of the host throughout the year, and the area over which suitable host material is planted within the pest risk analysis area. If no host material is available (as may be the case with a very host-specific pest for instance), we might designate that as negligible. If host material is widely available, the pest is highly polyphagous and the area over which such hosts occurs throughout the year is very large, we could designate that as 'very high'.

The application of consistent and clear criteria is a critical component of any pest risk assessment model that is based on ratings. Any criteria developed for a particular risk element or event should be specific enough to provide sufficient guidance to an analyst to make consistent judgements, but flexible enough to fit a wide range of situations. In some cases, it may be useful to provide examples of particular pests that meet a given criterion to aid the analysts in making judgements. The European pest risk analysis project 'PRATIQUE' (Baker *et al.*, 2009) has adopted this approach, and its rating system is based on having well-defined criteria with examples for every rating (Schrader *et al.*, 2010). The results of PRATIQUE were used to revise the EPPO pest risk analysis Decision Support Scheme (EPPO, 2007), and develop their online risk analysis tool, 'CAPRA' (computer-assisted pest risk analysis).

CAPRA takes the analyst through a series of questions in the risk analysis, and each question includes various types of guidance or examples for how that question should be rated. For example, under 'Probability of entry of pest', there are several sub-questions. One question addresses how likely the pest is to be associated with the pathway at the point(s) of origin taking into account the biology of the pest, and provides the following points to consider in determining a rating:

- Is the pest in the life stage that would be associated with the commodities, containers, or conveyances?
- For plants do seeds or other propagules have access to commodities, containers or conveyances?
- Is seasonal timing appropriate for the pest to be associated with the pathway at origin?

By having specific criteria, and even examples, for each rating, we can provide guidance to the analyst as to how to rate different elements. In addition, if the criteria are readily available and understood, the criteria themselves increase transparency and provide additional analytical support to decision makers.

There are, however, drawbacks to having highly defined criteria. First, the sheer variety of pests, commodities, pathways and other factors to consider makes it impossible for criteria to fully cover all possibilities. It is inevitable that pests or situations will arise that do not 'fit in the box', and constrain the analyst by forcing them to select options from criteria that may not be the most appropriate. Second, providing specific criteria may keep the analyst from

fully exploring the available evidence if the inclination exists to simply find enough evidence to support a particular rating. As with all aspects of risk analysis, it is therefore critical that we focus on understanding the benefits and limitations of any particular approach we choose to use.

9.4.1. How are ratings combined?

Recall from earlier chapters that risk is a product of probability and consequences. Furthermore, the probability of an event occurring is usually the result of a sequence of smaller events – each of which can be individually rated. In order to complete the analysis, and summarize the information in a useful way for the audience and the decision makers, we are then faced with trying to combine different ratings into an end result. This is also the point where we start crossing the line from purely 'qualitative' approaches to semi-quantitative approaches – where we take qualitative information and 'quantify' it, enabling us to use mathematical methods to describe risk.

9.4.2. Adding scores

There are a variety of means for combining qualitative ratings. The simplest approach is to simply add ratings together to come up with a numerical score or rank. Recall that ratings may be assigned verbal descriptions (e.g. high, medium, low), or numerical values (e.g. 3, 2, 1). In cases where verbal descriptions are used, numerical values can be assigned to those verbal values (e.g. High = 3, Medium = 2 and Low = 1). For any given model, there would be a minimum possible score, a maximum and then a range of results in the middle. The NPPO of the USA (US Department of Agriculture) used a rating system for conducting pest risk analyses for the importation of fruits and vegetables for consumption. The model included several 'elements' for both likelihood of introduction and consequences of introduction. Each element was scored high, medium or low and the numeric rank of 3 (for high), 2 (for medium) or 1 (for low) assigned. The scores were then totalled together and overall risk was expressed as the 'pest risk potential' and ranked according to the final score as high, medium or low.

The advantage of such an approach is that it is simple to apply and understand. The logic behind the model is clear and transparent, and it can be broadly applied to a variety of situations. However, although this approach has been widely used and is generally accepted, it has some flaws.

The most important flaw to a strictly additive approach stems from the fact that likelihood of an event occurring results from a series of *dependent* events – each event in a chain of events must occur with some probability in order for the risk to be realized. If any one event has a zero probability of occurring, then the risk is negligible. However, the additive approach described above does not allow for dependence of events since a value of 'zero' or negligible can never result. Every element is treated as an independent event, and therefore the additive approach does not accurately reflect the mathematical relationship of events in risk. This approach could lead to an artificial inflation of the risk.

9.4.3. Matrices

A common approach to combining ratings is to use matrices (Altenbach, 1995; Yoe, 2012). In this approach a matrix can be used to combine ratings for two different elements – most commonly overall likelihood and consequences. A simple 3×3 matrix is shown in Fig. 9.1. In this instance, both consequences and probability each

have three possible ratings (high, medium, low), and the resulting overall risk rating may also be high, medium or low. If probability is high, and consequences are high, then the overall risk is rated high. If, on the other hand, the probability is low, but consequences are high, then the overall risk (according to this matrix) is medium.

Matrices can be expanded to include more options beyond high, medium and low. Australia uses a six-rating matrix (Fig. 9.2) for its import risk analyses (Biosecurity Australia, 2001). Australia goes one step further in their matrix, defining their ALOP as 'very low risk' (see Chapter 4 for a discussion of ALOP). Thus, the risk for any organism that falls below or at the level of 'very low risk' is considered to be acceptable, and anything that falls above that level is considered to pose an unacceptable risk (and therefore would require mitigation). Note that this matrix is applied by Australia in different sectors, and they refer to 'entry and exposure' – this is equivalent to entry and establishment (or introduction) in IPPC terms.

This approach demonstrates one of the main advantages to presenting information in a matrix format – the results are transparent and informative. In addition, results are easily compared from one analysis to another, facilitating greater consistency in decision making and risk management.

The two figures above combine ratings for probability (or likelihood) and consequences – often the last step in the risk assessment process. However, ratings earlier in the process, particularly for likelihood can be combined using the matrix approach as well. For instance, the likelihood of a pest finding suitable host material can be combined with the likelihood of the pest finding a suitable climate to provide an

Consequences of event		Probability of event		
	Rating	Low	Medium	High
	Low	Low	Low	Medium
	Medium	Low	Medium	High
	High	Medium	High	High

Fig. 9.1. Example of a simple 3×3 matrix.

Likelihood of entry and exposure						
High likelihood	Negligible risk	Very low risk	Low risk	Moderate risk	High risk	Extreme risk
Moderate likelihood	Negligible risk	Very low risk	Low risk	Moderate risk	High risk	Extreme risk
Low likelihood	Negligible risk	Negligible risk	Very low risk	Low risk	Moderate risk	High risk
Very low likelihood	Negligible risk	Negligible risk	Negligible risk	Very low risk	Low risk	Moderate risk
Extremely low likelihood	Negligible risk	Negligible risk	Negligible risk	Negligible risk	Very low risk	Low risk
Negligible likelihood	Negligible risk	Negligible risk	Negligible risk	Negligible risk	Negligible risk	Very low risk
Rating	Negligible impact	Very low impact	Low impact	Moderate impact	High impact	Extreme impact
Consequences of entry and exposure						

Fig. 9.2. Risk matrix used by Australia. (Biosecurity Australia, 2001.)

Likelihood of finding suitable host material	Likelihood of finding suitable climate			
	Rating	Negligible	Low	Medium	High
	Negligible	Neg.	Neg.	Neg.	Neg.
	Low	Neg.	Low	Medium	Medium
	Medium	Neg.	Low	Medium	High
	High	Neg.	Medium	Medium	High

Fig. 9.3. Likelihood of establishment.

overall rating of likelihood of establishment (Fig. 9.3).

If either likelihood (climate or host) is 'negligible' then the overall likelihood of establishment is also negligible. Mathematically, if we consider 'negligible' to be a value close to zero, we know that multiplying a number by zero results in a value of zero. Logically, if there is either no host material available, or if there is no suitable climate, then we can judge that the pest will not be able to establish. By using a matrix, we can clearly show the relationship and dependence of these two elements.

9.4.4. Combining matrices within a pest risk analysis

More detailed and complex models have been developed that enable us to combine a series of matrices together to develop an overall risk rating, based on a 'matrix models' approach. In Chapter 11 (Pest Risk Assessment) we explain the steps that should be analysed for a pest to be able to enter, establish and spread. Each one of these elements may be further broken down into sub-elements that can be analysed and rated.

The resulting ratings for each element or sub-element can be combined – two at a time – into matrices, and the results of one matrix can then subsequently be combined with results of a second matrix. For example, we can take the results of Fig. 9.3 (where we have already combined two elements) and combine that with a third element. The result of the new matrix could then be combined with yet another element and so forth, until all of the elements are combined together, resulting in a single overall rating.

9.4.5. Example of how multiple matrices can be combined

Figures 9.4a–c provide an abbreviated example of how results from two different matrices can be combined into a third matrix. In this simplified example, we are looking at the likelihood of entry (Fig. 9.4a) as a product of the likelihood of a pest being associated with the commodity at time of shipment and the likelihood of the pest surviving transit. For our hypothetical example, we have rated the likelihood of the pest being on the commodity at the time of shipment as 'high', and the likelihood of surviving transit as 'medium', giving a combined rating of 'high' according to this matrix (see highlighted cell).

Figure 9.4b shows the likelihood of establishment as a product of the pest finding suitable climate and host material. Our hypothetical rating for the likelihood of the pest finding a suitable climate is 'low', and the rating for it finding suitable host material is 'high', giving an overall rating of 'medium' for the likelihood of establishment according to this matrix (see highlighted cell).

Finally, Fig. 9.4c shows the likelihood of introduction as a product of the likelihoods of entry and establishment. The combined ratings from the previous two matrices (high likelihood of entry and medium likelihood of establishment), giving a final result of 'high' likelihood of introduction.

Although it may be possible to conduct such an analysis by hand, there are software packages that can provide the analysis once we input the appropriate values. Relatively simple models that have a limited number of steps can be handled manually, while more complex models with many steps are best handled using computer software. Bayesian belief networks (BBNs) are based on such an approach of combining ratings

		Likelihood of being on the commodity at time of shipment			
Likelihood of surviving transit	*Rating*	Negligible	Low	Medium	High
	Negligible	Neg.	Neg.	Neg.	Neg.
	Low	Neg.	Low	Medium	Medium
	Medium	Neg.	Low	Medium	*High*
	High	Neg.	Medium	Medium	High

Fig. 9.4a. Likelihood of entry.

		Likelihood of finding suitable climate			
Likelihood of finding suitable host material	*Rating*	Negligible	Low	Medium	High
	Negligible	Neg.	Neg.	Neg.	Neg.
	Low	Neg.	Low	Medium	Medium
	Medium	Neg.	Low	Medium	High
	High	Neg.	*Medium*	Medium	High

Fig. 9.4b. Likelihood of establishment.

		Likelihood of entry			
Likelihood of establishment	*Rating*	Negligible	Low	Medium	High
	Negligible	Neg.	Neg.	Neg.	Neg.
	Low	Neg.	Low	Medium	Medium
	Medium	Neg.	Low	Medium	*High*
	High	Neg.	Medium	Medium	High

Fig. 9.4c. Likelihood of introduction.

in a stepwise manner – and a BBN model has been developed for pest risk analysis under the European project PRATIQUE in cooperation with EPPO. The details of developing and using BBNs for pest risk analysis are beyond the scope of this text; however, more information on the application of BBNs can be found in McCann *et al.* (2006) and through the EPPO pest risk analysis tool CAPRA (www.eppo.org).

9.4.6. Converting verbal scores into numerical values in a matrix

Matrices are extremely useful for conveying information in a simple, transparent manner. However, it is important to understand that, in many cases, the combination of scores to produce a summary rating can be arbitrary (Altenbach, 1995). It is often left to the decision maker to interpret what a combination of 'high likelihood + low consequences' means compared with 'medium likelihood + medium consequences', and so forth. Converting ratings into numerical values (e.g. H = 3, M = 2, L = 1) can promote objectivity since the final 'score' is easily comparable with other scores.

Figures 9.5a and 9.5b show how qualitative ratings from Fig. 9.1 (our earlier example of a matrix) can be converted to numerical ratings. A 'low' likelihood of introduction ('1') multiplied by a medium consequences rating ('2') produces a score of '2'. The same can occur if the rating is for low consequences but medium likelihood. A 'high' likelihood of introduction ('3') multiplied by a 'high' consequences rating ('3') produces a score of '9'. The range of possible scores is then 1–9 (1 being the lowest and 9 being the highest).

What constitutes deciding whether a final combined score is 'low', 'medium' or 'high' risk is fairly arbitrary however. For example, we may decide that scores of 1–2 are low, 3–4 are medium and 5–9 are high; or we may decide that 1–3 are low, 4–6 are medium and 9 is high (in this model, final scores of 7 and 8 are not possible). The interpretation of the final score is ultimately a policy decision but interpretation of results should be consistently applied from one analysis to another.

The information we can gain from this approach is the *relative* risk of a hazard within the context of that analysis, or between different analyses that use the same model. Note, however, that if different models are used, values should not be compared between different analyses since they may not be using the same criteria for ratings (see earlier section on ratings).

Be mindful that the decision to convert qualitative ratings to numerical values can mislead readers into thinking that the analysis is a quantitative analysis since the

Consequences of event	Probability of event		
	Rating	Low	Medium	High
	Low	Low	Low	Medium
	Medium	Low	Medium	High
	High	Medium	High	High

Fig. 9.5a. Example of a simple matrix for combining scores – verbal values 'combined'.

Consequences of event	Probability of event		
	Rating	1	2	3
	1	1	2	3
	2	2	4	6
	3	3	6	9

Fig. 9.5b. Example of simple matrix for combining scores – verbal values converted to numerical scores and combined.

end result is expressed numerically, rather than qualitatively. This makes providing the basis and justification for the model itself, as well as the evidence behind the ratings all the more important. Furthermore, simple matrices are not able to directly incorporate uncertainty – therefore an accompanying discussion of uncertainty is absolutely essential to provide additional guidance for the interpretation of results.

The value and use of qualitative ratings combined into matrices is not without some controversy. Some risk analysts point out the arbitrary nature of ratings and how final scores are added together and caution that such approaches can be misleading. Others point out that, in situations where quality data are lacking, the qualitative approach (and resulting matrices) is our best option for providing a transparent analysis.

The authors of this text recommend that, no matter what approach is used, the most important factors are that the analysis is clearly documented, the model is transparent, based on the best scientific evidence available and incorporates and discusses uncertainty. This is consistent with the expectations of the relevant international agreements and standards, and effectively meets the needs of decision makers and trading partners, and provides the basis for further discussions.

9.5. Rating and Integrating Uncertainty

Just as specific elements of a pest risk analysis can be rated, uncertainty can also be rated. Individual elements can have uncertainty rated, or uncertainty can be given an overall rating at the end of the analysis. In practice, it is difficult to directly incorporate uncertainty into the ratings for elements, or into matrices. However, new methods are being developed that use a semi-quantitative approach and assign distributions to values derived from ratings, with the distributions incorporating the uncertainty. This approach has been adopted by the PRATIQUE project and the revised EPPO pest risk analysis Decision Support Scheme (EPPO, 2007; Baker et al., 2009).

However uncertainty is described (e.g. narrative, rating, quantified), it should be clearly communicated. This includes identify the sources and level of uncertainty, and gaps in knowledge where additional information could prove useful. Uncertainty is more fully explained in Chapter 16.

9.6. Qualitative Approaches

9.6.1. Narratives and decision sheets

Narratives and decisions sheets are usually the simplest approaches for conducting a pest risk analysis. A narrative analysis is usually a short report that outlines the specific issue of concern, and provides a verbal description of risk (including uncertainty). Even though a narrative report is usually a simplified type of pest risk analysis, it should still include a description of key components of a pest risk analysis. Specifically, the analysis should describe the likelihood that a pest will be able to enter, establish and spread, explain the potential economic consequences and describe any uncertainty in the analysis. The format for narratives can be extremely flexible and left to analysts to decide – but should be structured in such a way as to provide a logical flow.

Likelihoods for the entry, establishment and spread components are usually described in qualitative terms such as 'high', 'medium', 'low' or 'negligible', or simply 'likely' or 'unlikely'. However, the use of such descriptive terms is often left to the analyst's discretion (rather than specified ranking criteria), and the criteria for assigning the qualitative terms may be undefined and subject to interpretation. For example, one analyst may consider a moth that can lay 100 eggs in its lifetime to have a high level of fecundity (based on that analyst's background); another analyst with a different background may judge 100 eggs per moth to be a low level of fecundity. The same information in this case has been interpreted very differently by two analysts because no specific criteria exist for rating, in this case, the fecundity of the moth.

Consequences may also be described qualitatively (low, medium, high or acceptable/unacceptable), or quantitatively if impacts can be reasonably estimated. For example, we may say 'the potential impacts from the pest would be high because it attacks many important crop plants' or we may say 'the estimated trade impacts from the pest could exceed X $ value'. Chapter 7 and Chapter 11 provide more information on describing economic consequences.

A decision sheet is a type of narrative analysis that usually follows a specific format or template. Decision sheets are often used to provide quick, simple summaries to risk managers and decision makers regarding a particular pest or situation. Decision sheets are a means to provide an analysis in a highly structured format that makes it easy to find and interpret information and thus formulate recommendations for decisions. Box 9.1 shows an example of a report template for

---

Box 9.1. Example of a report template. (Adapted from the USDA PPQ New Pest Advisory Group report template; USDA, 2010.)

Report Template

Pest identification: Identify the pest as specifically as possible and indicate why an analysis is necessary

Current policy: Describe the current national policy regarding the pest.

Pest situation overview:

Exotic status: Indicate whether the pest is new to the pest risk analysis area, reintroduced, or an imminent threat with a pathway for introduction (specify the pathway).

Biology: Provide details on the biology of the pest needed to assess risk and develop scientifically sound recommendations (e.g. type of damage caused, reproductive patterns, reproductive potential, developmental thresholds, adaptability, lethal environmental conditions, associated plant pathogens or vectors, modes of transmission...).

Prevalence and global distribution: Provide a list of countries where the pest occurs.

Host range: Provide a list of hosts.

Potential distribution in the PRA area and spread: Based on the biology of the pest, its current prevalence and global distribution, and host range, write a short paragraph that describes where it could establish in the pest risk analysis area. Discuss methods of dispersal and likelihood of spread. Use risk maps when appropriate or if available.

Potential pathways of introduction: Provide information on any means (e.g. importation of a host, interstate trade, smuggling, a hurricane...) that would allow the introduction of the pest as defined by the IPPC. Interception data can be used as evidence.

Detection and control: Explain detection and control options that exist or that are being developed for this pest. Indicate whether these options are currently available in the pest risk analysis area.

Potential economic impacts: Explains potential economic impacts as described in Supplement No. 2 of the IPPC *Glossary of Phytosanitary Terms* (IPPC, 2010) and Section 2.3 of ISPM No. 11 (IPPC, 2004).

Trade implications: Explain anticipated trade implications. Is the presence of this pest likely to close foreign or domestic markets? Do any foreign countries regulate commodities for the presence of the pest? Do any foreign countries consider your organism a quarantine pest? Is there any indication that the pest may limit domestic movement of potentially infested or infected commodities?

Potential environmental impacts: Explain potential environmental impacts as described in Supplement No. 2 of the IPPC *Glossary of Phytosanitary Terms* (IPPC, 2010) and Section 2.3 of ISPM No. 11 (IPPC, 2004).

Current regulatory response and activities: Explain regulatory activity regarding the pest's introduction or potential introduction.

Need for new technology or knowledge: Describe areas of uncertainty. Provide an explanation of missing information that is essential to make scientifically sound recommendations or conduct a regulatory programme.

Recommended policy: Recommends an appropriate regulatory policy.

Recommendations for phytosanitary actions: Provide recommended phytosanitary actions.

analysing newly introduced pests and pests that have the potential to be introduced into a pest risk analysis area. Note that it includes the basic elements we should analyse in a pest risk analysis: likelihood of entry, establishment and spread, potential economic consequences and uncertainty.

There are several advantages to using a narrative approach to pest risk analysis. First, the requirements for information and quantitative data are minimal compared with other types of analyses and are particularly well suited for situations where available information is lacking. Second, these types of reports can usually be prepared relatively quickly – often such reports are used as preliminary analyses to decide whether more in-depth (and time-consuming) analyses should be conducted. Third, they do not require special expertise or other specialized resources (e.g. computer programs, etc.) and are therefore an accessible type of analysis that anyone can prepare. Last, narrative reports are easy to understand and use.

Narrative reports are extremely useful, but have some disadvantages. First, because they are descriptive, rather than quantitative, results can be difficult to compare from one analysis to another and are therefore not suited for comparisons or rankings. Second, they may not be very precise since they are based on general information, and therefore can only provide rough estimations and summaries of risk. Third, narrative analyses are not explicit with regard to the appropriate level of protection, so the basis for a decision for risk management may be less transparent. Last, because narrative reports can have variable formats and information, there may be a lack of consistency from one report to another, although using decision sheets and templates can help to reduce problems from inconsistency and help to increase transparency.

9.6.2. Generic guidelines

A more structured approach than simple narratives and decision sheets is to use defined guidelines. Like ISPMs, guidelines provide general guidance on how to analyse particular risks. This approach is commonly used for routine pest risk analyses, such as those done for the importation of specific commodities.

By using guidelines, the NPPO ensures a consistent approach to routine analyses, and allows for comparison between analyses. In addition, following guidelines can make the work of the analyst much easier by providing a framework for the analysis. As with ratings however, guidelines usually need to be fairly generic so that they are broadly applicable to a range of situations. This is one of the main disadvantages to using guidelines. Another disadvantage is that analysts may become overly dependent on using guidelines and become less creative and analytical in their work.

Examples of generic guidelines used by different NPPOs are available online on the internet. The USDA Plant Protection and Quarantine Service (USDA, 2000) and the Australian Import Risk Analysis process (Biosecurity Australia, 2001) have been used for years for routine pest risk analyses for the importation of commodities by the respective countries. Both of these approaches are consistent with the requirements of the SPS Agreement, the IPPC and the relevant international standards. Generic guidelines provide a framework of questions or elements that should be addressed in the analysis, and incorporate the use of ratings for elements. As discussed earlier in this chapter, guidance is provided on criteria for rating each element, as well as for the interpretation of final results. Figure 9.6 provides an example of a generic model for pest risk analysis, including elements that can be rated.

Guidelines are structured to cover essential elements identified in ISPM No. 11, but vary in specific approaches. This is because NPPOs may address various elements using different criteria, depending on their specific situation. For instance, the USA contains a wide range of climate zones, from sub-tropical to cold temperate areas – therefore any elements related to climate and host plants need to account for the

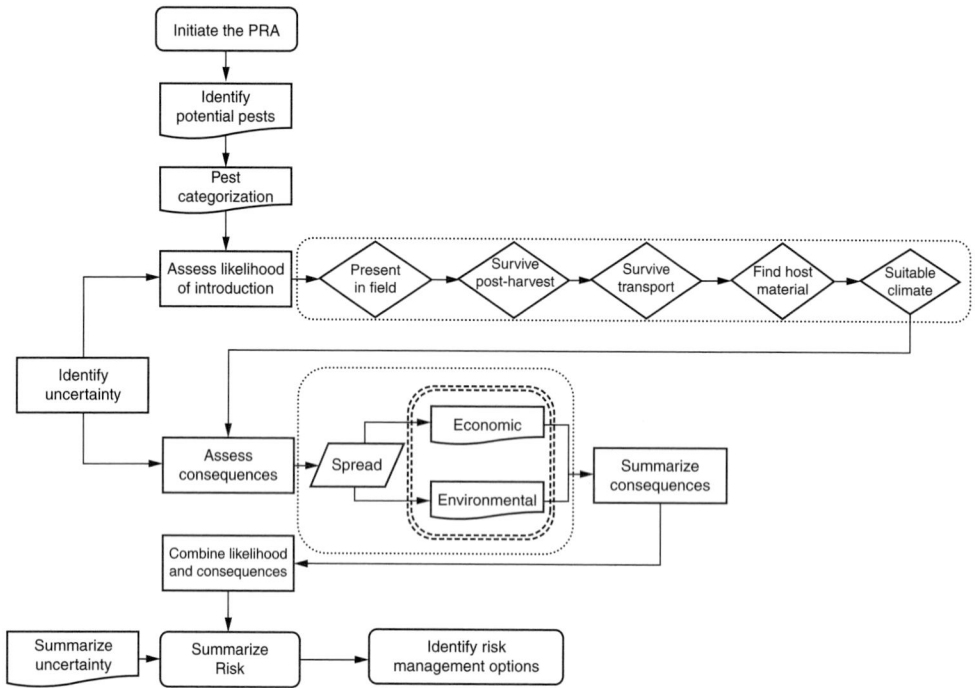

Fig. 9.6. Example of a generic model for conducting a commodity pest risk assessment.

entire range of possibilities. Other countries, for instance Caribbean countries, are sub-tropical to tropical, so the criteria used to judge elements related to climate and host plants will look very different from the criteria used by the USA.

In general, generic guidelines address the following:

Probability (or likelihood) the pest will:	**Consequences**
Occur on a commodity in the field	Direct economic
Survive post-harvest handling	Indirect economic
Survive transport from one area to another	Environmental
Escape detection	Other
Find suitable climate	Social
Find suitable host(s)	Political
Reproduce and spread	Aesthetic

As discussed in the earlier section in this chapter on ratings, results can be combined in a variety of ways. Ratings can be added, multiplied or assembled into matrices to communicate a final result. Likelihood and consequences may be communicated as a single value (as in the case of a matrix for likelihood and consequences, see Fig. 9.5a and b for an example); or simply communicated as the likelihood and associated consequences (e.g. 'low likelihood with high consequences'). If this approach is used, then uncertainty is discussed either throughout the analysis (i.e. with each element), as a summary at the end or some combination of those.

Some guidelines use a 'semi-quantitative' approach. Ratings are either converted to numeric values according to rating level (e.g. 1, 2, 3, 4, 5, etc.) (as with USDA's commodity pest risk analysis guidelines) or converted to numeric probabilities (the approach used by the NPPO of Australia). If ratings are converted into numeric probabilities then the resulting probabilities can be combined using a quantitative method (discussed in the next section). This is a valid approach, even given a lack of specific quantitative data that might go into the analysis; however, all assumptions must be clearly stated and the methods used to conduct the analysis must be fully transparent.

9.6.3. Schemes and keys

Guidelines provide a general framework for an analysis, but are often very flexible in how the specific analysis is conducted. The most structured approaches follow schemes or keys that guide an analyst through a series of very specific questions to assemble the analyst's responses to produce a result or answer.

The advantage of using highly specified schemes is that the scheme itself provides the analysis, ensuring that (given that all questions are answered) all key elements are thoroughly addressed. In addition, because the approach is highly structured, it increases consistency between analyses and allows for easy comparison from one analysis to another – this ability to compare results between different analyses is often extremely useful for prioritizing work (MacLeod and Baker, 2003). The disadvantages are that such approaches lack flexibility, and pests or situations that do not readily fit into the scheme may prove difficult to analyse. As with guidelines, the sheer range of pests or situations that must be dealt with in plant quarantine makes it difficult for even the most thorough schemes to accommodate every possibility. Lastly, schemes may be useful particularly for inexperienced analysts, but may frustrate more experienced analysts who feel constrained by the structure of the scheme.

One of the best-known schemes used for pest risk analysis is the EPPO pest risk analysis Decision Support Scheme mentioned above. The scheme consists of over 80 questions that address key components of a pest risk analysis – likelihood of entry (including analysis of all relevant pathways for a pest), likelihood of establishment and spread, and evaluation of the economic and environmental consequences of a pest introduction. The questions in the scheme adhere very closely to major points identified in ISPM No. 11 and go into detail for each aspect of pest introduction and consequences. Each question is assigned a rating – the rating scale in the scheme having five possible scores. In addition, the analyst should provide a score for uncertainty for each question, as well as provide relevant scientific and technical evidence to support the rating for the question and the related uncertainty.

Furthermore, the pest risk analysis scheme has been made into the online tool CAPRA. The online tool allows analysts to enter information about pests, following the pest risk analysis scheme and answering specific questions. The analyst provides ratings for each element (and rates uncertainty for each element), following the guidance provided by the scheme. The CAPRA tool then automatically summarizes the overall risk and uncertainty for that organism.

Table 9.3 provides a comparison of assessing likelihood of entry of a pest (one part of assessing likelihood of introduction of a pest) using a generalized approach (used by the US to analyse commodity imports) and a specified scheme that is used by EPPO to analyse specific pests.

9.7. Prioritization Models

Prioritization models are a special application of pest risk analysis that typically use specific characteristics to rank pests. They usually represent only a portion of a full pest risk analysis, and rather than being applied to a single pest, a prioritization model is applied to a group of pests (or to compare similar situations, such as ports of entry) to estimate which pest or situation is most 'risky' and which is least 'risky'. They are often applied for specific programmes, such as national surveillance programmes, or in making determinations regarding resources. As a hypothetical example, for instance we may wish to rank a group of pests (Pest X, Y and Z) for which we are considering eradication programmes to determine which of the pests we would be most likely to succeed in eradicating:

- *Pest X* A species of snail that would be relatively inexpensive to eradicate, but has a very high reproductive rate and a wide host range, and though it does not

Table 9.3. Evaluating likelihood of entry – comparison of specific and general approaches. (From USDA, 2000; EPPO, 2007.)

EPPO (pest risk analysis scheme for pests)	USA (revised pest risk analysis process for commodity imports)
List the relevant pathways (single pest identified with multiple pathways)	List quarantine pests (single pathway identified with multiple pests)
Select from the relevant pathways, using expert judgement, those which appear most important.	(Pathway identified through the import request for a specified commodity)
How likely is the pest to be associated with the pathway at the point(s) of origin taking into account the biology of the pest?	Likelihood of pest population on the harvested commodity
How likely is the pest to be associated with the pathway at the point(s) of origin taking into account *current management* conditions?	Likelihood of surviving post-harvest processing before shipment
Consider the volume of movement along the pathway (for periods when the pest is likely to be associated with it): how likely is it that this volume will support entry?	(Volume is defined by the import request and therefore not analysed)
Consider the frequency of movement along the pathway (for periods when the pest is likely to be associated with it): how likely is it that this frequency will support entry?	
How likely is the pest to survive during transport or storage?	Likelihood of surviving transport and storage of the consignment
How likely is the pest to multiply/increase in prevalence during transport or storage?	
Under current inspection procedures how likely is the pest to enter the pest risk analysis area undetected?	
How likely is the pest to be able to transfer from the pathway to a suitable host or habitat?	(Considered under 'establishment' in next part of PRA process)
Do other pathways need to be considered?	(Pathway is defined by commodity request)
Describe the overall probability of entry taking into account the risk presented by different pathways and estimate the overall likelihood of entry into the pest risk analysis area for this pest (comment on the key issues that lead to this conclusion).	Describe conclusions and uncertainty

disperse naturally, it is capable of dispersal through human assistance.

- *Pest Y* A species of rust that would be extremely costly to eradicate, is capable of reproducing quickly, has a narrow host range, but can spread easily through wind dispersal.
- *Pest Z* A species of beetle that would be costly to eradicate, has a moderately high rate of reproduction and a moderate range of hosts, but which exhibits a poor ability to disperse naturally or through human assistance.

As an NPPO, we may only have enough resources to focus our efforts on some of the

pests – therefore we need to prioritize which pests we will target first. The first step to developing a prioritization model is selecting appropriate characteristics that will be used to evaluate a given situation. In the case of eradication programmes, the characteristics we select might be:

- *Characteristic 1* Projected cost of eradication (higher costs increase difficulty of eradication);
- *Characteristic 2* Rate of reproduction (high rate of reproduction would increase difficulty of eradication);
- *Characteristic 3* Host range (wide host range would increase difficulty of eradication);

- *Characteristic 4* Ability of the pest to spread, either naturally or through human assistance (high rate of spread would increase difficulty of eradication).

We can then evaluate each pest according to the characteristics above – often this is done using rating systems such as described earlier in this chapter (Table 9.3). Note that for each characteristic above, we would want to provide guidance on what constitutes particular ratings mean (i.e. defined criteria for ratings). For instance, for Characteristic 3 (host range), we can specify that a 'Low' score means the pest attacks only on a few species in a single plant family, a 'Medium' means the pest attacks several species in at least two plant families and a 'High' means the pest attacks multiple species in multiple plant families.

In this hypothetical example, higher cumulative scores mean that a pest would be more difficult to eradicate (based on cost, host range, rate of reproduction and rate of spread) and lower scores mean a pest would be easier to eradicate, relative to other pests ranked in this evaluation. Pest Z had the lowest score, meaning it has a lower predicted cost for eradication, and lower rates of reproduction, host range and spread (and therefore a higher chance of success *relative to the other pests in this evaluation*). This structure allows us to pick which pest we have the best chance of successfully eradicating, based on the characteristics we have selected. Note that this model does not predict chance of success of eradication (it simply ranks the pests within this evaluation in order of chance of success), nor can we compare the results of

this evaluation to another evaluation unless the same methods are used.

This highlights both the strengths and the potential shortcomings of prioritizing pests based on specific criteria. One of the strengths of prioritizing using this method is that it is highly flexible and relatively easy to apply. We can simply select a few key characteristics and apply them to a range of pests. In addition, we can weight certain characteristics – further refining even a relatively simple model. In the example above for instance, we may decide that the projected cost of eradication is the most important characteristic and we could 'weight' that characteristic to count for more.

We can also add new characteristics if we want to further refine our results (although adding additional criteria may or may not improve the analysis). For instance, in the example above, we may decide to add a fifth characteristic – such as availability of cost-effective treatments — to further inform our decision. There are computer programs that can be used for more complex models where many characteristics need to be evaluated (see discussion on analytical hierarch process below for an example of more complex prioritization models). The results are usually clear, and easy to interpret and communicate. Thus, it represents a potentially powerful tool for prioritizing resources.

However, there can be disadvantages to using such models. Because the rankings are dependent on the characteristics we select, this means that we need to carefully decide on characteristics that are adequately defined and applicable to the situation. For instance, in prioritizing pests for eradication, using distribution of pests in other countries as a characteristic would not contribute to estimating how likely we will be able to eradicate the pest. This may be useful information in general, but would not necessarily inform on the likelihood of eradication. Poorly selected characteristics can lead to a model that fails to inform decision making, or even be outright misleading.

There are different types of prioritization models. Multi-criteria decision making (or MDCM) is often used for prioritizing (Yoe, 2012). The Analytical Hierarch Process

Table 9.4. Ranking pests according to specified criteria.

Characteristic	Pest X	Pest Y	Pest Z
1	L (1)	H (3)	M (2)
2	H (3)	H (3)	M (2)
3	H (3)	L (1)	M (2)
4	M (2)	H (3)	L (1)
Cumulative score	9	10	7
Overall rank	2	3	1

(or AHP) (see Box 9.2) is an application of MDCM that has been used to prioritize pests – the UDSA has used AHP for prioritizing pests in the CAPS program (see Chapter 8 for information on the CAPS program), and the State of Victoria in Australia has used AHP to evaluate and rank weeds.

Example of AHP applied for pest risk analysis: Victorian Weed Risk Assessment

The Victorian Weed Risk Assessment method was developed by the Department of Primary Industries in the State of Victoria, Australia as a way to prioritize resources

---

Box 9.2. Analytical Hierarchy Process.

The Analytical Hierarch Process (AHP) is a structured method for organizing and considering information. It uses a structure of making pair-wise comparisons between different 'objectives' (or groups, or characteristics) to organize and weight those objectives in a hierarchy. Objectives or groups that are higher in the hierarchy are 'weighted' and will influence a given decision more than objectives or groups that are ranked lower in the hierarchy. The hierarchy is determined through expert judgement – experts are consulted to provide their opinion on the relative importance of a given set of objectives related to a specific problem.

According to Saaty (2008), there are four basic steps involved in developing an analytical hierarchy:

1. Define the problem and determine the goal of the decision that needs to be made.
2. Organize the hierarchy beginning with the overall goal of the decision (at the top level) working down through broad objectives (intermediate levels) through to the lowest objectives (the lowest levels).
3. Construct a pair-wise comparison matrix with each objective listed in the matrix.
4. Using expert judgement, weight each objective compared with every other object (e.g. the pair-wise comparison) to obtain the weighted values of each objective, where more important objectives are scored higher, and the less important objective in the pair-wise comparison is ranked the reciprocal value.

Saaty uses a scale of 1–9 for order of importance. An objective (or group) that is the most important would be ranked '9' (based on expert judgement), and the second objective in the pair-wise comparison would be given a reciprocal score of '1/9'. Note that an objective compared to itself would score '1' since that denotes equal importance.

A simple hypothetical matrix for deciding what form of transportation is best to take to work, where scores have been assigned through expert judgement, might look like the matrix below (note the scores are purely hypothetical and provided for example only). One way to calculate the relative ranking (approximately) is to sum the scores for each row, then divide by the total score for the matrix.

	Car	Bus	Bike
Car	1	5	9
Bus	1/5	1	4
Bike	1/9	1/4	1
Relative rank	0.69	0.24	0.07

The results of this ranking indicate that *relative* to the other forms of transportation in the analysis, experts judged that taking a car to work is almost three times better than taking a bus, and almost 10 times better than taking a bike.

Note that in this example, we did not include any sub-objectives, such as comfort, time, cost, etc. (and each one of those objectives can be ranked using the same process just described). We could make the ranking process more complex (and perhaps more informative) if we were to add another level of objectives to our hierarchy. Note also that we have relied on expert judgement – but have not specified exactly what the experts should consider in determining rank, meaning that our process could be lacking in transparency unless we have provided more guidance to the experts we consulted, and clearly documented how experts reached their decisions.

related to weed management (Weiss and McLaren, 2002). They needed a way to determine which weeds were the most important – working with limited data and high levels of uncertainty. They decided to use an AHP model, and began by identifying specific criteria that makes a plant 'weedy'. They determined that in addition to 'invasiveness', other characteristics such as impacts (economic, social, environmental) and the plant's potential distribution relative to its actual distribution also were important. They constructed a hierarchy based on these criteria (Fig. 9.7).

Through the AHP process, the team weighted the relative importance of these major groups:

- Invasiveness: 0.12;
- Distribution: 0.32;
- Impact: 0.56.

This means that impact was judged to be approximately four times more important than invasiveness, and almost twice as important as distribution in determining whether a plant was a more serious weed or not. Thus, in performing the evaluation of each weed, 'impact' was given greater weight in the evaluations compared with the other characteristics of invasiveness and distribution.

Each of these groups then had several sub-groups that were also ranked. For instance, invasiveness was given four subgroups: establishment potential, growth/competitive ability, reproductive strategy and dispersal ability. Each one of these subgroups was then assigned further subgroups. Within each grouping, the individual components were also ranked through AHP. This same process was used for distribution and impacts, to develop the overall model. Experts then rated weeds according to the groups and sub-groups, and the hierarchical model provided final scores for all of the weeds evaluated.

9.8. Tailored Approaches to Pest Risk Analysis

In some cases, NPPOs may choose to tailor a pest risk analysis to a particular situation. The advantage of such an approach is that it allows analysts to be flexible in how they analyse a pest or pathway. Greater flexibility in the analysis encourages analysts to seek out the optimal approach for each analysis, taking greater advantage of the most appropriate tools available. For instance, if a pest is already known to have significant consequences, then the analysis can focus primarily on the likelihood of introduction and spread, allowing the analyst to devote greater time and resources to that aspect of the analysis. Furthermore, it allows the analysis to adapt to the available information – something that is not possible if following a specific scheme. Most nonroutine pest risk analyses are – to some extent – tailored to specific situations, and almost all quantitative analyses (Chapter 10) are tailored to specific problems.

There are disadvantages to using tailored approaches to pest risk analysis. First, because each model is unique to the problem

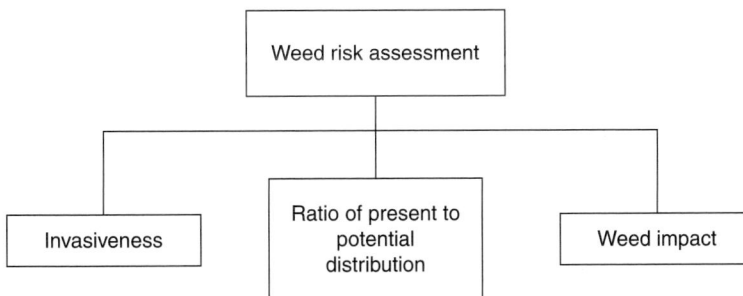

Fig. 9.7. AHP Model used for the Victorian Weed Risk Assessment method showing components evaluated (and weighted). (From Weiss and McLaren, 2002.)

it is trying to solve and because different methods and criteria may be used in making judgements, it is not possible to make comparisons between pest risk analyses. Second, tailored approaches to pest risk analysis can lead to a lack of consistency in decision making, and this could be problematic particularly in trade-sensitive situations where consistency is both necessary and required. Third, substantial effort may be needed to develop new models or apply new methods – thus, tailored approaches are best applied in unique or specific situations, but are not practical for routine situations such as commodity import pest risk analyses.

Tailored approaches to pest risk analysis are particularly well suited for some types of pathway pest risk analyses. Pathway pest risk analyses are highly variable – and the processes used to analyse pathways range from relatively simple descriptions to complex mathematical models, or even combinations of methods within a given analysis. Furthermore, pathway pest risk analyses may address certain parts of pathways, rather than a full pathway encompassing entry, establishment and spread. For instance, a country may want to know whether a particular border is particularly vulnerable to pest entry and the analysis may focus only on that aspect of risk. In some cases, specific pests may be identified within a pathway, but in other cases the pathway itself may be the focus of the analysis. Thus the methods used to conduct a pathway pest risk analysis are highly dependent on the scope and purpose of doing the pest risk analysis.

Usually pathway PRAs describe the pathway that is being analysed, including information on potential hazards associated with the pathway. Describing the pathway involves specifying the 'start' and 'end' points of the analysis (e.g. at what point in the pathway does the analysis begin and end), as well as key events or elements that occur throughout the pathway.

For instance, in the case of the movement of military equipment, we may wish to describe the equipment that was used (what equipment, and when, where and how it was used), if and how it was handled overseas, and any other events that could occur that would affect the presence of pests with that equipment. The pathway would start at a point in a foreign location, and end post-entry. However, if the pathway pest risk analysis was examining the potential for spread of a new pest after it has been introduced, then the pathway PRA would not include an analysis of 'entry' (since entry has already occurred) – it would focus instead on events related to 'establishment' and 'spread'. In this case, the pest risk analysis would start after pest entry, and end with spread.

Depending on the type of analysis that is being done, the most common hazards that are analysed are quarantine pests. In the case of the New Zealand analysis on vehicles and machinery (MAF, 2007), for instance, the hazards were identified as pests that were found associated with that particular pathway. In other cases, pathway pest risk analyses may not focus on specific pests, but on other aspects of the pathway in question – for instance, the amount and variety of cargo arriving at a particular port or border, or the amount of international flights (including cargo and passengers) arriving at an airport.

9.9. Summary of Qualitative Methods

There are a range of qualitative methods available to the risk analyst – these methods are often practical and relatively easy to implement, especially in situations where quantitative information is lacking. In conducting any risk analysis – qualitative or quantitative – we should:

- Apply criteria, judgements and decisions consistently;
- Be transparent in assumptions that are made in the analysis;
- Be transparent about our uncertainty.

Each method – narratives, guidelines and schemes – has advantages and disadvantages, and no one method will work ideally in every situation. Therefore, the key is to understand the technical basis for the methods, and then be able to pick and choose the most appropriate method(s) for any given situation. Whether using a scheme, a guideline or a novel approach, the analysis should be based on the best available scientific information. The judgements and assumptions that are made in the analysis should be transparent, and the uncertainty should be communicated.

References

Altenbach, T.J. (1995) *A Comparison of Risk Assessment Techniques from Qualitative to Quantitative*. ASME Pressure and Piping Conference, Honolulu, Hawaii. Lawrence Livermore National Library, Washington, DC.

Baker, R.H.A., Battisti, A., Bremmer, J., Kenis, M., Mumford, J., Petter, F., Schrader, G., Bacher, S., De Barro, P., Hulme, P.E., Karadjova, O., Lansink, A.O., Pruvost, O., Pyek, P., Roques, A., Baranchikov, Y. and Sun, J.H. (2009) PRATIQUE: a research project to enhance pest risk. *EPPO Bulletin* 39, 97–93.

Biosecurity Australia (2001) *Guidelines for Import Risk Analysis*. Agriculture, Fisheries and Forestry Australia/ Biosecurity Australia, Canberra.

EPPO (2007) *Guidelines on Pest Risk Analysis: Decision-support Scheme for Quarantine Pests*. EPPO Standard PM 5 / 3(3) EPPO, Paris (FR). Available online at: http://archives.eppo.org/EPPOStandards/PM5_PRA/ PRA_scheme_2007.doc, accessed 15 December 2011.

Hamm, R.M. (1991) Selection of verbal probabilities: a solution for some problems of verbal probability expressions. *Organizational Behavior and Human Decision Processes* 48, 193–223.

IPPC (2004) International Standards for Phytosanitary Measures, Publication No. 11: *Pest Risk Analysis for Quarantine Pests Including Analysis of Environmental Risks and Living Modified Organisms*. Secretariat of the International Plant Protection Convention (IPPC), Food and Agriculture Organization of the United Nations, Rome.

IPCC (2007) *Climate Change 2007: Synthesis Report, Contribution of Working Groups I, II and III to the Fourth Assessment Report of the Intergovernmental Panel on Climate Change*; Pachauri, R.K. and Reisinger, A. (eds), IPCC, Geneva.

IPPC (2010) International Standards for Phytosanitary Measures, Publication No. 5: *Glossary of Phytosanitary Terms*. Secretariat of the International Plant Protection Convention (IPPC), Food and Agriculture Organization of the United Nations, Rome.

Kent, S. (1964) *Words of Estimative Probability: Studies in Intelligence*. Center for the Study of Intelligence, Washington, DC, pp. 49–64.

Kesselman, R.F. (2008) Verbal probability expressions in national intelligence estimates. A comprehensive analysis of trends from the fifties through post 9/11. MSc thesis. Institute for Intelligence Studies, Mercyhurst College, Erie, PA.

MacLeod, A. and Baker, R.H.A. (2003) The EPPO pest risk assessment scheme: assigning descriptions to scores for the questions on entry and establishment. *EPPO Bulletin* 33, 313–320.

MAF (2007) *Import Risk Analysis: Vehicles and Machinery*. Ministry of Agriculture and Forestry, Wellington.

McCann, R.K., Marcot, B.G. and Ellis, R. (2006) Bayesian belief networks: applications in ecology and natural resource management. *Canadian Journal of Forest Research* 36, 12, 3053–3062.

Patt, A.G. and Schrag, D.P. (2003) Using specific language to describe risk and probability. *Climatic Change* 61, 17–30.

Saaty, T.L. (2008) Decision making with the analytic hierarchy process. *International Journal of Services Sciences* 1, 1, 83–98.

Schrader, G., MacLeod, A., Mittinty, M., Brunel, S., Kaminski, K., Kehlenbeck, H., Petter, F. and Baker, R.H.A. (2010) Enhancements of pest risk analysis techniques: enhancing techniques for standardising and summarising pest risk assessments – review of best practice in enhancing consistency. *EPPO Bulletin* 40, 1, 107–120.

USDA (2000) *Guidelines for Pathway-initiated Pest Risk Assessments*, Version 5.02. Plant Protection and Quarantine, Animal and Plant Health Inspection Service, United States Department of Agriculture, Washington, DC.

USDA (2010) *New Pest Advisory Group Report Template*. Plant Protection and Quarantine, Animal and Plant Health Inspection Service, United States Department of Agriculture, Washington, DC.

Weiss, J. and McLaren, D. (2002) Victoria's pest plant prioritisation process. In: Jacob H.S., Dodd, J. and Moore, J.H. (eds) *Proceedings of the 13th Australian Weeds Conference Plant Protection Society of Western Australia*, Perth, pp. 509–512.

Yoe, C. (2012) *Principles of Risk Analysis: Decision Making Under Uncertainty*. CRC Press, Boca Raton, Florida.

10 Quantitative Methods

Robert Griffin

10.1. Introduction

In Chapter 9 we learned about different qualitative and semi-quantitative methods for doing pest risk analysis. This chapter will focus on quantitative methods – providing a general overview of the different types of methods that are used, the reasons for doing quantitative analysis and the applications for quantitative analysis. A number of comprehensive references are available with detailed information on both the theory and practice of a wide range of proven quantitative methods available to analysts. Two excellent choices are Vose (2005) and Yoe (2012). Software programs are also available to assist analysts with modelling and calculations (e.g. @Risk).

This chapter complements these resources with a brief introduction to the basic concepts and methodologies as well as potential applications for quantitative methods in pest risk analysis. Although this text has separated the methods into qualitative, semi-quantitative and quantitative, in reality most pest risk analyses are some combination of these methods. The most important factors driving which method (or combination of methods) we choose are necessity, availability of data, resources available (including expertise and time) and the purpose of the analysis.

10.2. Quantitative Approaches – Issues and Applications

Nunn (1997) highlights the issue of qualitative, semi-quantitative and quantitative methods in his review of quarantine risk analysis. In most cases, we lack suitable data for carrying out full quantitative analyses – in those cases, qualitative and semi-quantitative approaches are appropriate. Furthermore, when quantitative approaches are used, they are typically resource-intense – requiring specialized expertise, data and computing resources, and often require more time than qualitative approaches. Given situations where data and time are simply not available, quantitative approaches may not be possible. Within the phytosanitary community, there are some who argue that quantitative approaches are more objective, as they are based on numerical data rather than expert judgement. Others argue that this may not necessarily be the case, since in many instances even quantitative approaches will rely on estimates based on expert judgement, and the models developed for quantitative approaches may also have subjective elements. Still others point to the fact that quantitative approaches are more explicit in presenting risk estimates and are therefore more transparent.

The issue of quantitative approaches has also been highlighted in two different dispute settlement cases (see Chapter 19). In the case of Australia and salmon imports, the findings specifically noted that qualitative methods are sufficient for demonstrating risk. More recently, in the Australia/ New Zealand apples case, the findings of the dispute noted that the quantitative methods used overestimated risk – the use of expert judgement to estimate probabilities and the model using normal distributions for estimates resulting in higher probabilities for certain events both contributed to the high risk estimate.

10.3. Applications of Quantitative Methods

These issues aside, quantitative methods can be a powerful tool to have as part of our analytical toolbox, and there are some advantages to using quantitative approaches in certain situations. Some of the applications and uses for quantitative analyses include:

- Gaining understanding;
- Cost- and time-efficient analyses;
- Addressing complex phenomena;
- Synthesizing knowledge/set priorities/ communicate;
- Testing hypotheses;
- Suggesting strategies.

Quantitative methods use numerical values from data or expert opinion to represent elements of the pest risk analysis that have a mathematical relationship and can therefore be combined in calculations based on formulae or models to achieve a result that quantitatively characterizes the risk or some aspect of the risk, including effects that mitigate risk.

In Chapter 9 we noted that it is difficult to label pest risk analyses as purely qualitative or quantitative, adding that the majority of pest risk analyses are primarily qualitative but certain aspects of analyses also lend themselves to quantitative methods. An obvious example is the economic consequences of pest introduction, which must be expressed in economic terms according to the IPPC (2010).

Methods for economic analysis are discussed in Chapters 7 and 11.

Broadly speaking, quantitative methods for pest risk analysis may be categorized as tools for risk assessment and tools for risk management, where the objective of the former is to characterize the unmitigated and mitigated risk, while the latter is primarily concerned with quantitatively representing the effects of measures, procedures or conditions on reducing the risk. 'Semi-quantitative' methods mentioned in Chapter 9 use numerical values that are subjectively assigned to an element of the pest risk analysis for relative comparisons but not representing actual values (e.g. ratings). Quantitative methods use actual values (probabilities, quantities, amounts, money, etc.) although these values are sometimes based on subjective estimates, as may be the case when values are derived from expert opinion. The key difference with quantitative methods is that subjectively derived values used in quantitative analyses are linked to real-world units that are meaningful in a mathematical sense and carry through in calculations.

In Chapter 14 (risk management) we discuss some of the quantitative aspects of risk-based inspection and probit 9 as a required response for treatments. These are central concepts in risk management that are supported by quantitative analyses associated with the need to understand *the prevalence of pests* that either escape detection or survive treatment, respectively. Our ability to calculate these values and scale these analyses (e.g. calculate probit 8 or other efficacy) provide a level of precision and confidence that would be difficult to achieve using only qualitative methods, even though the ultimate conclusion is likely to be expressed qualitatively (e.g. 'highly effective').

In Chapter 16 we discuss the analysis of uncertainty and mention that quantitative and 'semi-quantitative' methodologies can be useful for modelling variability. Quantitative models are also important for evaluating mitigation measures ranging from treatment efficacy to the likelihood of a mated pair (Liquido *et al.*, 1995). The key point here is that quantitative methods are sometimes the most appropriate tool and in other cases they are essential to pest risk analysis. Analysts need to be familiar with both qualitative and quantitative methods, and aim to use the most appropriate analytical methods for the audience, situation and available data.

Chapter 16 also makes reference to the usefulness of quantitative methods for developing and evaluating systems approaches. This point is raised to contrast a risk assessment based on probability and consequences with one that focuses mainly on the effects of events 'from farm to fork', which cause attrition in pest prevalence (see Fig. 10.1). This methodology fits well with the scenario analysis that is discussed below because it involves modelling related events in a scenario and assigning values (or probabilities) to each.

Other quantitative approaches used in risk management focus on the potential for *pest establishment* given a particular set of conditions. Included among these are:

• Probability of one or more mating pairs surviving a commercial shipment (Landolt *et al.*, 1984);

• Probability of one or more mating pairs (Vail *et al.*, 1993);
• Maximum pest limit (Baker *et al.*, 1990).

Although these techniques have been developed for pests with a specific type of biology (mainly but not exclusively fruit flies), adaptations of these and similar techniques can be used in risk assessment to determine the probability of having a threshold prevalence, or for risk management to determine the efficacy of measures designed to ensure that pest prevalence is below a threshold level. The concept is equivalent in either case, but the use of methodologies such as these requires a suitable pest and situation, and assumes threshold prevalence has been established via research or policy, or both.

10.4. Describing Scenarios – Linear Event Trees and Deterministic Models

The underlying objective of quantitative methods applied to risk management is some form of measurement associated with pests

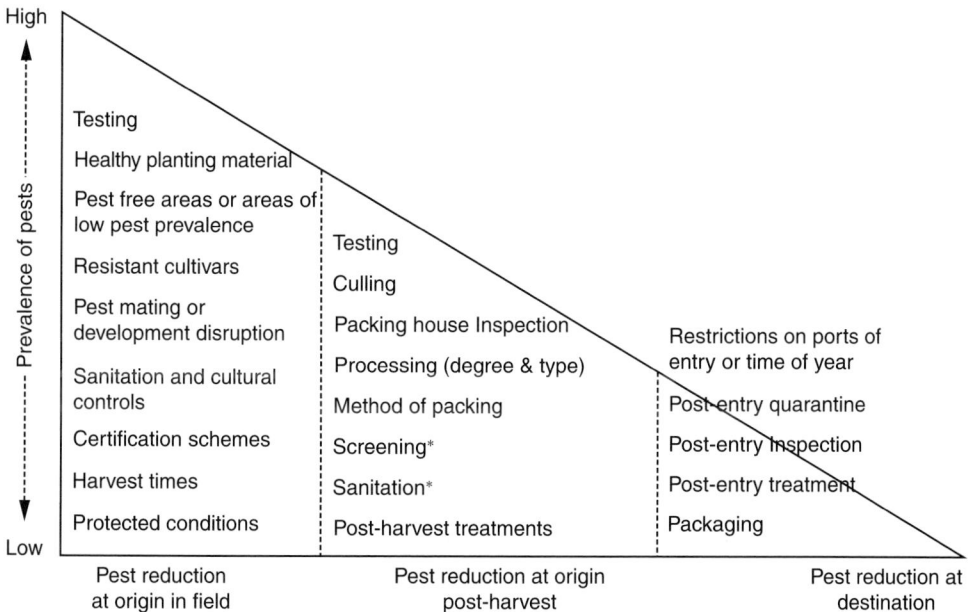

Fig. 10.1. Examples of measures leading to pest attrition in a pathway for 'farm to fork'.

*Measures are examples and representative but not all inclusive of causes of pest attrition in the pathway

that continue (survive) along a pathway (scenario). Obviously, pests that are avoided, removed or treated do not present a risk. The pests that proliferate, escape detection and survive treatment are the ones that increase the likelihood of introduction and thus present the risk. This means that in order to understand the effects of attrition or risk management in a quantitative context, we must:

- Understand the scenario of concern;
- Estimate the initial volume and pest load;
- Estimate the effects of specific conditions or events.

We begin by identifying a starting point and estimating the magnitude of the problem at its beginning and then follow changes through critical events leading to the adverse outcome using data we have on the effects.

Imagine for instance a simple hypothetical scenario that begins with 30,000 'Bilbo'

fruit being exported with an anticipated infestation rate of 5% for a quarantine pathogen of citrus. Assume that inspection procedures are 60% effective for removing infected fruit. Assume also that market distribution data shows 95% of the Bilbo fruit will go to areas where no citrus is present, and fruit that is distributed in citrus growing areas is highly unlikely to be exposed to citrus in the field under conditions for disease transmission (conservative expert estimate of 99.99% will be consumed). The scenario we have described and the data associated with it can be represented as follows in Fig. 10.2.

The result is a simple linear scenario that allows us to calculate a predicted rate of outbreaks per year by multiplying point estimates. The scenario is based on the existing or 'unmitigated' condition, which includes the expected or normal events. To understand the effects of a mitigated scenario,

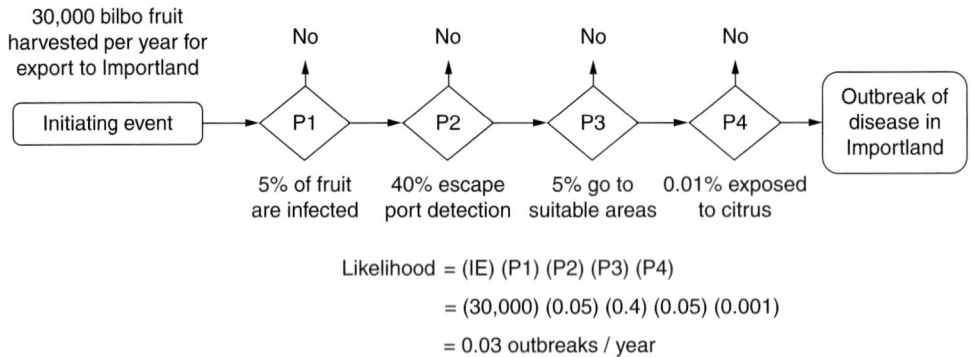

Fig. 10.2. Estimate of the likelihood of establishment for the *unmitigated* risk using point estimates and a linear event tree.

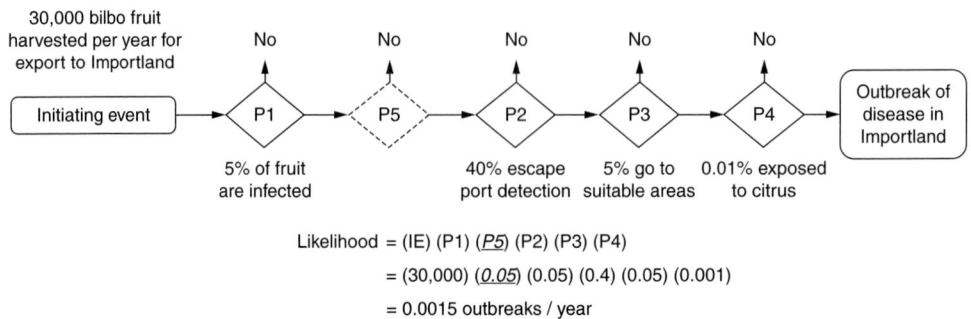

Fig. 10.3. Estimate of the likelihood of establishment for the *mitigated* risk using point estimates and a linear event tree.

we would add an event such as a treatment and corresponding efficacy data to the scenario as shown in Fig. 10.3. Note that where inspection is 60% effective, we use the 40% that escapes for our calculation. Likewise, where the treatment is 95% effective, we use the 5% that survive. This is consistent with our intention to track the undesirable aspects of attrition and mitigation events in order to calculate an outcome that can be related to risk. Note also that if any event results in no (zero) pests surviving or escaping, we can have no subsequent event and do not reach the adverse outcome, i.e. there is no risk.

The example provides a simple demonstration of a *linear event tree* that is a simulation or deterministic model (i.e. using single point values) of the expected scenario, identifying significant events which are related in a linear fashion. This example corresponds to the left side of the schematic pathway figure we first saw in Chapter 8, which represents the series of events leading to establishment (see Fig. 10.4). The calculation for unmitigated risk is useful for risk assessment purposes because it provides an estimate for the likelihood of introduction. The calculation for the mitigated scenario is useful for risk management because it shows the mitigating effect of a treatment. Additional measures could be added and input estimates adjusted to explore the results of different data and other scenarios, including more complex measures such as systems approaches.

Additional events can also be added to represent different scenarios or to address a different problem. Fig. 10.5 is an example

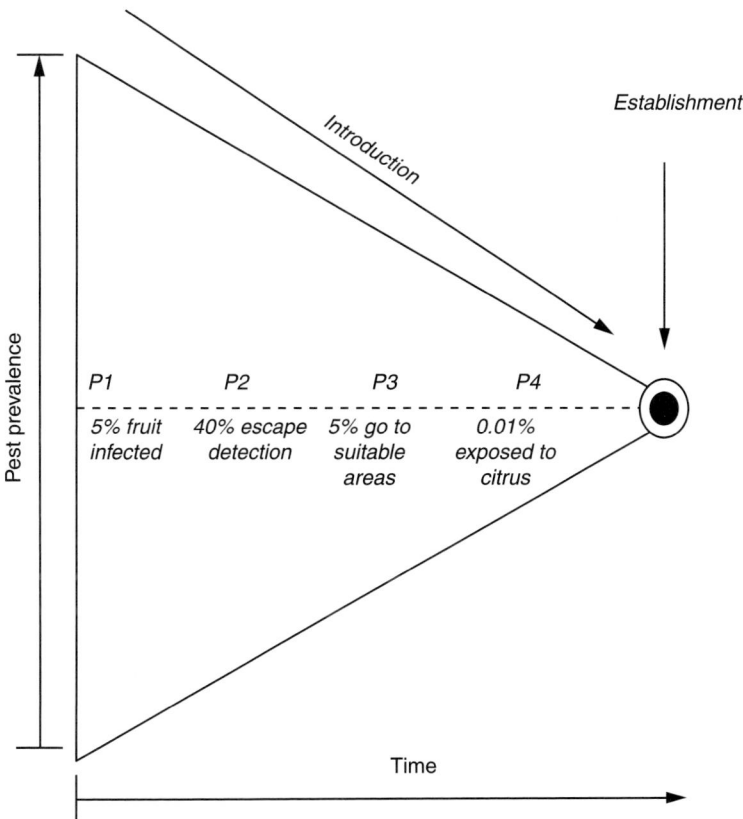

Fig. 10.4. Graphic representation of the pathway for introducing infested 'Bilbo' fruit, showing probabilities associated with events.

```
┌─────────────────────────┐        V1. No. of flights from a selected region
│    Number of flights    │
└─────────────────────────┘
             │              P1. Probability of pests entering aircraft hold
             ▼
┌─────────────────────────┐        V2. No. of pests per flight
│    Pests in cargo hold   │
└─────────────────────────┘
             │              P2. Probability of transit survival (1–percentage mortality)
             ▼
┌─────────────────────────┐        V3. No. of surviving pests
│    Pests survive transit │
└─────────────────────────┘
             │              P3. Probability of being adult (able to disperse)
             ▼
┌─────────────────────────┐        V4. No. of surviving adult pests
│       Adult pests        │
└─────────────────────────┘
             │              P4. Probability of being female
             ▼
┌─────────────────────────┐        V5. No. of mated adults pests (all females)
│ Mated pests (all females)│
└─────────────────────────┘
             │              P5. Probability of being a high risk pest
             ▼
┌─────────────────────────┐        V6. No. of high risk, mated adult pests
│     High risk pests      │            (or, pests in particular females)
└─────────────────────────┘
             │              P6. Probability of neither detecting nor controlling
             ▼
┌──────────────────────────────────┐  V7. No. of undetected and uncontrolled pests
│ Mated pests not detected or controlled │
└──────────────────────────────────┘
             │              P7. Probability of finding host plant
             ▼
┌─────────────────────────┐        V8. No. of mated pests on plant hosts (sites)
│   Mated pests find host  │
└─────────────────────────┘
             │              P8. Probability of birth and survival (1–percentage mortality)
             ▼
┌──────────────────────────────────┐  V9. No. of sites at which pests establish
│ Sites with pests surviving to next season │
└──────────────────────────────────┘
```

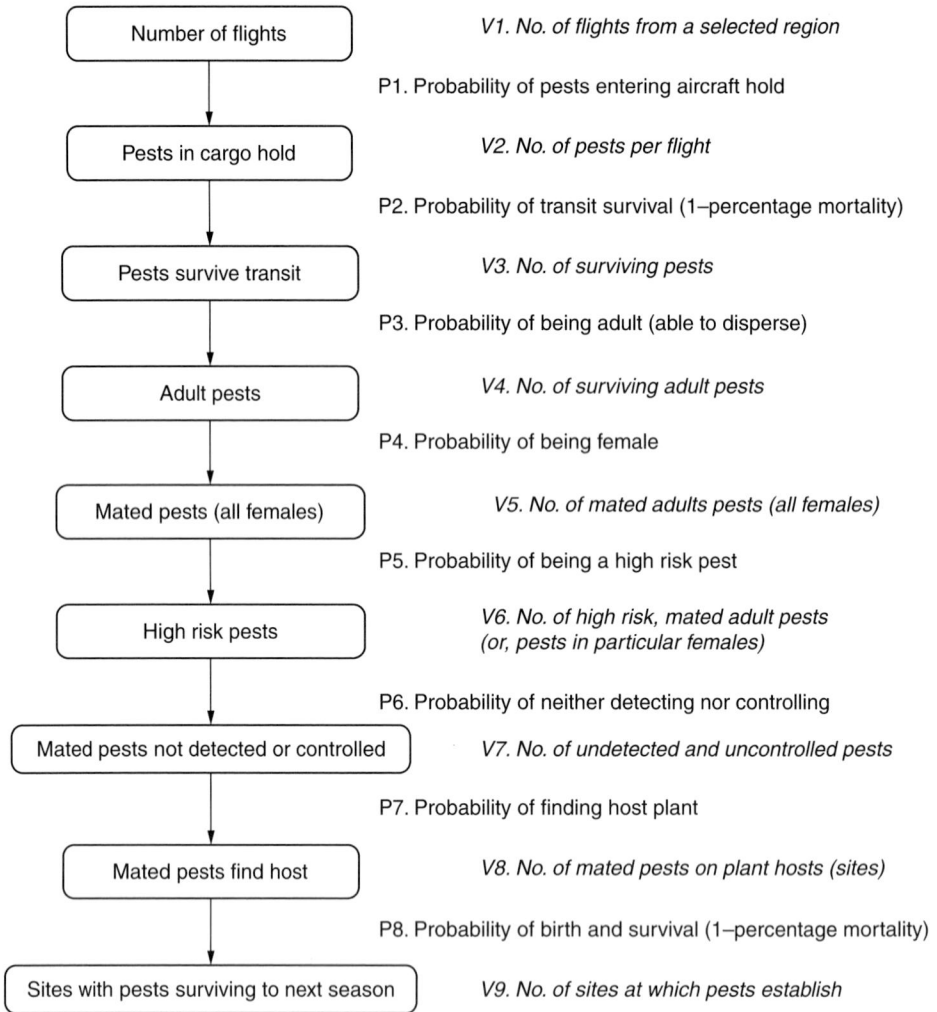

Fig. 10.5. Linear scenario describing pathway for pest outbreaks resulting from shipments of infested air cargo. (From USDA, 2004.)

of a slightly larger linear scenario aimed at estimating the number of pest outbreaks that result from cargo aircraft infested with insect pests. Note that the scenario includes events related to the biology (reproduction) of the pest, which is important to this scenario but was not a factor included in the previous example.

The difference in events between these two examples serves to emphasize that a key aspect of scenario analysis is first establishing the model. Neither example describes every detailed condition or event, nor are they oversimplified to the point of ignoring important events. Most importantly, the events are not 'invented' to correspond to available data. Note for instance that in the first scenario, there were relatively 'hard' data for the infestation rate, efficacy of inspection and market distribution, but an expert estimate was required for the exposure event. The exposure event is considered a critical last step in the Bilbo fruit scenarios, but the comparable event in the air cargo pest scenario (finding a host plant) is followed by additional events associated with the survival of subsequent generations of the insect. Rather than fit the scenarios to the data, the data are fitted to scenarios that are thoughtfully constructed of distinct events that are significant for the purpose they are intended to demonstrate, the pests they address, and accurately describe the situation being considered.

No model is perfect, but establishing a reasonable scenario is a critical step worthy of special attention and broad expert collaboration. Once a scenario is agreed, parameters can be set for inputting data and shortcomings in data will quickly emerge.

10.5. Probabilistic Models

As noted in Chapter 16, quantitative methods can provide powerful tools for analysing uncertainty. The relationship of uncertainty to probability is also noted. Looking closely at the Bilbo examples, we can see where questions might arise around the precision associated with the data. Point estimates (deterministic values) are used in our examples, but it is unlikely that these represent exact values, especially the expert estimate. What is more likely is that each value represents the mode in a range or distribution that is the uncertainty around the estimate. For instance, the 5% of fruit estimated to be infected is more accurately represented by 5% ±2%, i.e. ranging from 3–7% with a mode of 5%. The same is probably true of most other data, except in cases where the data point is not an estimate of the mode but actually a precise observation, in which case there is no distribution.

Distributions are graphically represented as curves known as probability density functions (PDFs). PDFs take different forms, but in all cases, the area under the curve represents a value of 1, or the full range of possible points for a particular value. The best known PDF is the normal distribution or 'Bell' curve known for its symmetrical bell shape (see Fig. 10.6).

A bell curve would represent the distribution of possible values for infected Bilbo

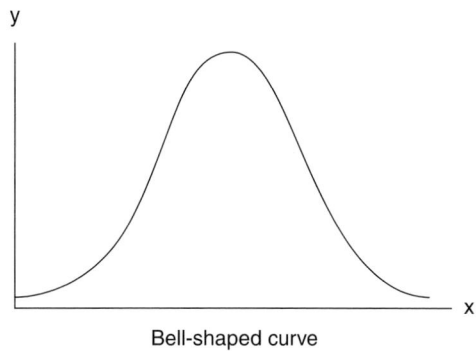

Bell-shaped curve

Fig. 10.6. A normal distribution. (From Yoe, 2012.)

fruit if they were distributed 'normally' across the range from 3– 7% with a mode of 5%, i.e. the most likely value is 5% but the real value could be as low as 3% (not lower) or as high as 7% (not higher) with an equally low probability for either. If there were a higher probability for the real value to be near 3%, the curve would be 'skewed' to the left with a longer 'tail' to the right because the probability for values to the right would be proportionally lower.

Other distributions that are useful for probabilistic risk analysis in pest risk analysis are the uniform distribution and the triangular distribution. The uniform distribution is used when a lower and upper limit can be established, but either there is no difference in probabilities for values between these limits, or the probabilities are unknown. The triangular distribution also requires a higher and lower limit but adds a most likely value, the mode (see Fig. 10.7). A similar-looking distribution, the beta-PERT is arguably the most common distribution for pest risk analysis (see Vose, 2005 for detailed discussion).

The great advantage of using PDFs in stochastic modelling rather than point estimates (in deterministic modelling) is that we are able to express the true nature of the data. Uncertainty becomes explicit in the size and shape of the curve. The disadvantage of using PDFs is that we are unable to multiply them as easily as we did the point estimates. This is where the software programs become important.

In order to combine PDFs, random points within each distribution are multiplied

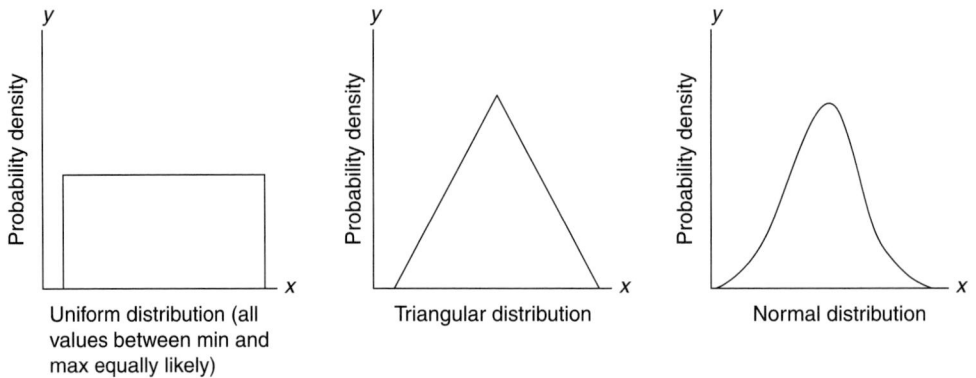

Fig. 10.7. Common PDF curves. (From Yoe, 2012.)

together. This is done using Monte Carlo methods that rely on repeated random sampling from each distribution to compute the results as a new distribution. Monte Carlo simulations are used when we cannot compute an exact result with a deterministic algorithm. Since thousands of calculations (iterations) are required to derive a result from combining two distributions, Monte Carlo simulations are not feasible without a computer and appropriate software. Fortunately, most ordinary computers are capable of running such simulations and a number of software products are available for this purpose. Many software packages also provide additional tools to assist the analyst with creating scenarios, testing assumptions, performing sensitivity analysis and exporting graphics.

The hard work associated with scenario analysis is building and populating the model. Once the model is established, running it with different inputs and assumptions is a simple matter of adjusting the parameters. Aside from making uncertainty very explicit, a scenario analysis is also an excellent way to test assumptions and understand how, and how much, changes in the data or even changes in the scenario affect the results.

To demonstrate a scenario analysis in practice, let's look at a slightly more complex example. Figure 10.8 describes four scenarios (A–D) determined by experts to be the primary pathways for possible new outbreaks of Pine shoot beetle, *Tomicus piniperda* (L.), resulting from the movement

of logs from infested areas. In addition, each scenario was analysed according to seasons corresponding with the insect's activities (summer, autumn, winter, early spring, and late spring). This created a total of 20 sub-scenarios. However, the summer sub-scenarios were determined by experts to have a negligible risk after the first event because insects would be feeding in shoots and therefore would not be associated with logs during this period.

In each scenario, the most likely probability of any in the sequence of events occurring was represented by a point estimate (mode value) and surrounded by an estimate of the lowest and highest probability in a triangular distribution. Expert consensus concerning the evidence determined that if point estimates were wrong, they would not be above the high value or below the low value. Thus, the high and low estimates for the frequency of each event create a triangular distribution around the point estimate and define the entire range of possible values. Experts were encouraged to estimate conservatively in order to ensure that the actual probability lies within the area of the curve defined by the estimates. A point estimate alone was used when the evidence indicated a very high degree of certainty. Estimates and continuing calculations of probability were terminated when any event resulted in the elimination of the pest risk.

Computer simulation was used to graphically represent the distributions for each event and to calculate the product of all

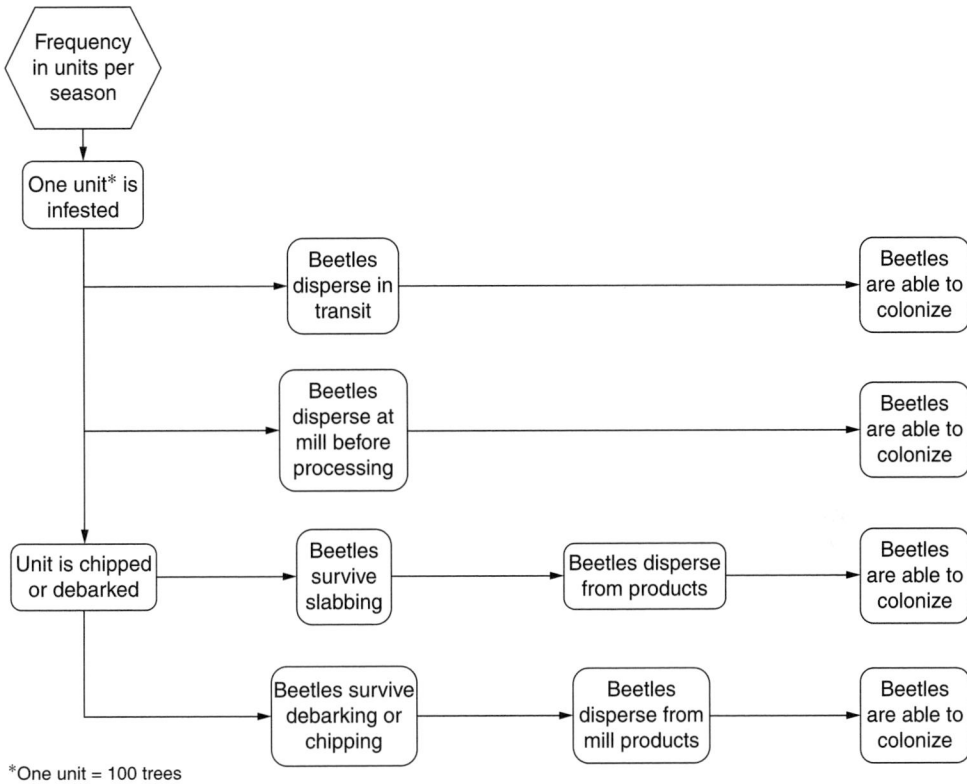

Fig. 10.8. Pine shoot beetle spread scenarios. (Griffin and Miller, 1994)

events for a scenario. Curves were generated using 3000–9000 iterations (trials with random numbers). Figures 10.9 and 10.10 show the data for Scenarios A and C, respectively. Figures 10.11 and 10.12 show the PDF results for Scenarios A and C. It should be noted that the values on the *x* axis of each graph are the log of the actual values (i.e. frequencies are shown on a log scale). The log scale is used in this example to produce a more readable graph that facilitates interpretation. Note also that each graph is marked with a dotted line indicating the expected value (mean). The log value for the mean is noted in the upper left corner of each graph.

By combining the curves for each event in a scenario pathway, an overall estimate of the risk and associated uncertainty were developed for scenarios describing the situation(s) as they would be without the addition of mitigation. This facilitated the identification of

high risk scenarios and events. It also provided the background for evaluating the application and value of mitigation schemes applied to specific scenarios and events.

The mean, mode and 95% probability level were extracted from the PDFs for Table 10.1. A summary like this helps to better contrast the scenarios. In this case, Scenario C is in stark contrast to other scenarios. Another presentation of the data is shown in Fig. 10.13 where mode values show again that Scenario C is clearly the pathway where regulatory strategies need to focus. Table 10.2 ranks pathway point estimates for yet another presentation of the results.

In this example, a substantial amount of the data was based on expert estimates. Fewer points came from 'hard' data. This is noteworthy because many analysts have the impression that the data-intensive nature of probabilistic methods demands hard data.

Frequency in units per season (AF) → Probability that one unit is infested (A1) → Probability that beetles disperse in transit (A2) → Probability that beetles are able to colonize (A3)

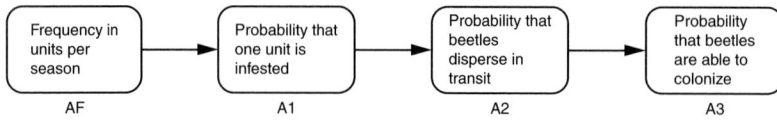

Combined probability estimates and point products for scenario A:

Event	Summer point	Low	Autumn point	High	Low	Winter point	High	Low	Early spring point	High	Low	Late spring point	High
AF	225		225			225			112.5			112.5	
A1	0	0.01	0.45	0.7	0.05	0.5	0.8	0.25	0.6	0.9	0.05	0.2	0.4
A2		0.0002	0.02	0.05	0.0001	0.0075	0.03	0.05	0.125	0.25	0.05	0.125	0.25
A3		1.0E–06	1.0E–05	0.0001	1.0E–06	1.0E–05	0.0001	1.0E–05	0.0001	0.001	1.0E–06	1.0E–05	0.0001
Point product			2.0E–05			8.4E–06			8.4E–04			2.8E–05	

Fig. 10.9. Data for Scenario A. (Griffin and Miller, 1994)

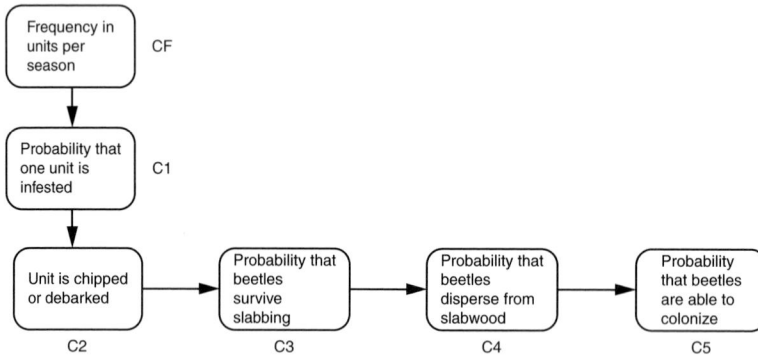

Frequency in units per season (CF) ↓ Probability that one unit is infested (C1) ↓ Unit is chipped or debarked (C2) → Probability that beetles survive slabbing (C3) → Probability that beetles disperse from slabwood (C4) → Probability that beetles are able to colonize (C5)

Combined probability estimates and point products for scenario C:

Event	Summer point	Low	Autumn point	High	Low	Winter point	High	Low	Early spring point	High	Low	Late spring point	High
CF	225		225			225			112.5			112.5	
C1	0	0.01	0.45	0.7	0.05	0.5	0.8	0.25	0.6	0.9	0.05	0.2	0.4
C2		0.5			0.5			0.5			0.5		
C3		0.9	0.95	0.99	0.9	0.95	0.99	0.9	0.95	0.99	0.9	0.95	0.99
C4		0.1	0.5	0.9	0.1	0.5	0.9	0.3	0.8	0.95	0.3	0.8	0.95
C5		0.0001	0.00225	0.05	0.0001	0.00225	0.05	0.001	0.0225	0.5	0.0001	0.00225	0.05
Point product			5.4E–02			6.0E–02			5.8E–01			1.9E–02	

Fig. 10.10. Data for Scenario C. (Griffin and Miller, 1994)

Of course it would be desirable to have actual data for every event but, in this case, expert judgement provided the bulk of the information used to populate the model. This worked well for two reasons: first, the experts made conservative estimates to ensure that the results covered every possibility; and second, because the results for one scenario were so distinctly different from the others

that any variability not accounted for in the conservative estimates of experts would be unlikely to significantly affect the outcome. Under conditions where the results are much closer, a higher level of resolution may be required and the broad conservative estimates of experts would be less useful.

This example also demonstrates the 'tree' aspect of scenario analysis, as 'branches'

Scenario A – Autumn

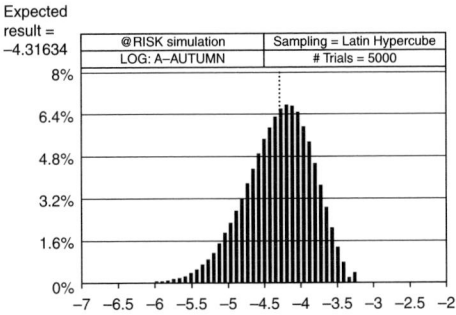

Expected
result =
−4.31634

@RISK simulation	Sampling = Latin Hypercube
LOG: A–AUTUMN	# Trials = 5000

Scenario A – Winter

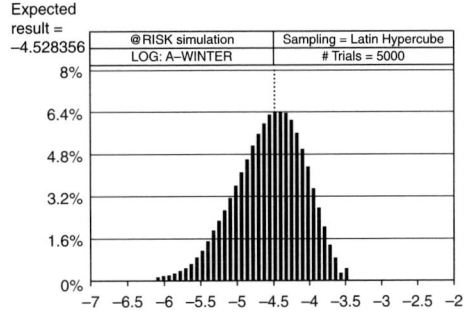

Expected
result =
−4.528356

@RISK simulation	Sampling = Latin Hypercube
LOG: A–WINTER	# Trials = 5000

Scenario A – Early spring

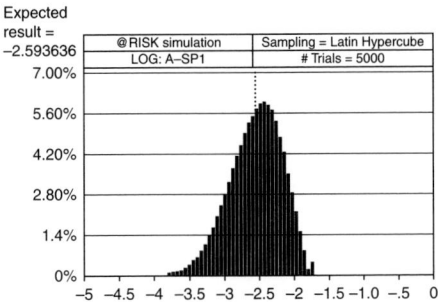

Expected
result =
−2.593636

@RISK simulation	Sampling = Latin Hypercube
LOG: A–SP1	# Trials = 5000

Scenario A – Late spring

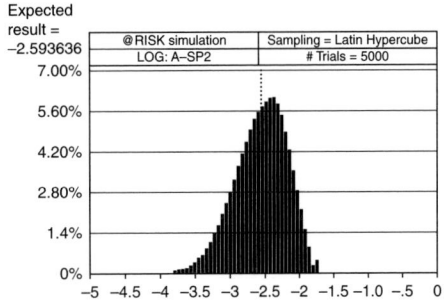

Expected
result =
−2.593636

@RISK simulation	Sampling = Latin Hypercube
LOG: A–SP2	# Trials = 5000

Scenario A – All seasons

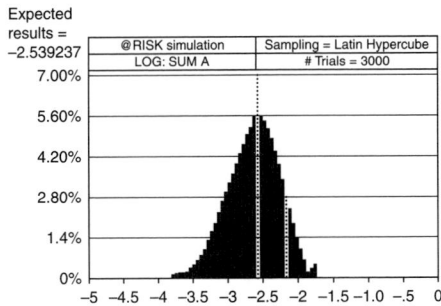

Expected
results =
−2.539237

@RISK simulation	Sampling = Latin Hypercube
LOG: SUM A	# Trials = 3000

Fig. 10.11. PDF results for Scenario A. (Griffin and Miller, 1994)

represent different scenarios, including a yes–no split that forms Scenarios C and D. In addition, each scenario has sub-scenarios for the seasons. The design is fundamentally a linear event tree similar to the Bilbo fruit example but with the additional complexity needed to address the different pathways and seasons.

10.6. Fault Trees – Working Backwards

We mentioned above that the Bilbo fruit example corresponds to the events found on the left side of the 'bow tie'. The Pine shoot beetle example is focused on pathways for pest spread that correspond to the right side of the 'bow tie'. In both examples, the direction

Scenario C – Autumn

Expected result = −0.6409544		
@RISK Simulation LOG: C–AUTUMN	Sampling = Latin Hypercube # Trials = 5000	

Scenario C – Winter

Expected result = −0.5664466		
@RISK Simulation LOG: C–FALL	Sampling = Latin Hypercube # Trials = 5000	

Scenario C – Early spring

Expected result = 0.4181073		
@RISK Simulation LOG: C–SP1	Sampling = Latin Hypercube # Trials = 5000	

Scenario C – Late spring

Expected result = −1.026544		
@RISK Simulation LOG: C–SP2	Sampling = Latin Hypercube # Trials = 5000	

Scenario C – All seasons

Expected result = 0.5787764		
@RISK Simulation LOG: C–SUM C	Sampling = Latin Hypercube # Trials = 3000	

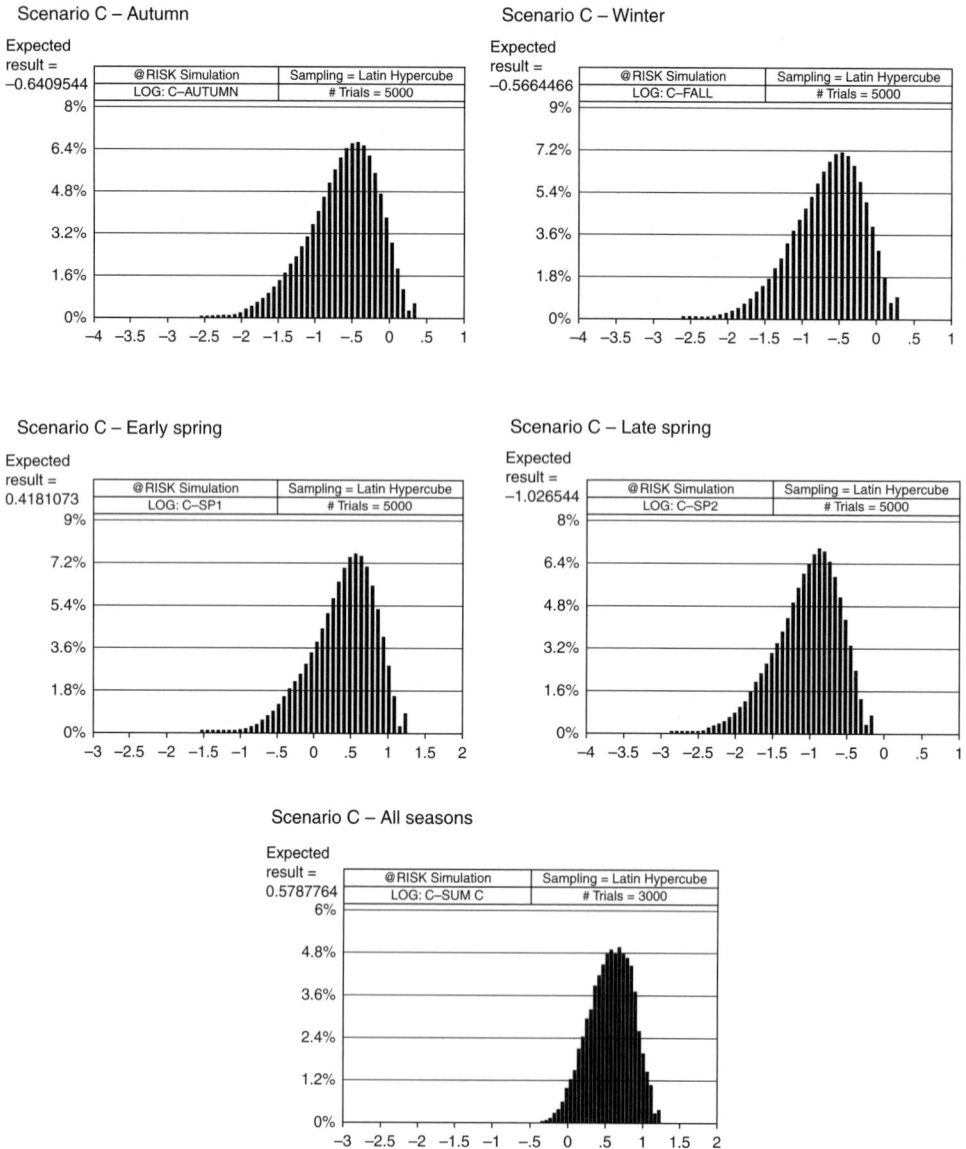

Fig. 10.12. PDF results for Scenario C. (Griffin and Miller, 1994)

of the analysis is from left to right. It is possible however for the direction to be reversed. This type of analysis might be used for trace-back and would employ the same principles for the analysis except that the 'tree' would begin from an outcome and work back to an initiating event in a design known as a 'fault tree'. Figure 10.14 provides an example of a fault tree. Although few examples of this application exist in the phytosanitary world, such analyses could be used more often to understand which of several possible initiating events is most likely responsible for an adverse outcome (e.g. pest establishment). The great advantage of a fault tree is that the analyst is more likely to have real-world data to populate the model when the analysis is based on events that have already occurred.

Table 10.1. Summary calculations for all scenarios. (Griffin and Miller, 1994)

Dispersal from:		Autumn	Winter	Early spring	Late spring	New outbreaks year	Years/ outbreak
Scenario A transit	Mean	7.9E–05	4.7E–05	3.4E–03	1.3E–04	3.7E–03	271
	Mode	2.0E–05	8.4E–06	8.4E–04	2.8E–05	9.0E–04	1,110
	95%	2.2E–04	1.4E–04	8.8E–03	3.3E–04	9.5E–03	105
Scenario B the mill	Mean	1.7E–07	2.0E–06	3.7E–03	1.4E–04	3.8E–03	264
	Mode	5.1E–08	5.6E–07	6.8E–04	2.3E–05	7.0E–04	1,432
	95%	4.8E–07	5.3E–06	9.6E–03	3.6E–04	9.9E–03	101
Scenario C slabwood	Mean	3.6E–01	4.2E–01	3.7E+00	1.4E–01	4.6E+00	0.2
	Mode	5.4E–02	6.0E–02	5.8E–01	1.9E–01	8.8E–01	1.1
	95%	1.0E+00	1.2E+00	9.2E+00	3.7E–01	1.2E+01	0.1
Scenario D mill by-products	Mean	9.4E–08	1.1E–07	9.9E–06	2.3E–07	1.0E–05	97,290
	Mode	1.3E–08	1.4E–08	1.2E–06	2.0E–08	1.2E–06	802,118
	95%	2.8E–07	3.4E–7	2.9E–05	7.1E–07	3.0E–05	33,078
All scenarios by season	Mean	3.6E–01	4.2E–01	3.7E+00	1.4E–01	4.6E+00	0.2
	Mode	5.4E–02	6.0E–02	5.8E–01	1.9E–01	8.9E–01	1.1
	95%	1.0E+00	1.2E+00	9.2E+00	3.7A–01	1.2E+01	0.1

10.7. Good Practices for Quantitative Modelling

In this chapter we've introduced some quantitative methods used in pest risk analysis. Because these methods are not commonly applied, there is very little specific guidance available to pest risk analysts on when, where and how to apply such methods. Likewise, there are no 'guidelines' or 'schemes' that use fully quantitative methods. Most quantitative analyses are unique – 'tailor made' pest risk analyses done to address specific and non-routine problems. While there are no specific guidelines for quantitative analysis, there are good practices that can be followed when doing a quantitative pest risk analysis. Burmaster and Anderson (1994) provide a list of 14 principles for the use of Monte Carlo techniques in ecological modelling, but the principles can be broadly applied to most quantitative analysis used in pest risk analysis. The principles are summarized as:

- Show all formulae used in the risk assessment.
- Calculate and present the point estimates of exposure and risk that are generated following the current deterministic risk assessment guidelines from the appropriate regulatory agency.
- Present the results of sensitivity analysis for the deterministic calculations to identify inputs suitable for probabilistic treatment.
- Restrict the use of probabilistic methods to pathways of regulatory importance to save time, money and resources.
- Provide detailed information on the input distributions selected.
- Show to the extent possible how input distributions capture variability and uncertainty.
- Use measured data for input distributions to the extent possible, making sure the data are relevant to the situation.
- Discuss the data and report goodness-of-fit statistics for parametric distributions for input variables.
- Discuss the presence or absence of moderate to strong correlations between or among the input variables.
- Provide detailed information and graphs for each output distribution.
- Perform probabilistic sensitivity analyses for all key inputs in such a way as to distinguish the effects of variability from the effects of uncertainty.

Scenario		Autumn	Winter	Early spring	Late spring	Summer
A:	Transit	2.3E–03	9.5E–04	9.5E–02	3.2E–03	0.1
B:	The mill	5.8E–06	6.3E–05	7.6E–02	2.5E–03	0.1
C:	Slabwood	6.1	6.8	65.2	21.7	99.8
D:	Mill by-products	1.5E–06	1.6E–06	1.4E–04	2.2E–06	00

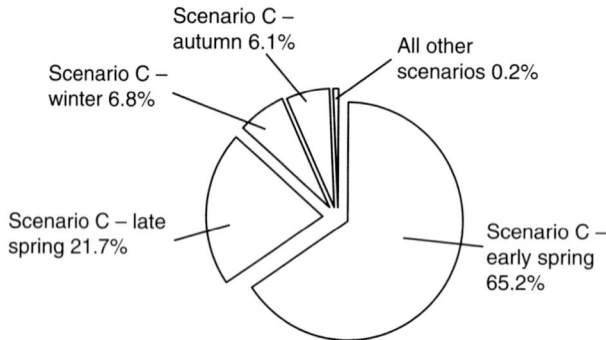

Fig. 10.13. Percentage risk by scenario using mode values. (Griffin and Miller, 1994)

Table 10.2. Pathway ranking based on point estimate results. (Griffin and Miller, 1994)

	Product of point estimate	Scenario
Lowest probability	0.000000013 (1.3 ×× 10^{-8})	D – Autumn dispersal from chips and bark at the mill site
	0.000000014 (1.4 × 10^{-8})	D – Winter dispersal from chips and bark at the mill site
	0.000000020 (2.0 × 10^{-8})	D – Late spring dispersal from chips and bark at the mill site
	0.000000051 (5.1 × 10^{-8})	B – Autumn dispersal from logs at the mill before processing
	0.00000056 (5.6 × 10^{-7})	B – Winter dispersal from logs at the mill before processing
	0.0000012 (1.2 × 10^{-6})	D – Early spring dispersal from chips and bark at the mill site
	0.0000084 (8.4 × 10^{-6})	A – Winter dispersal from logs in transit
	0.00002 (2.0 × 10^{-5})	A – Autumn dispersal from logs in transit
	0.000023 (2.3 × 10^{-5})	B – Late spring dispersal from logs at the mill before processing
	0.000028 (2.8 × 10^{-5})	A – Late spring dispersal from logs in transit
	0.000675 (6.75 × 10^{-4})	B – Early spring dispersal from logs at the mill before processing
	0.00084 (8.4 × 10^{-4})	A – Early spring dispersal from logs in transit
	0.019	C – Late spring dispersal from slabwood and rough-cut lumber at the mill site
Highest probability	0.054	C – Autumn dispersal from slabwood and rough-cut lumber at the mill site
	0.06	C – Winter dispersal from slabwood and rough-cut lumber at the mill site
	0.57	C – Early spring dispersal from slabwood and rough-cut lumber at the mill site

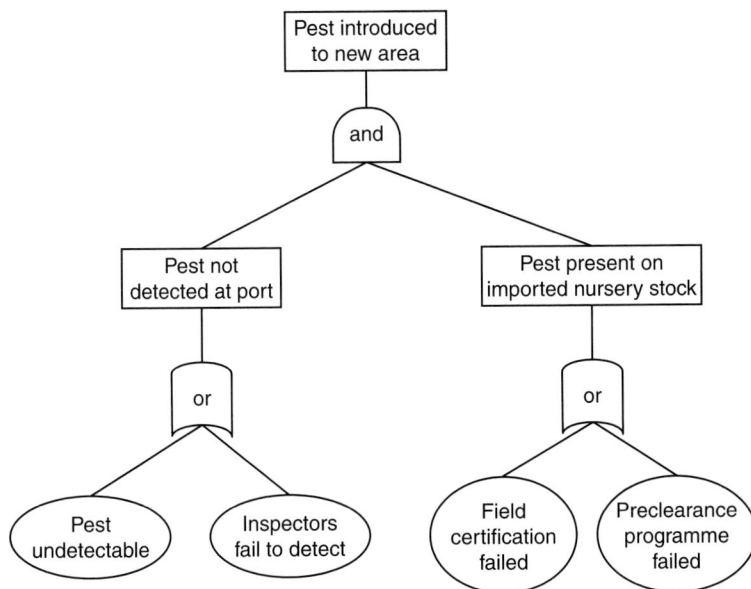

Fig. 10.14. Simple fault tree for a pest introduction on nursery stock.

- Present the name and statistical quality of the random number generator used.
- Discuss the limitation of the methods and interpretation of the results.

10.8. Summary

All forms of probabilistic scenario analysis provide a high level of transparency associated with uncertainty. Although the type, shape and range of distribution curves are very informative in this regard, an explanation of the nature of the uncertainty and its importance to the scenario is needed for a full understanding. Thus the PDFs do not stand alone but require interpretation and explanation to be understood. Likewise, an explanation of the model, the rationale behind selected events, and the type and source of data inputs is essential, particularly for readers who are unfamiliar with the methodology or uncomfortable with mathematical methods. We noted previously that there are a number of excellent technical references to assist analysts with these applications, but it is ultimately the responsibility of the analyst to make the analysis understandable for the intended audience and defendable to trading partners.

References

Baker, R.T., Cowley, J.M., Harte, D.S. and Frampton, E.R. (1990) Development of a maximum pest limit for fruit flies (Diptera: Tephritidae) in produce imported into New Zealand. *Journal of Economic Entomology* 83, 13–17.

Burmaster, D.E. and Anderson, P.D. (1994) Principles of good practice for the use of Monte Carlo techniques in human health and ecological risk assessments. *Risk Analysis* 14, 4, 477–481.

Griffin, R. and Miller, C.E. (1994) *Scenario Analysis for the Risk of Pine Shoot Beetle Outbreaks Resulting from the Movement of Pine Logs from Regulated Areas.* USDA APHIS, Washington, DC. Copy on file with USDA-APHIS-PPQ-CPHST-PERAL, Raleigh, NC.

IPPC (2010) International Standards for Phytosanitary Measures, Publication No. 5: *Glossary of Phytosanitary Terms*. Secretariat of the International Plant Protection Convention (IPPC), Food and Agriculture Organization of the United Nations, Rome.

Landolt, P.J., Chambers, D.L. and Chew, V. (1984) Alternative to the use of probit-9 mortality as a criterion for quarantine treatments of fruit fly-infested fruit. *Journal of Economic Entomology* 77, 285–287.

Liquido, N.J., Griffin, R.L. and Vick, K.W. (1995) Quarantine security for commodities: current approaches and potential strategies. *USDA Publ. Ser.* 1996–2004, USDA, Washington, DC.

Nunn, M. (1997) Quarantine risk analysis. *Australian Journal of Agricultural and Resource Economics* 41, 4, 559–578.

USDA (2004) Quantitative analysis of insect pest risks from the international cargo aircraft pathway to Miami. On file with USDA – APHIS – PPQ – CPHST – PERAL, Raleigh, North Carolina.

Vail, P.V., Tebbetts, J.S., Mackey, B.E. and Curtis, C.E. (1993) Quarantine treatments: a biological approach to decision-making for selected hosts of codling moth (Lepidoptera: Tortricidae). *Journal of Economic Entomology* 86, 70–75.

Vose, D. (2005) *Risk Analysis: a Quantitative Guide*. John Wiley & Sons Ltd, Chichester.

Yoe, C. (2012) *Principles of Risk Analysis: Decision Making Under Uncertainty*. CRC Press, Boca Raton, Florida.

11 Pest Risk Assessment

Christina Devorshak and Alison Neeley

11.1. Introduction

In previous chapters, we've provided the basic building blocks for how to conduct pest risk analysis. This includes understanding the legal and regulatory environment we work in, the terminology we use, the types of information we need and the various types of methods used to analyse risk.

In Chapter 2, we learned that risk is a product of the probability of an adverse event and the magnitude of the consequences should they even occur. In addition we learned that the probability of an event occurring is typically the product of a series of dependent events – in other words, if any of the individual events have a zero probability of occurring, then the overall probability becomes zero as well. Finally, we learned that consequences are usually additive, and that under the SPS Agreement and the IPPC, we must look at economic consequences in pest risk analysis.

The next sections of this text will describe the components of a pest risk analysis, specifically the pest risk assessment and pest risk management stages. This chapter describes the pest risk assessment (as a component of pest risk analysis). The first half of this chapter will provide an overview initiating a pest risk analysis and describing the likelihood of introduction, and the second half of this chapter will provide options to describe economic consequences. Pest risk management is described in Chapters 13 and 14.

ISPM No. 2 and ISPM No. 11 state that certain aspects of pest risk analysis are common to all stages. This includes:

- Determining the scope of the analysis;
- Information gathering;
- Risk communication (described in Chapter 15);
- Documentation and transparency.

Throughout the earlier chapters, we have emphasized the need for transparency, the scientific basis for risk analysis, the importance of describing uncertainty and the need to be clear about assumptions and judgements. Box 11.1 highlights ten general principles for conducting pest risk analysis – these principles were the results of an expert workshop conducted with the aim of providing guidance to pest risk analysts. These principles are useful to consider here (and to keep in mind), as we lead into the discussion on the process of conducting pest risk analysis.

Box 11.1. General principles for conducting pest risk analysis. (Summarized from Gray *et al.*, 1998.)

1. Analyse the appropriate attributes – Identify the problem as specifically as possible, and analyse the appropriate factors.

2. The scope of the analysis should be relevant to the decision – the complexity of the analysis should meet the needs of the decision that needs to be made.

3. Use the analysis, not just the result – the analysis is not just a trigger for a decision, but should be used to inform the decision.

4. Consider human factors – specifically, the effects of uncertainty (due to error and lack of knowledge) should be accounted for in the pest risk analysis.

5. Stimulate and be receptive to new information – pest risk analysis may identify information gaps (e.g. uncertainty), and this represents an opportunity for new information to be presented that will help the analysis.

6. Work with scientists – risk analysis is not science, but is based on scientific information; therefore consulting scientists and other experts is essential.

7. Use peer review – this helps to ensure scientific credibility.

8. Ensure complete and transparent documentation – this includes documenting assumptions, data sources, uncertainty and results.

9. Make judgements explicit – the use of expert judgement is often essential, but where it is used, it should be explicitly noted.

10. Make the pest risk analysis process open – include interested parties, be communicative and transparent.

11.2. General Overview of Pest Risk Analysis in the IPPC According to ISPM No. 2 and ISPM No. 11

There are many different approaches to pest risk analysis that are used in different countries; however, those approaches incorporate the requirements laid out in the SPS Agreement, the IPPC and the related standards. All the methods and techniques for pest risk analysis in this book are consistent with the requirements of the IPPC. The remainder of this book will provide more detailed explanations of how pest risk analysis can be conducted.

Figure 11.1 provides an overview of the pest risk analysis process. According to ISPMs No. 2 and No. 11, pest risk analysis includes three basic stages:

- **Stage 1.** Initiation;
- **Stage 2.** Pest risk assessment;
  - ○ Pest categorization;
  - ○ Likelihood of introduction and spread;
  - ○ Assessment of economic consequences;

- **Stage 3.** Pest risk management (Chapters 13 and 14).

11.3. Stage 1: Initiation

The first step in pest risk assessment ('Stage 1' of pest risk analysis according to the IPPC) is the initiation stage. According to ISPM No. 2 (*Framework for Pest Risk Analysis*) 'initiation is the identification of organisms and pathways that may be considered for pest risk assessment in relation to the identified PRA area'. Figure 11.1 shows initiation is the starting point of the pest risk assessment process.

The decision to undertake a pest risk assessment may be prompted for different reasons (see also Chapter 8 on applications of pest risk analysis). Many pest risk assessments are focused on single organisms – either organisms that have been recently introduced into an area, or organisms that have been identified as potential pests which have not yet been introduced. One of the most common reasons to initiate a pest risk assessment is a request by a trading

Pest risk analysis process overview

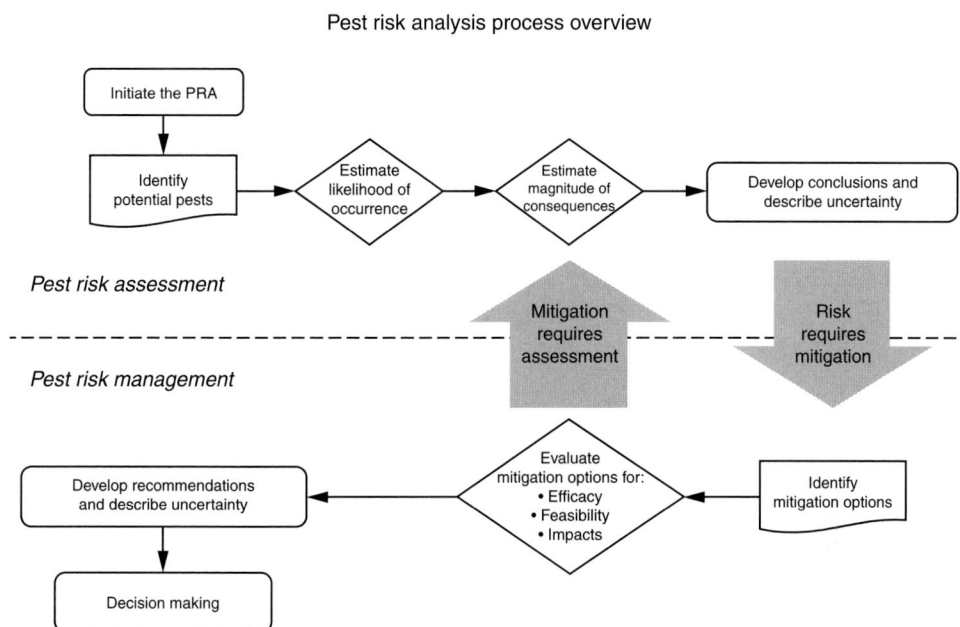

Fig. 11.1. Pest risk analysis process overview.

partner to export a particular commodity to an importing country. Other types of pathway risk assessments may be conducted, for instance to examine risks associated with smuggling or natural spread of pests.

Other reasons for initiating a pest risk assessment include changes to policies or operational approaches. This may include changes to laws or regulations, implementation of domestic programmes (such as official control), or changes to the phytosanitary status of a country. Operational concerns could include the re-allocation of resources for domestic surveillance or port-of-entry inspections.

Initiation involves determining whether an organism is a pest and defining the pest risk analysis area. In the initiation stage, the identity of the organism(s) of concern should be described (e.g. taxonomic status), as well as the characteristics that indicate the organism is a plant pest. This includes the organism's feeding habits, its ability to be a vector or other types of effects on plant health. The most common taxonomic level that is considered for a pest is species; however,

taxonomic groups below (e.g. biotypes or strains) or above (e.g. genus level) species level may be considered if evidence suggests such an approach is necessary and technically justified. For instance, many pathogens have different strains or 'pathovars' that are more or less virulent; and some species of whiteflies have different 'biotypes' that exhibit differing levels of resistance to insecticides.

Note that a variety of different types of organisms may come under consideration in the initiation stage. In addition to organisms traditionally understood to be pests (for example, fruit flies), organisms not traditionally viewed as pests may be examined in Stage 1. This includes:

• Biological control and other beneficial organisms (see also Chapter 17);
• Plants as pests (e.g. weeds, see also Chapter 18);
• Living modified organisms and invasive alien species (see also Chapter 20);
• Organisms new to science;
• Organisms imported for research and other similar purposes.

In the initiation stage, we define the pest risk analysis area. This involves describing the area for which the risk assessment will be applied based on the ability of the pest to enter, establish and spread. Recall from Chapter 5 that the pest risk analysis area is the area for which the pest risk analysis is conducted, while the endangered area is the area in which the pest may become established. In either case, however, the area in question may apply to a country, several countries or a part of a country. Providing maps in this part of the analysis can be useful to make clear exactly what area(s) is under consideration in the analysis (see Chapter 12 for more information on mapping in pest risk analysis).

The initiation stage is also where the scope of the analysis is described. The scope describes the concern or issue the analysis covers, the purpose of the pest risk analysis and the relevant background or expectations with regard to the analysis. The scope should be described as specifically as possible, since the scope will define and drive all of the subsequent steps of the analysis.

For instance, a request may be made to examine the risks associated with the importation of *Citrus* spp. In this case, we may specify that the scope of the analysis applies only to *Citrus* spp. fruit, a specific species of *Citrus* or *Citrus* plants. Each one of those analyses would result in very different pest profiles and different levels of risk. Thus, a well-defined scope provides focus to the analyst and sets the stage for a pest risk analysis that meets the needs of its audience.

All the information from the initiation stage should be clearly documented as part of the overall pest risk assessment, including whether any previous relevant assessments have been conducted. In the case of a pathway risk analysis (such as for a commodity import), initiation also requires that the relevant pathway(s) be identified and described, as well as the listing of pests that are likely to be associated with the pathway(s). In such cases, the pest risk assessment may then be composed of the combined assessments for each pest associated with the pathway.

11.4. Stage 2 Risk Assessment: Pest Categorization

After the initiation stage is completed, and a determination is made that an organism (or organisms in the case of a pathway assessment) should be assessed further, the first part of risk assessment is called 'pest categorization'. Pest categorization is 'the process for determining whether a pest has or has not the characteristics of a quarantine pest or those of a regulated non-quarantine pest' (e.g. a regulated pest).

The pest categorization step is a unique aspect of pest risk analysis compared with risk analysis conducted for other purposes (e.g. animal health or food safety). The reason for this step is that a determination must be made as to whether the pest(s) are considered to be regulated by the country doing the pest risk analysis. If the pests occur in that country already, and are not managed or controlled, then those pests should not be considered regulated pests.

The pest categorization step is usually a relatively simple assessment as to whether the pests identified in the initiation stage meet the criteria for being regulated pests. Usually, this means briefly examining evidence to make a judgement as to whether a pest occurs in the pest risk analysis area or not (for quarantine pests), whether the pest has the potential to cause economic harm, and whether the pest would be able to survive in the pest risk analysis area. If evidence indicates a pest meets these criteria, the pest risk assessment continues to the next step. If, however, evidence indicates that the pest is not likely to cause unacceptable economic harm, or that the pest is unlikely to become established in the pest risk analysis area, then the pest risk assessment can be stopped at this point. See also the section on economic assessment of consequences later in this chapter for more on assessing the potential for economic harm in the pest categorization stage.

In the case of a pathway analysis, the pest categorization step also examines whether pest(s) identified in the initiation stage are likely to be associated with the specified pathway. For instance, if the specified pathway is citrus fruit, only those pests likely to be associated with the fruit should be further analysed since pests associated with other parts of the plant are generally not expected to follow the pathway. Pests that are not associated with imported plant parts in this case need not be analysed further. Exceptions may arise; however further analysis of such pests should be technically justified.

Pests that are categorized as regulated pests, and most often as quarantine pests, and that are likely to follow a pathway for entry, are subject to further assessment. As with the initiation stage, all of the evidence used to support judgements made in the pest categorization stage should be clearly documented. For the purposes of this chapter, we will focus on the assessment of quarantine pests since regulated non-quarantine pests are rarely assessed in practice.

11.5. Stage 2 Risk Assessment Continued: Likelihood of Introduction (Assessing Entry and Establishment)

In the next step of the pest risk assessment stage, pests that were categorized as regulated pests are assessed for the likelihood of introduction (recall that introduction includes entry and establishment). Recall from Chapter 5 that the term 'introduction' refers to the entry and establishment of a pest. In order for a pest to enter an area, a series of events must occur in sequence, with some reasonable probability. Likewise, after a pest has entered a new area, several conditions (e.g. climate, host availability, etc.) have to be met and additional events must occur with some probability. Before we discuss these elements in detail however, we can conceptualize what has to happen for a pest to be introduced and spread in a new area.

A useful exercise in this stage of the pest risk analysis is to 'map' the pathway for introduction (and spread). This allows the analyst (and the audience) to visualize the events that must occur for pest introduction – it makes the analysis more transparent and is useful for identifying both key events in the pathway and areas of uncertainty. Furthermore, such an approach lends itself to quantifying events in the pathway – for both deterministic and stochastic modelling. New Zealand uses this approach in its routine import risk analyses (MAF, 2006) (even where such analyses are qualitative or semi-quantitative).

An example of such a pathway description is shown in Fig. 11.2, showing the pathway for the importation of fresh pear (*Pyrus* spp.) fruit from China into New Zealand. The visual depiction of the pathway focuses the analysis on key points where risk can be introduced (e.g. infestation by a pest) or where risk may be reduced (e.g. removal of infested fruit or treatment). Such a depiction also lends itself to later discussions on risk management.

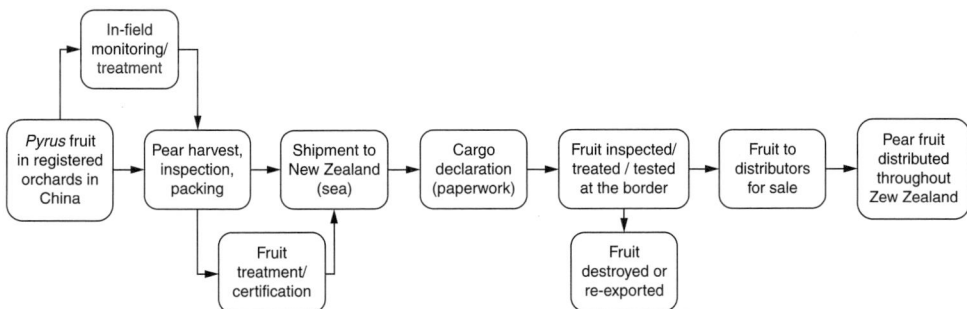

Fig. 11.2. Example of a visual pathway description: potential *Pyrus* fresh fruit pathway from China to New Zealand. (Excerpted from MAF, 2009.)

11.5.1. Defining the events for likelihood of introduction (and spread)

ISPM No. 11 provides a detailed description of the elements that should be analysed, including the likelihood or probability the pest(s) will:

- Be associated with pathways that could lead to introduction;
- Survive existing pest management procedures;
- Remain with the pathway at origin;
- Escape detection (either pre- or post-entry);
- Survive transit or movement to the pest risk analysis area;
- Find a favorable location in the pest risk analysis area;
- Find suitable host material in the pest risk analysis area;
- Overcome biotic and abiotic resistance (e.g. find suitable environment);
- Be able to reproduce and spread.

Each of these elements should be considered to determine the likelihood of introduction for a pest. Recall that 'introduction' includes both 'entry' and 'establishment' – thus, we often see the assessment of likelihood broken down into those two components, where entry and establishment are analysed separately and then combined. Some pest risk analyses consider 'spread' in connection with introduction (e.g. the analysis covers entry, establishment and spread together in that order); in other cases, spread is considered together with consequences, since degree of spread can directly influence the severity of consequences. Lastly, some pest risk analysis models consider these elements as individual questions to be addressed; other models may 'lump' some of these elements together and still other models may 'split' elements into sub-elements to be answered (see also Table 8.3 in Chapter 8 for examples of how questions for elements can be 'lumped' or 'split').

In some cases, we may have quantitative data, for instance for the prevalence of the pest in a particular pathway at origin, or its ability to survive a particular treatment. In these cases, we can assess likelihood of introduction using quantitative methods outlined in Chapter 10. More often than not, however, we lack specific data on pest prevalence or probabilities related to specific events. In these cases, we may opt to use a narrative approach to describing risk, or use a rating system such as those described in Chapter 9.

Whatever method we decide to use, however, we must be clear about the model we are using (e.g. what events in the pathway(s) are we considering, the assumptions we are making in the analysis, and the sources and level of uncertainty in our analysis) and how we are relating events and other factors in the analysis. It is also useful to link our pathway description (such as might be provided a 'pathway map' like Fig. 11.2) back to the model we are using to analyse risk.

11.5.2. Example of rating events for likelihood of introduction

As a hypothetical example, let's say we were assessing risks associated with particular species of fruit, and during the pest categorization stage, determined that a species of fruit fly is known to occur in the exporting country, feeds on that species of fruit (internally) and does not occur in the importing country. We decide to analyse this pest further, looking first at the likelihood of introduction. Figure 9.5 in Chapter 9 provided a simple model for assessing the likelihood of introduction in a commodity pest risk analysis. The events leading to introduction, according to that model are:

- Pest is present in field;
- Pest survives post-harvest handling;
- Pest survives transit;
- Pest finds suitable host material post-entry;
- Pest finds suitable climate post-entry.

We can assign ratings to each one of these elements, based on pre-defined criteria for each element (see also Chapter 9 for more on criteria and combining ratings). In addition to assigning ratings to each one of these elements, we should provide supporting

Table 11.1. Hypothetical example of ratings for a fruit fly on imported fruit.

Element	Risk rating[a]	Uncertainty[a,c]	Supporting evidence
Pest is present in field	H (3)	C	*In this section we would provide evidence (e.g. information from journals, databases, compendia or other sources) that supports a particular rating for each element. This may include data or information on a pest's prevalence in the field, host range, geographic range, fecundity, virulence, dispersal ability or any other biological or technical information relevant to particular elements. We may also provide a verbal explanation of uncertainty here.*
Pest survives post-harvest handling	M (2)	C	
Pest survives transit	M (2)	MC	
Pest finds suitable host material post-entry	M (2)	MU	
Pest finds suitable climate post-entry	M (2)	MU	
Summary/combined ratings	M (11)[b]	*Describe overall uncertainty*	*In this section we would summarize the likelihood of introduction and discuss overall uncertainty*

[a]These ratings are purely hypothetical and for example only.
[b]In this case, we have added the ratings for each element to come up with a final risk score, but other methods for combining ratings could be used (see Chapter 9 for more information on combining ratings). Recall that the relationship between specific events is multiplicative, but for ease of use, some models simply add ratings.
[c]C, MC, MU and U are ratings for uncertainty as Certain, Moderately Certain, Moderately Uncertain and Uncertain. This is described in more detail in Chapter 16.

evidence in the form of a short discussion for each element, and we should describe any uncertainty for that particular element. Table 11.1 provides an example of a hypothetical pest subject to an analysis.

Note that this model is a relatively simple model for assessing introduction, and is only looking at unmitigated risk. We have not included, for instance, likelihood of the pest being detected at entry, or the effect of a specific phytosanitary treatment (e.g. fumigation) in the assessment at this stage. However, we could also decide to analyse mitigated risk, for instance if specific conditions were defined in the initiation of the pest risk analysis. We could also return to this model during the risk management stage and look at the effect of including specific mitigations on the overall risk (see also Chapters 13 and 14).

Likewise, we may decide that a more detailed model is needed. If we want a

more detailed model, we can split various elements into sub-elements, or include additional events in a model (similar to the approach that EPPO uses in its pest risk analysis Scheme – see EPPO, 2007). For instance, in the model above, we consider the element 'survive post-harvest handling procedures' as a single element. However, if we wished to develop a more detailed model, we might include a few elements to cover this one point, including likelihoods that:

- The pest remains associated with the commodity at harvest;
- Infested fruit is not culled during post-harvest handling;
- The pest survives post-harvest handling (e.g. washing, waxing, etc.);
- The pest survives processing and packaging;
- The pest survives storage.

The decision to use more or less detailed models is largely a matter of preference, and there are advantages and disadvantages to each approach (see Chapter 9, Section 9.6.3. for advantages and disadvantages of using highly specified schemes). The important points are that the analysis is transparent and consistent, the model and the evidence are scientifically and technically sound, and the analysis is clearly documented.

11.6. Assessing Economic Consequences

The next part of the pest risk assessment, after estimating likelihood of introduction, is to look at likelihood of spread, and the potential for economic consequences. Because consequences are directly related to the likelihood and extent of spread, these two points are considered together in this section. Recall that Chapter 7 provided an overview on the role of economics in pest risk analysis, and that under both the IPPC and the SPS Agreement, we are required to demonstrate the potential for economic harm.

Although it may be relatively easy to identify, describe and estimate the total impacts of a past pest incursion, making predictions about the potential impacts of a particular organism or group of organisms that have *not* yet been introduced is infinitely more challenging. Species often behave very differently in novel environments than they do in their native range. For example, some organisms that were not major pests in their native range have caused major harm to cultivated plants or the natural environment when they were introduced to a new area free from natural predators and parasites. Others, however, have not had a significant impact in the new area, even though these same organisms were pests in their native range.

It is estimated that approximately 50,000 non-indigenous species have been introduced into the USA over the past 200 years; less than 15% of those cause any kind of significant economic damage (Evans, 2003). Past experiences with introductions of non-indigenous organisms suggest two things: (i) the actual impacts that result from a pest introduction depend on the very complex interaction of many variables, not all of which are known or understood; and (ii) because of this complexity, it is very difficult to predict potential impacts with any kind of precision. In general, these variables fall into three broad categories:

1. *The biology of the pest*: its natural dispersal ability and its inherent ability to adapt to novel environments;
2. *The suitability of the environment*: climatic conditions, availability of suitable host material (in the proper phenological stage), presence and abundance of natural enemies, widespread presence of suitable vectors;
3. *Agricultural and natural systems*: types of crops and livestock produced, production systems used, control methods used as part of routine agricultural management, continuity of host material, etc. (FAO, 2001).

11.6.1. Impacts

Types of impacts

The full range of economic impacts that occur as a result of a biological incursion can be broadly classified into two categories: direct and indirect impacts. A pest or disease can directly impact a host or group of hosts by reducing its value. Indirect effects, on the other hand, result from the presence of the pest, but are not specific to the pest–host dynamics and therefore, extend well beyond the primary impacts to affected hosts (Evans, 2003). Examples of indirect effects include shifts in consumer demand, changes in relative prices of inputs, loss of environmental amenities, displacement of native species, loss of international markets, costs for research and changes to related markets such as energy and tourism.

According to the IPPC, the scope of both direct and indirect 'economic impacts' includes impacts to *all* plant resources, including those that are commercial (such as crops), cultivated (such as horticultural species) and found in the environment (IPPC, 2004). In the past the IPPC has been incorrectly characterized as commercially focused; to remedy this impression, an appendix to the glossary was created to clarify what is meant by 'economic consequences.' According to the appendix, 'potential economic damage' includes all types of impacts including environmental and social concerns.

Table 11.2 lists some examples of direct and indirect consequences that can occur as a result of a new pest incursion.

Magnitude of impacts

Just because a particular pest has the *potential* to cause damage, or is known to be 'injurious to plants or plant products' (IPPC, 2010), does not mean it will necessarily cause *economically significant* damage upon introduction into a new area. The magnitude of damage that a pest will cause depends upon some of the same variables that influence its ability to establish, but also depends on how quickly the pest can spread over time and how large the area being analysed is. Therefore, before we can begin to estimate the magnitude of consequences that would likely result from the introduction (and establishment) of a given pest, we must first define our *time horizon* and then estimate the pest's *spread potential* over that period of time.

TIME HORIZON. The impacts that result from pest introductions are not constant; they are expressed over time as the pest spreads throughout the pest risk analysis area. A time horizon is simply a fixed, defined point of time in the future, over which we estimate impacts. Just as it is critical to define the exact pest risk analysis area being analysed, it is equally critical to clearly define the time horizon being considered in the analysis. A time horizon can be chosen to be the point of time in the future where, given the pest's spread potential, it will likely reach its maximum distribution in the pest risk analysis area, or it can be a fixed length of time chosen arbitrarily (e.g. 1 year, 10 years, 50 years, etc.). The time period selected can greatly influence the outcome of an analysis, because the greater the time horizon, the greater the estimated total impacts generally are.

SPREAD POTENTIAL. Spread potential has two components:

1. the pest's predicted *rate* of spread: or in other words, how fast the population of the pest will expand beyond its initial point(s) of introduction;

Table 11.2. Examples of potential consequences from a biological incursion.

	Direct consequences	Indirect consequences
Market effects 'Commercial consequences'	Yield loss (due to lost production or loss of marketable commodity) Reduction in value of host (due to quality loss or diverted market)	Market loss (domestic and export) Increased production costs Regulatory costs (e.g. inspection, phytosanitary requirements) Long-term changes in demand Costs for research Costs of restoring environment Impact on other industries (e.g. tourism, energy, etc.)
Non-market effects Impacts that cannot be measured in the market place	Reduction of ecologically significant species Impacts on threatened and endangered species	Displacement of native species Social effect of control measures Impacts on plant communities Reduction in aesthetic beauty

2. the endangered area: which is simply the total portion of the pest risk analysis area where the pest is likely to survive.

Spread potential is important because the establishment of a pest usually does not result in immediate significant impacts, but as the pest spreads beyond its initial point of introduction, impacts will begin to increase, often exponentially. The spread potential of a pest is an important element in determining how quickly the pest may impact the environment, the size of the area it impacts and how easily the pest can be contained.

Over the course of many years we have developed the following hypotheses with regard to pests' spread potential:

- *Adaptability of pest*: pests that possess special abilities that enable them to survive adverse conditions are more likely to become naturalized in a new environment. Examples of such adaptations include:
- Ability to reproduce and develop rapidly;
- Ability to feed on a variety of different foods or go without food for long periods of time;
- Altered growth or form to suit a particular habitat, or ability to become dormant until the climatic conditions favour development (such as through hibernation, aestivation, cryptobiosis, diapause, production of sclerotia, cysts, spores, pupae, etc.);
- Ability to survive wide ranges in abiotic conditions such as temperature, humidity, rainfall, UV radiation, soil type, chemistry, etc.
- *Endangered area*: pests can be expected to survive and maintain their normal behaviour in new areas with environmental conditions comparable with those in their native range.
- *Host range*: pests that have a broad host range with many hosts in the pest risk analysis area will have a greater spread potential than pests that only have few hosts, or feed on hosts that have a restricted distribution in the pest risk analysis area.

- *Reproductive potential*: pests that have a high biotic potential (e.g. high female fecundity; ability to reproduce parthenogenically) are more likely to be able to disperse rapidly.
- *Dispersal potential*: pests that are highly mobile, or that can 'hitchhike' easily have a greater likelihood of dispersing rapidly. Other variables that influence dispersal potential include: suitability of the natural and/or managed environment (including routine controls), presence of natural enemies, the necessity and widespread presence of vectors of the pest in the pest risk analysis area, and presence of natural barriers in the pest risk analysis area.

11.6.2. Estimating potential consequences of introduction

Once it has been determined that there is some probability that a particular plant pest or disease will be introduced into the pest risk analysis area, the next step in the risk analysis process is to determine whether and to what extent the pest will adversely affect the pest risk analysis area. Consequence evaluation can take many forms, from simple qualitative narratives to complex economic models. True economic assessment requires a considerable amount of biological and non-biological information, and generally involves considerable time, expense and expertise.

In many cases, the data necessary for such analyses simply do not exist, especially in cases where there has not been a previous outbreak or there is no disease history. Complications may also arise from the uncertainty of the scientific evidence about the probability of entry and establishment, rate of spread and the extent of damage (Evans, 2003). Even in the best of cases, available data are usually based on extrapolation from similar pest situations, and hence there is a great deal of uncertainty involved; therefore, is very important to realize that consequence evaluation is rarely capable of producing precise results. Luckily, in many cases precise results are not necessary. ISPM No. 11 embraces a broad range of approaches to economic consequence analysis in pest risk analysis.

In the next section we discuss three separate approaches to consequence evaluation:

1. Pest categorization;
2. Determining the value at risk;
3. Economic analysis.

Pest categorization

Pest categorization, according to ISPM No. 11, is the process of determining whether a pest meets the definition of *quarantine pest*, and in this case, whether the pest is of 'potential economic importance'. To be of potential economic importance, three criteria must be met: (i) there must be a potential for the pest to be introduced into the pest risk analysis area; (ii) there must be a potential for the pest to spread after establishment; and (iii) the pest must be able to cause identifiable, harmful impacts to economically, environmentally or socially important hosts in the pest risk analysis area. See also Section 11.4 of this chapter for more discussion of the pest categorization stage.

When there is sufficient evidence or general agreement that the introduction of a pest will have unacceptable economic consequences in the pest risk analysis area, consequence evaluation can stop at pest categorization. In such cases, the risk assessment will focus primarily on probability of introduction and spread. Note that this abbreviated approach assumes a certain tolerance for pest damage. In other words, some pests may have 'acceptable' economic consequences, if the total impact of those consequences does not meet a certain threshold.

Determining the value at risk

Determining the value at risk is the method described in Section 2.3 of ISPM No. 11: *Assessment of Potential Economic*

Pest categorization – determining whether the pest meets a threshold	
This approach is appropriate when:	This approach is *not* appropriate when:
There is general agreement on the type and magnitude of likely consequences	When the level of consequences is in question
There are sufficient data to demonstrate impacts will exceed the threshold for being considered 'unacceptable'	For pest prioritization When the level of consequences is needed for risk management decisions (e.g. evaluation of strength of measures; deciding appropriate management strategies, etc.)
There is little chance of challenge or disagreement	
The magnitude of impacts is not necessary for decision making in the near term.	

Consequences. The purpose of this method is to identify the hypothetical total potential consequences that could occur as a result of the pest's introduction. Note that this approach often assumes a hypothetical 'worst case scenario' where the pest has been introduced everywhere it can survive in the pest risk analysis area and is fully expressing all potential economic consequences.

Steps to determining value at risk

1. Identify all potential direct pest effects (market and non-market). Consider all hosts in the pest risk analysis area, and the type and magnitude of damage caused by the pest.
2. Identify potential indirect pest effects, especially potential trade and environmental impacts.
3. Based on biology of the pest, determine the area where the pest is likely to establish.
4. Based on where the pest can establish and the distribution of the pest's potential hosts in the pest risk analysis area, determine the *endangered area*, in other words, the portion of the pest risk analysis area

where ecological factors favour the establishment of the pest AND where the presence of that pest will result in economically, environmentally or socially important loss.

5. Estimate the total area or amount of the hosts in the endangered area and quantify their value.

6. Estimate the value at risk (based on the value of the hosts at risk and the potential identified direct and indirect pest effects).

In determining value at risk, consider:

- How the pest being analysed has behaved in the area where it currently occurs.
- How a similar pest has behaved in the area being analysed.

It is important to note that the value at risk represents the total *potential* economic impact but it *does not* represent likely or actual economic impact. The worst case scenario is rarely ever realized (at least within a realistic time horizon) because pests are not likely to become established everywhere they can survive all at once, are not likely to cause consistent damage (some hosts will be more resistant than others, some pests more virulent than others), and serious pests are likely to be controlled (either by humans or by adaptive natural predators) before they reach their full pest potential. Additionally, as mentioned previously, costs will not be constant over time; the rate of spread over time greatly influences actual economic costs of an incursion.

Economic analysis

As mentioned previously, economic analysis generally requires a relatively large amount of resources, expertise and data. Even when using economic analytical techniques, it is not always necessary (or even possible) to do a complete economic analysis. For example, because environmental and other non-market impacts consequences are often very difficult and controversial to monetize, it may be possible to focus the assessment on market impacts that are more easily quantified, and then address non-market impacts

Determining value at risk as an approach to estimating consequences	
This approach is appropriate:	This approach is *not* appropriate:
When you need to compare the potential risk of pests within an analysis	Actual level of economic consequences is needed to evaluate the strength of measures (trading partners may view basing measures on 'worst-case' scenario to be precautionary)
As a method for prioritizing pests that have not yet been introduced	
When the results of pest categorization are in dispute and further evidence is needed to demonstrate a particular organism is likely to be a pest in the pest risk analysis area	Level of consequences is needed for risk management decisions or in assessing the cost–benefit of exclusion or control

qualitatively. In some cases, it may not be necessary to analyse all market impacts. If you just need to show the consequences are 'high' or greater than a certain amount, you may only need to focus on quantifying one or two important consequences. The rest can be addressed qualitatively.

So how do you decide when detailed economic analysis is needed? Consider conducting full economic analysis any time you need to have an accurate understanding of what the true magnitude of consequences is likely to be, you have enough data to conduct the analysis, and resources are available (including time, personnel and expertise).

There are many economic models and techniques, but in terms of phytosanitary risk assessment, the most common are:

- Partial budgeting;
- Partial equilibrium analysis;
- Bioeconomic modelling.

Because these methods generally require an economist and other technical expertise, a detailed description of how to conduct such analyses is not within the scope of this book; however each method is briefly described below.

Economic analysis	
This approach is appropriate when:	This approach is *not* appropriate when:
The estimated value of the economic consequences is highly controversial, political or called into question	Simpler analyses would be just as effective for decision making
It is needed for risk management decisions, such as evaluating how strong phytosanitary measures should be	You do not have enough data for the analysis to be meaningful
	Economic analysis is not feasible due to lack of resources including expertise, personnel and time
You need to know the best way to allocate limited resources	
You are conducting policy analysis	

Partial budgeting

Partial budgeting models represent the simplest and consequently most commonly used technique. Although many economists consider partial budgeting to be a form of 'accounting' rather than economic analysis *per se*, partial budgeting models do use economic tools such as marginal analysis and are generally adequate if the economic effects are likely to be relatively narrow expected or limited to producers. Unlike pest categorization and determining the value at risk, partial budgeting models can be used to estimate *actual* magnitude of potential impact, rather than the *comparative* magnitude of potential impact.

Partial budgeting models are used to evaluate the changes in resource uses (that are not fixed) that would occur as a result of a change in the system. Impacts are typically assessed in terms of changes in revenues and costs. For example, a partial budgeting model might be used to describe the reduced revenue and additional cost that would be incurred by producers as a result of the introduction of a particular pest. In conducting a partial budgeting analysis, it is very important to clearly describe the scope of the analysis including the time

horizon being considered, the area being considered and the sector (e.g. all domestic producers, domestic producers in the endangered area, domestic producers + public sector, etc.).

Partial equilibrium models

Partial equilibrium models estimate changes in social welfare based on changes in prices and quantities demanded and supplied of a given product. Usually, both producer *and* consumer impacts are considered. These models require knowledge of demand and supply functions, elasticities and price changes. As such, they are data-intensive, and therefore rarely done in phytosanitary risk assessment, where there is generally not enough time or data to complete them.

Partial equilibrium models differ from general equilibrium models in that they examine one market sector in isolation from other sectors in the economy. The market examined is treated as independent from other markets. Hence, one disadvantage of partial equilibrium models is that by holding prices and quantities of other goods at fixed values, the possibility that events in this market affect other markets' equilibrium prices and quantities are ignored. For example, although in reality, energy prices are closely linked to the maize market (because of ethanol production), in a partial equilibrium model that looks at the effect an introduced pest might have on maize, the price of energy would be held constant. In reality, a pest that impacts maize could also have a big impact on the energy industry as well.

On the other hand, general equilibrium models, which study how equilibrium is determined in all markets simultaneously are incredibly labour- and data-intensive, and are generally not done in the plant health world.

Bioeconomic models

Bioeconomic models combine economic and biological or epidemiological information to assess the effects of a change on a particular system. They involve close integration of important biophysical information and ecological processes with economic decision behaviour. Such models help policy analysts anticipate how interventions in one part of a complex system may result in responses from other parts of the system.

11.6.3. Example 1 – economic consequences of the European grapevine moth

The European grapevine moth, *Lobesia botrana* Denis and Schiffermüller, was detected in California for the first time in 2009. Native to the Palearctic regions of Europe, northern Africa and Asia, the pest had spread to central Africa and South America. After conducting an extensive literature review, pest risk analysts in the USA determined that although this moth had been reportedly associated with a large variety of hosts, grapevine was really the only major host crop in which damage was important. Little to no information was found regarding damage to other crops. Therefore, analysts focused their attention on the potential damage this moth would cause to grapes in the USA. The risk map shown in Plate 1 was developed by combining the areas of the USA with similar climates to where the moth is currently distributed and the areas of the USA where grapes are grown.

The analysts also found evidence in the literature that in some parts of the world, up to 80% of grapevine crops were lost when the pest was present, but not mitigated. Further, they found evidence that much of the damage caused by the pest throughout its current distribution was a result of secondary infections by various fungi that infected the feeding sites. Based on this evidence, and value of grapes in the states most at risk, risk analysts estimated that left unmitigated and allowed to spread everywhere in the USA it is capable of surviving, $2.4 billion of the (approximately $35 billion) grape and wine industry of the USA would be at risk.

If little information is available online on a specific pest, we can also look at similar pests that have been introduced previously to the pest risk analysis area or to other

climatically similar areas. Pests that are in the same genus or family can provide an indication of how the pest of concern might behave. The approach should be used with caution, however, since species can behave very differently. Be sure to explicitly state your assumptions and uncertainty.

11.6.4. Example 2 – economic consequences of wood boring beetles

As trade has increased, so has the risk that exotic wood boring beetles will be introduced into novel environments. Wood boring beetles are often carried in packing material, palettes and dunnage associated with the movement of commodities. Many of these beetles are harmless in their native areas but upon being introduced into new environments become devastating to the new forest ecosystems. Because so little information is known about many of these beetles, risk analysts around the world often have to look at how similar and closely related species have behaved when they have become introduced. For example, there are almost 15,000 identified species of metallic wood boring beetles in the family Buprestidae. Very little is known about most of these species; however, unless known to be otherwise, members of this family are often assumed to be very risky since several members of this family including the emerald ash borer beetle have been known to attack live trees in novel areas, causing a great deal of environmental and economic damage.

Using the information for where the pest occurs, the type of pest it is, and comparing that with the pest risk analysis area, we are able to assess the potential for economic importance using evidence from the literature and expert judgement. With more time and resources we can develop more complex models and use mathematical models and economic techniques. If a pest has no potential economic importance in the pest risk analysis area, then it does not satisfy the definition of a quarantine pest (nor a regulated non-quarantine pest) and you can stop the pest risk analysis.

After the pest risk assessment stage is completed, a determination is made as to whether the risk is considered to be acceptable or not. If the level of risk is acceptable, then no specific pest risk management measures are required. If the risk is not acceptable, then risk management measures may be needed in order to meet the appropriate level of protection.

11.6.5. Dealing with data gaps and uncertainty

Extrapolation

By its very nature, assessing potential consequences of introduction involves making predictions about something that has not occurred. Therefore, analysts often have to extrapolate the needed information from similar situations. The first step of this process is to examine the areas and other regions where the pest already occurs and note whether the pest causes major, minor or no damage and how often the pest causes damage (infrequently or frequently). If the area where the pest occurs is similar to the pest risk analysis area, then we can use that information to compare with abiotic and biotic conditions in the pest risk analysis area. We take particular note of how the pest behaves in novel environments (e.g. areas where it is not native). When examining damage potential, it is important to note whether the damage being considered is mitigated or unmitigated. All assumptions or uncertainty need to be explicitly stated.

Importance of describing uncertainty in economic analysis

Uncertainty is described in greater detail in Chapter 16. As with any other part of the analysis, uncertainty is a part of the estimation of economic consequences. For economic analysis to have any credence the underlying assumptions and data must be accurate, transparent, repeatable and defendable; therefore, it is important to clearly state basic assumptions, methods and data. Be careful of over-precision. Estimation of the probability of introduction of a pest and of its economic consequences involves many uncertainties. Reasonably accurate results are more beneficial than precise results that have no validity. Always provide a discussion of uncertainties associated with estimates.

11.7. Documenting the Results of the Pest Risk Assessment Stage

The final part of the pest risk assessment stage is documenting results. This means that results of both assessments of likelihood of introduction and spread and consequences are clearly documented. The results of the two parts of the assessment (introduction and consequences) may be combined together (for instance in a matrix) to produce an overall result (along with a discussion of the results), or reported separately in a narrative fashion. If quantitative methods are used, then the results may be reported as probabilistic statements, usually accompanied by a narrative discussion. In addition, relevant resources and references should be cited, and any assumptions, uncertainties and the model(s) should be clearly described.

11.8. Summary

Pest risk analyses follow the same basic steps they are addressing organisms, pathways, commodities or other situations. Those steps are:

- **Stage 1.** Initiation – where we identify organisms and pathways that may be considered for pest risk assessment in relation to the identified PRA area, and we describing the scope and purpose of the PRA;
- **Stage 2.** Pest risk assessment including:
  ○ Pest categorization;
  ○ Assessing likelihood;
  ○ Estimating consequences;
- **Stage 3.** Pest risk management.

Stage 1 initiation and Stage 2 pest risk assessment were detailed in this chapter. In assessing likelihood of introduction and spread, we assess the likelihood the pest will:

- Be associated with pathways that could lead to introduction;
- Survive existing pest management procedures;
- Remain with the pathway at origin;
- Escape detection (either pre- or post-entry);
- Survive transit or movement to the PRA area;
- Find a favourable location in the pest risk analysis area;
- Find suitable host material in the pest risk analysis area;

- Overcome biotic and abiotic resistance (e.g. find suitable environment);
- Be able to reproduce and spread.

Estimating consequences includes looking at direct and indirect consequences, and both market and non-market effects. The magnitude of consequences is dependent on both the time horizon for the pest risk analysis and the ability of the pest to spread. A range of methods are used to estimate consequences – and the method selected depends on the availability of information, and the specific application of the pest risk analysis. In all cases – for likelihood of introduction and spread, as well as estimating consequences – assumptions and uncertainty should be clearly described.

References

EPPO (2007) *Guidelines on Pest Risk Analysis: Decision-support Scheme for Quarantine Pests*. EPPO Standard PM 5 / 3(3) EPPO, Paris. Available online at: http://archives.eppo.org/EPPOStandards/PM5_PRA/PRA_scheme_2007.doc, accessed 15 December 2011.

Evans, E.A. (2003) Economic dimensions of invasive species. *Choices: The Magazine of Food, Farm and Resource Issues* second quarter, 5–10.

FAO (2001) *The State of Food and Agriculture 2001*. Food and Agriculture Organization of the United Nations, Rome.

Gray, G.M., Allen, J.C., Burmaster, D.E., Gage, S.H., Hammitt, J.K., Kaplan, S., Keeney, R.L., Morse, J.G., Warner North, D., Nyrop, J.P., Alina Stahevitch, O. and Williams, R. (1991) Principles for conduct of pest risk analyses: report of an expert workshop. *Risk Analysis* 18, 6, 773–780.

IPPC (2004) International Standards for Phytosanitary Measures, Publication No. 11: *Pest Risk Analysis for Quarantine Pests Including Analysis of Environmental Risks and Living Modified Organisms*. Secretariat of the International Plant Protection Convention (IPPC), Food and Agriculture Organization of the United Nations, Rome.

IPPC (2007) International Standards for Phytosanitary Measures, Publication No. 2: *Framework for Pest Risk Analysis*. Secretariat of the International Plant Protection Convention (IPPC), Food and Agriculture Organization of the United Nations, Rome.

IPPC (2010) International Standards for Phytosanitary Measures, Publication No. 5: *Glossary of Phytosanitary Terms*. Secretariat of the International Plant Protection Convention (IPPC), Food and Agriculture Organization of the United Nations, Rome.

MAF (2006) *Biosecurity New Zealand Risk Analysis Procedures*. Ministry of Agriculture and Forestry, Wellington.

MAF (2009) *Import Risk Analysis: Pears (Pyrus bretschneideri, Pyrus pyrifolia, and Pyrus sp. nr. communis) Fresh Fruit from China*. Ministry of Agriculture and Forestry, Wellington.

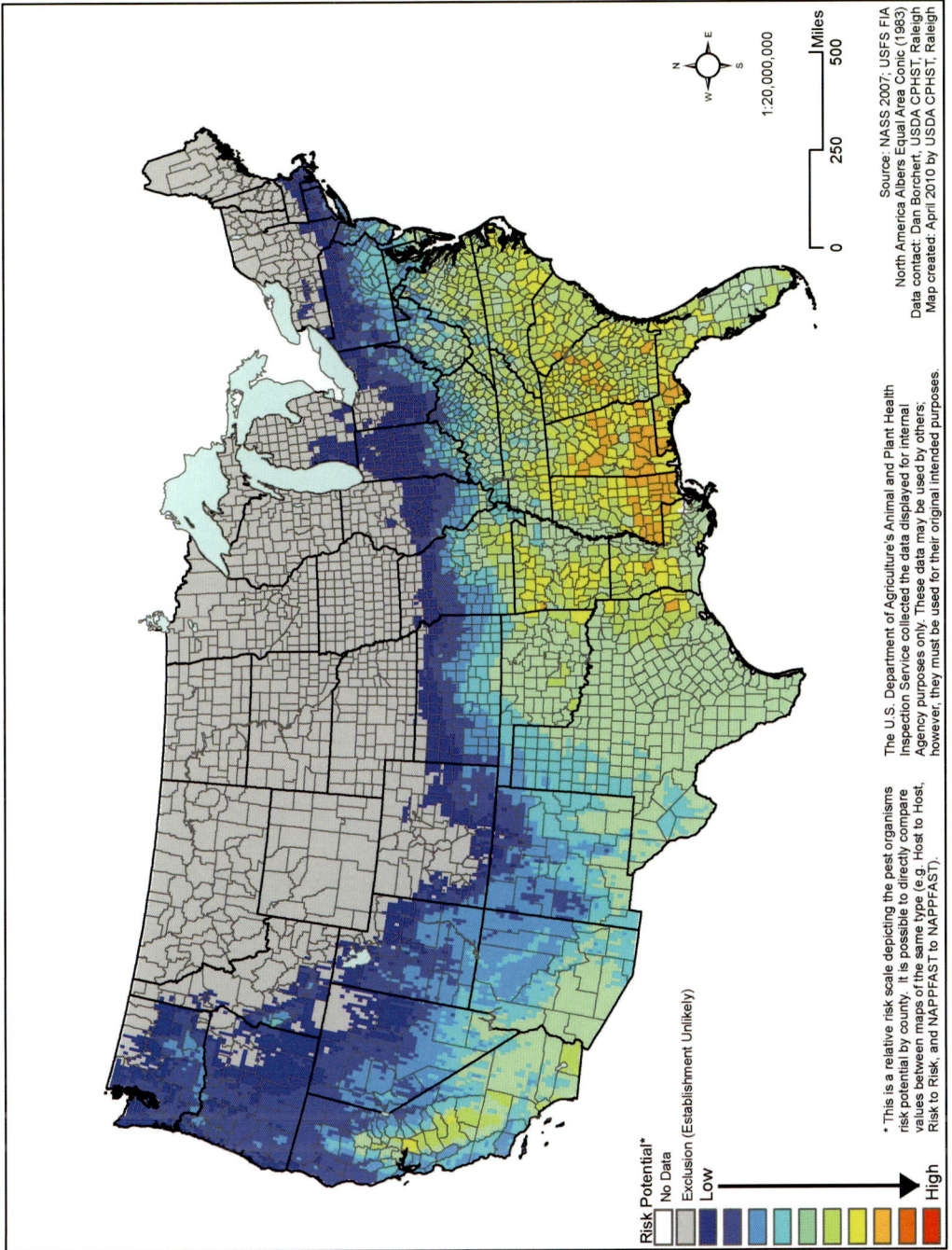

Plate 1. The risk map shows the areas of the USA that the risk analysts concluded were most at risk due to the introduction of *Lobesia botrana* (NASS, 2009).

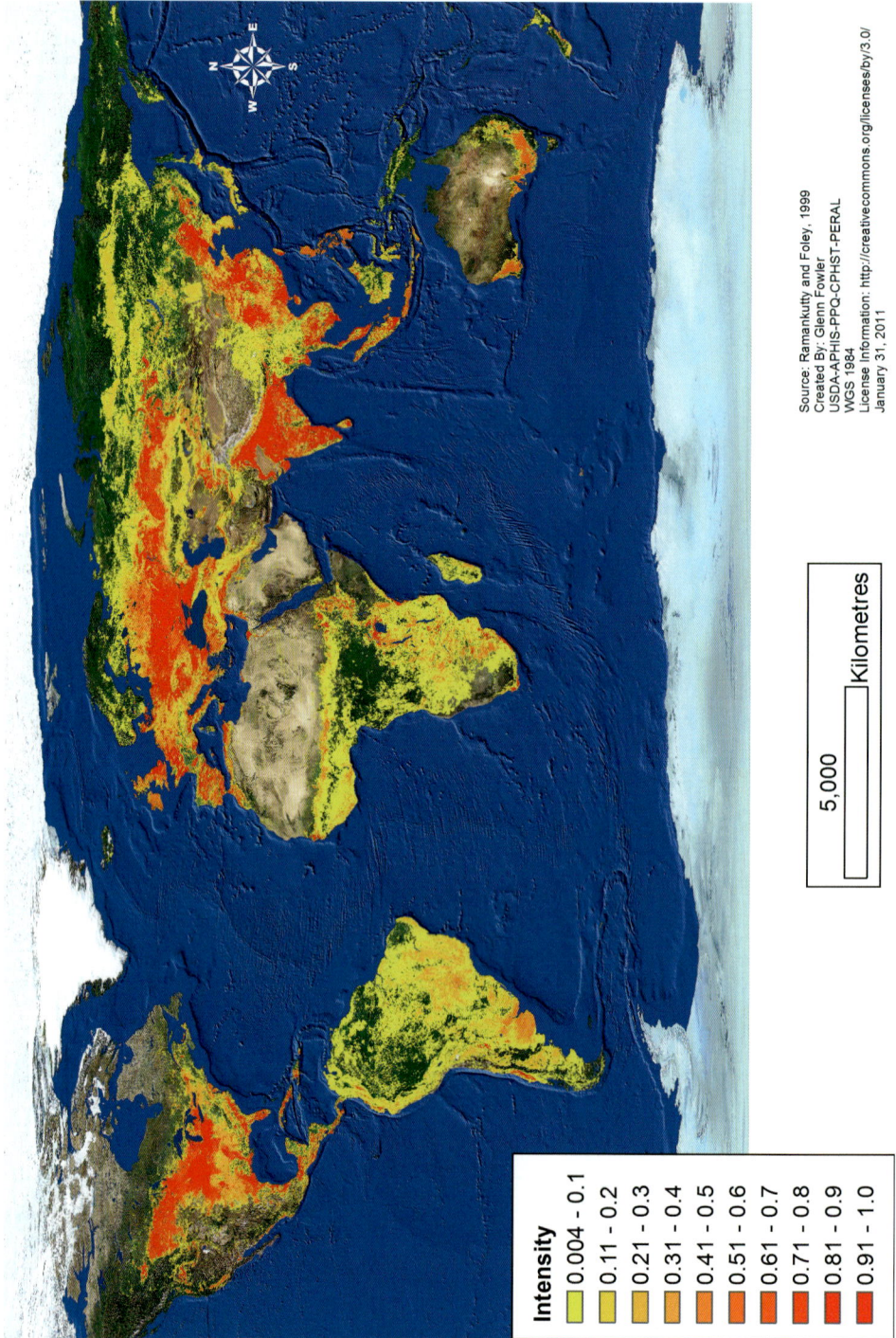

5,000 ☐ Kilometres

Intensity
- 0.004 - 0.1
- 0.11 - 0.2
- 0.21 - 0.3
- 0.31 - 0.4
- 0.41 - 0.5
- 0.51 - 0.6
- 0.61 - 0.7
- 0.71 - 0.8
- 0.81 - 0.9
- 0.91 - 1.0

Plate 2. An example of raster data: cropland intensity reported as a fraction of each grid cell (Ramankutty and Foley, 1999).

3

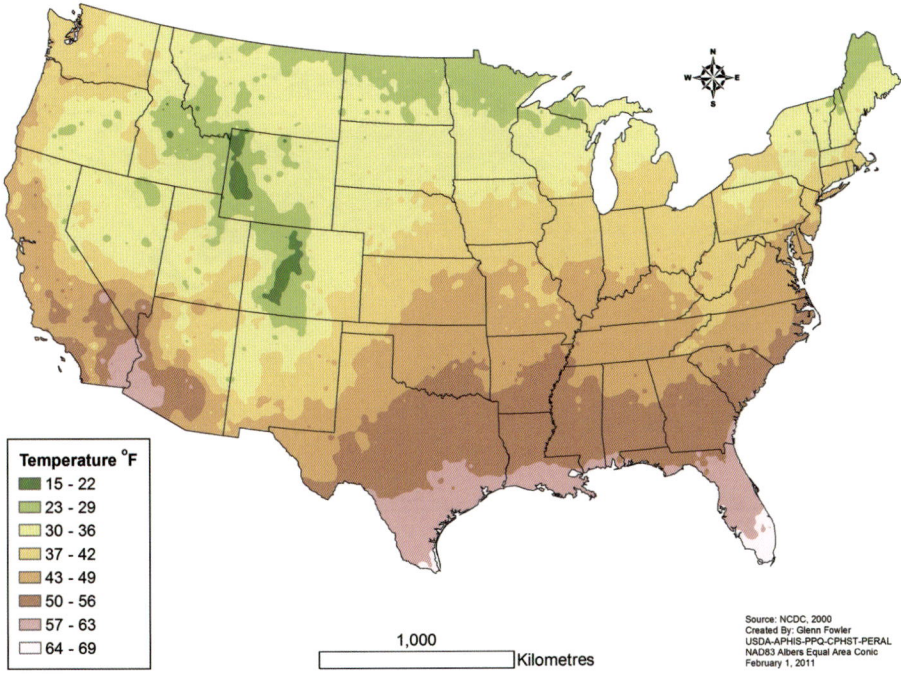

4

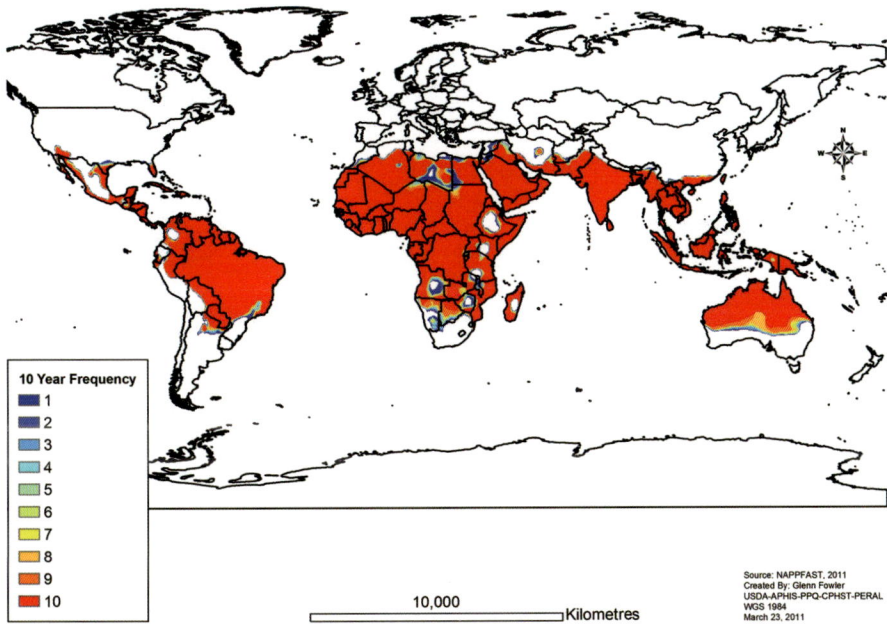

Plate 3. Interpolated mean annual minimum temperature map based on weather data from 1961–1990 (NCDC, 2000).

Plate 4. NAPPFAST probability map for areas where a hypothetical organism could complete four generations based on climate data from 2001 to 2010 (NAPPFAST, 2011).

5

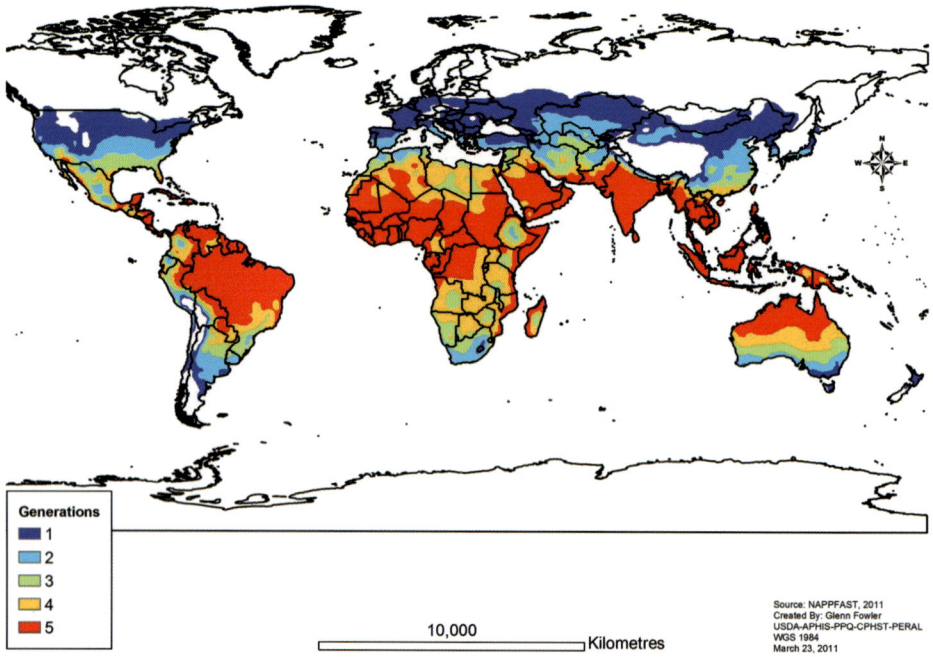

Generations
1
2
3
4
5

10,000 Kilometres

Source: NAPPFAST, 2011
Created By: Glenn Fowler
USDA-APHIS-PPQ-CPHST-PERAL
WGS 1984
March 23, 2011

6

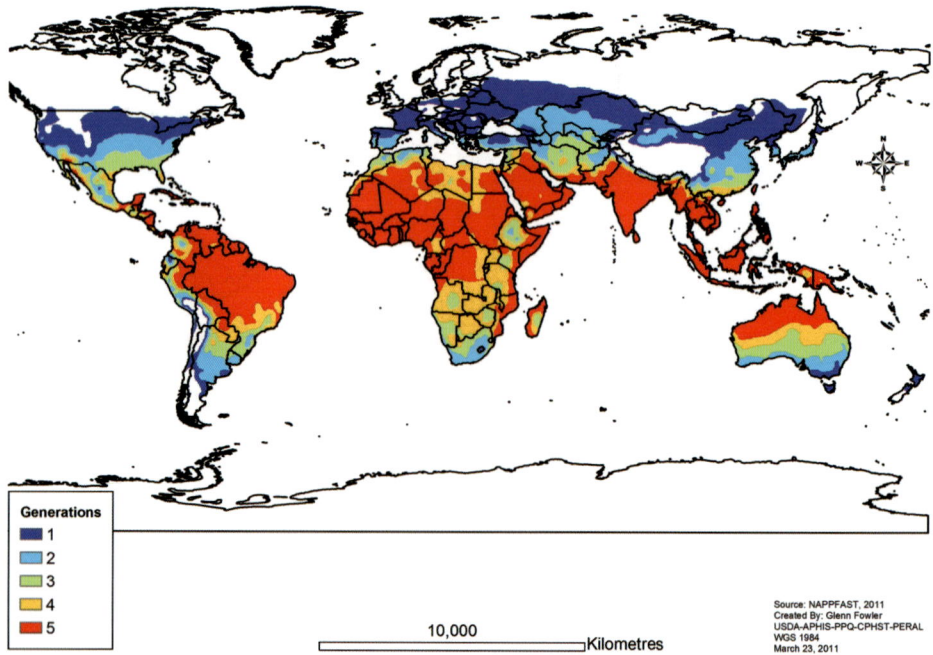

Generations
1
2
3
4
5

10,000 Kilometres

Source: NAPPFAST, 2011
Created By: Glenn Fowler
USDA-APHIS-PPQ-CPHST-PERAL
WGS 1984
March 23, 2011

Plate 5. NAPPFAST average history map for the number of generations a hypothetical organism could complete based on climate data from 2001 to 2010 (NAPPFAST, 2011).

Plate 6. NAPPFAST yearly map for the number of generations a hypothetical organism could complete based on climate data from 2010 (NAPPFAST, 2011).

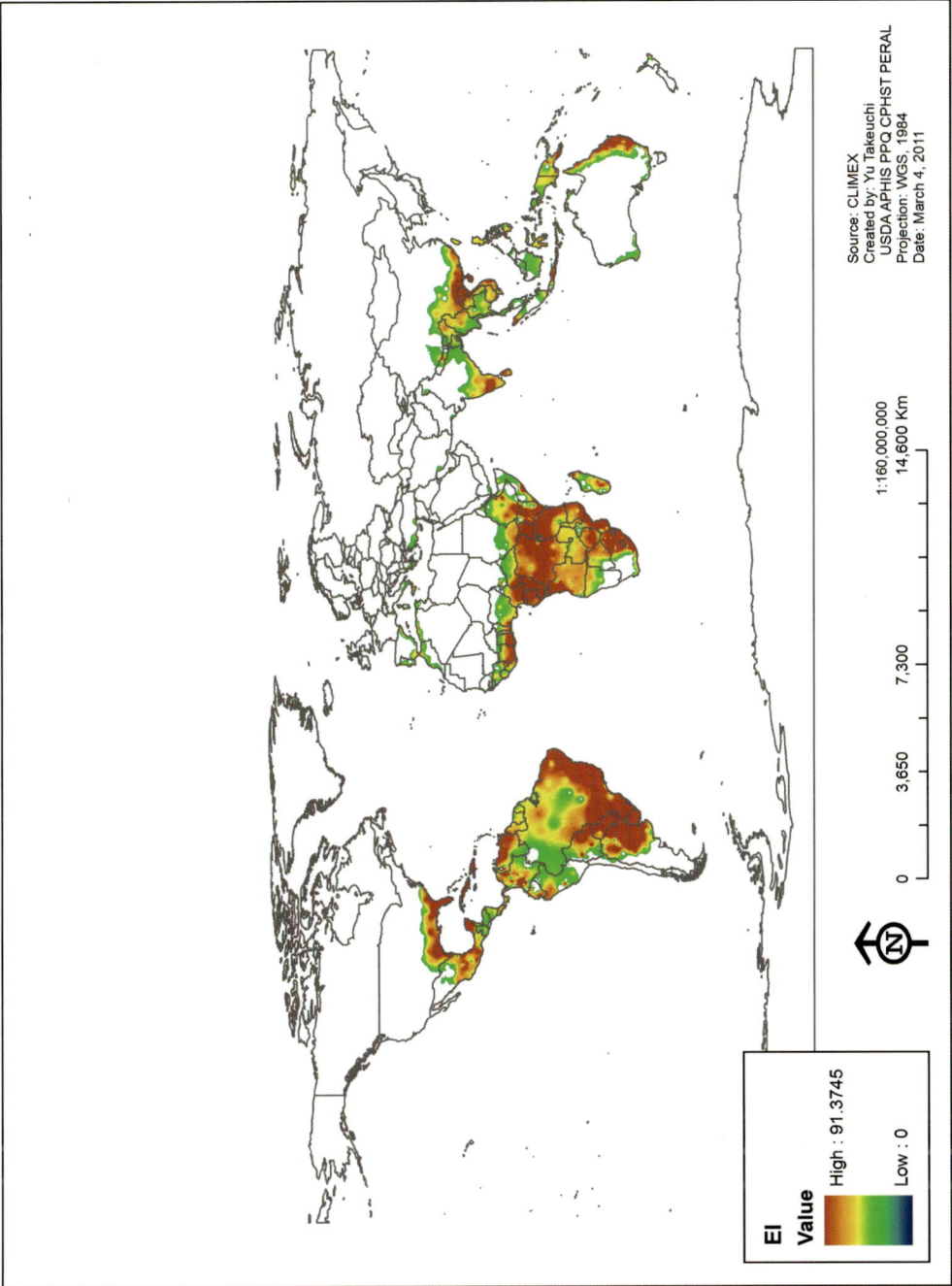

Plate 7. *Ceratitis capitata* climatologically suitable areas described by interpolated EI values in raster format, *Ceratitis capitata* suitable areas indicated by CLIMEX EI values (CLIMEX, 2011).

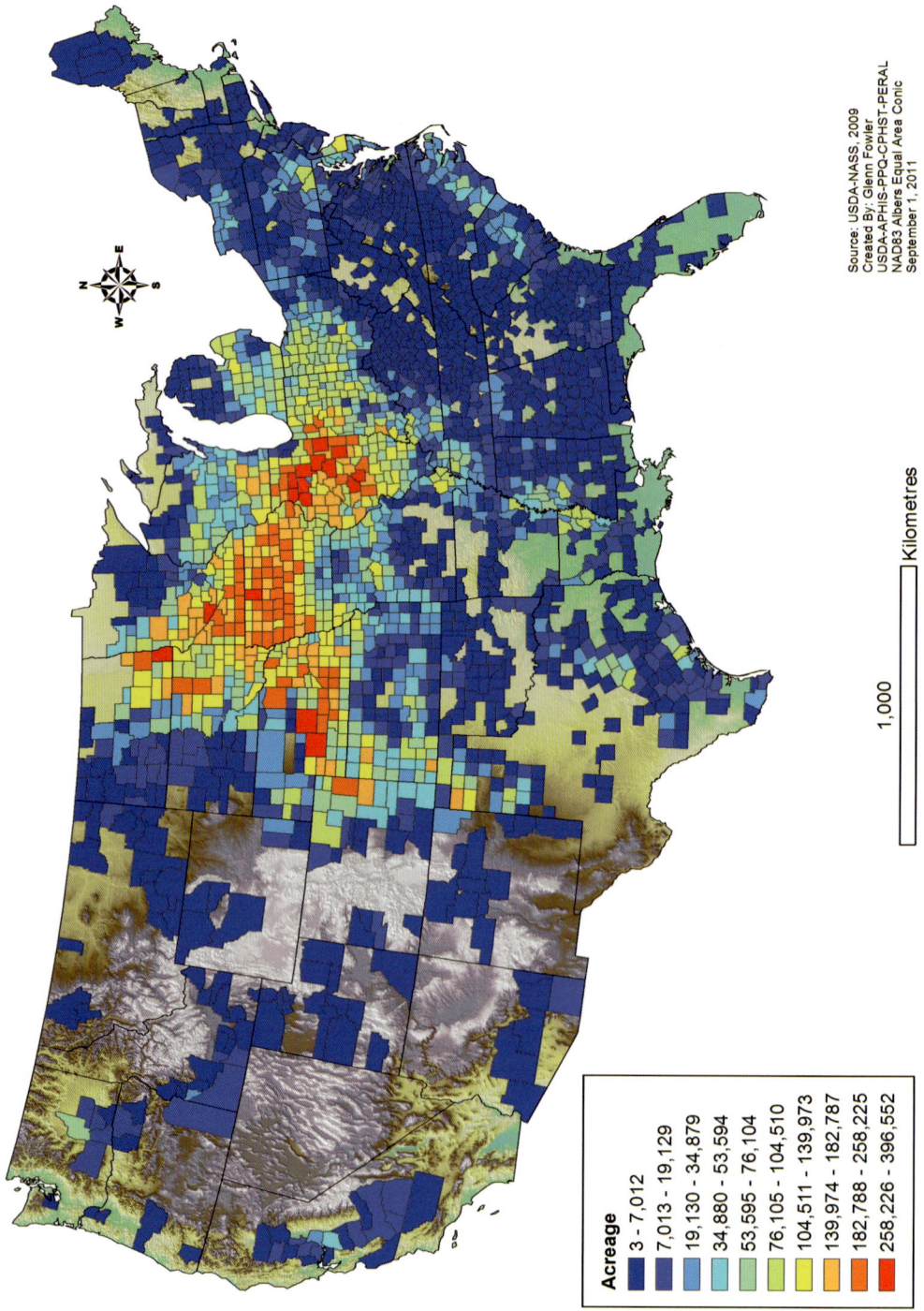

Source: USDA-NASS, 2009
Created By: Glenn Fowler
USDA-APHIS-PPQ-CPHST-PERAL
NAD83 Albers Equal Area Conic
September 1, 2011

Kilometres

1,000

Acreage

- 3 - 7,012
- 7,013 - 19,129
- 19,130 - 34,879
- 34,880 - 53,594
- 53,595 - 76,104
- 76,105 - 104,510
- 104,511 - 139,973
- 139,974 - 182,787
- 182,788 - 258,225
- 258,226 - 396,552

Plate 8. US harvested acres of corn for grain per county (USDA-NASS, 2009).

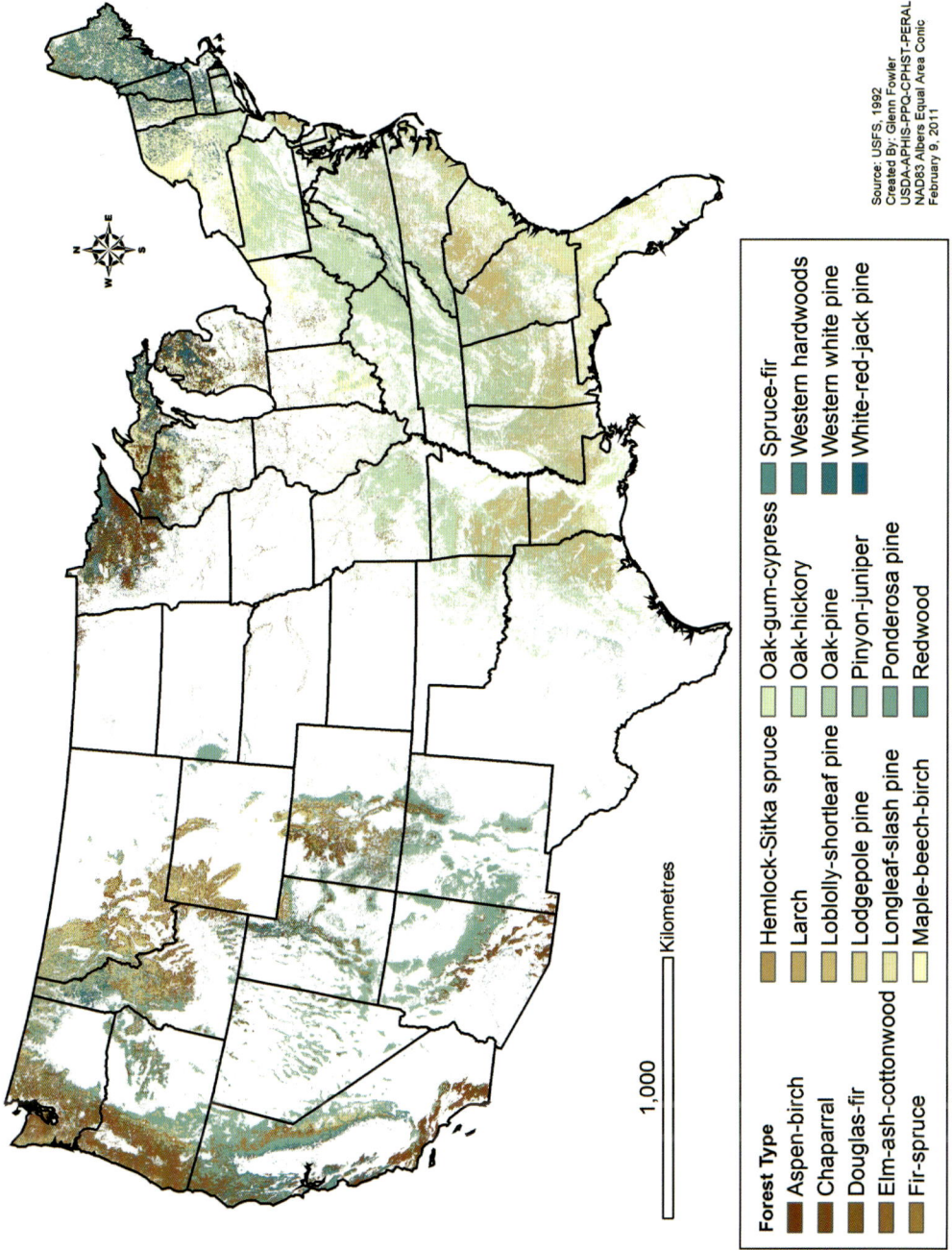

Plate 9. US forest types visualized at a 1-km resolution (USFS, 1992).

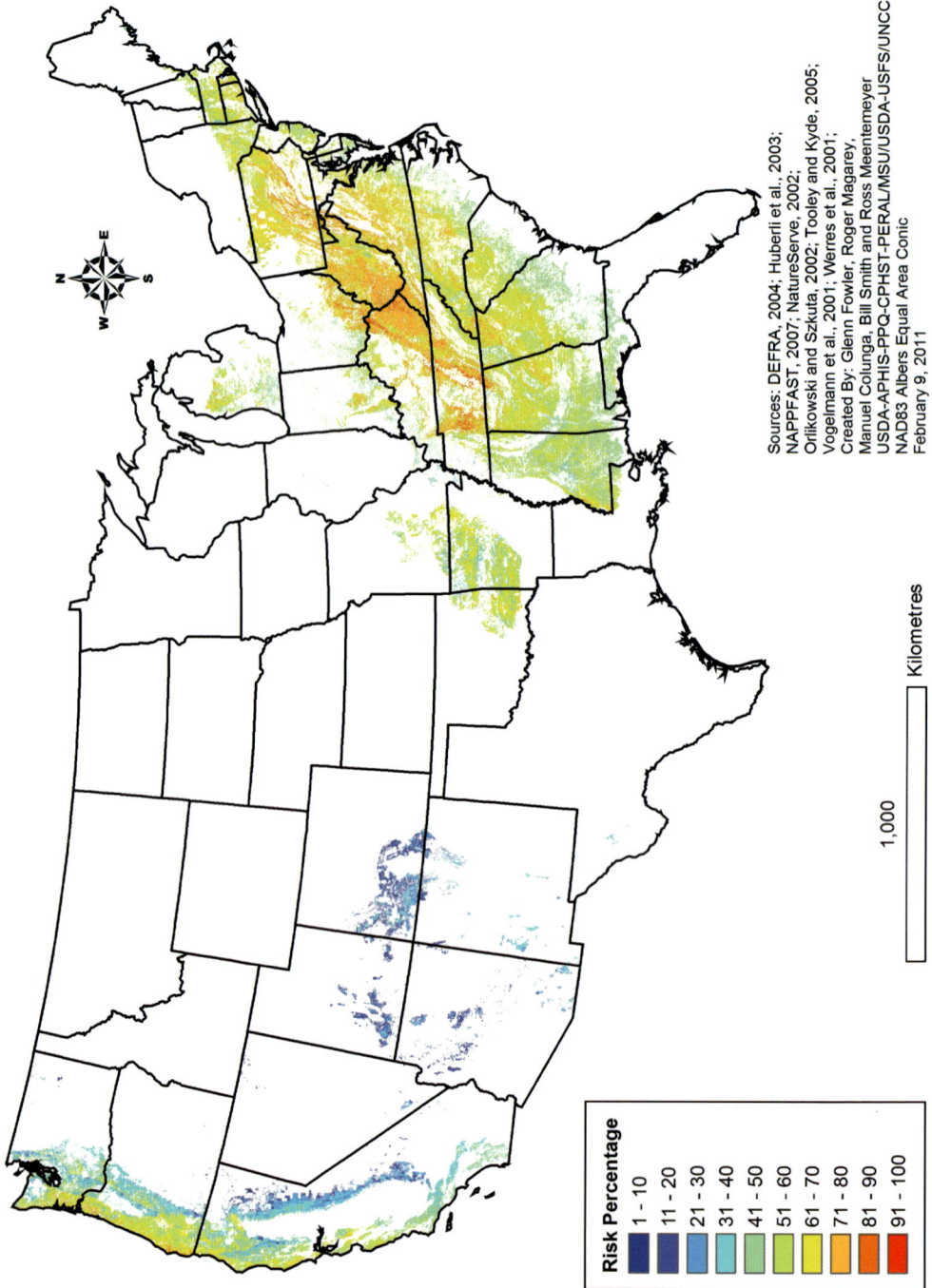

Plate 10. Risk map for areas at risk for infection by *Phytophthora ramorum* based on climate, understory hosts, overstory hosts and lethal cold temperature limit (Vogelmann *et al.*, 2001; Werres *et al.*, 2001; NatureServe, 2002; Orlikowski and Szkuta, 2002; Hüberli *et al.*, 2003; DEFRA, 2004; Tooley and Kyde, 2005; NAPPFAST, 2007b).

Sources: DEFRA, 2004; Huberli et al., 2003; NAPPFAST, 2007; NatureServe, 2002; Orlikowski and Szkuta, 2002; Tooley and Kyde, 2005; Vogelmann et al., 2001; Werres et al., 2001; Created By: Glenn Fowler, Roger Magarey, Manuel Colunga, Bill Smith and Ross Meentemeyer USDA-APHIS-PPQ-CPHST-PERAL/MSU/USDA-USFS/UNCC NAD83 Albers Equal Area Conic February 9, 2011

Kilometres

1,000

Risk Percentage

1 – 10
11 – 20
21 – 30
31 – 40
41 – 50
51 – 60
61 – 70
71 – 80
81 – 90
91 – 100

11

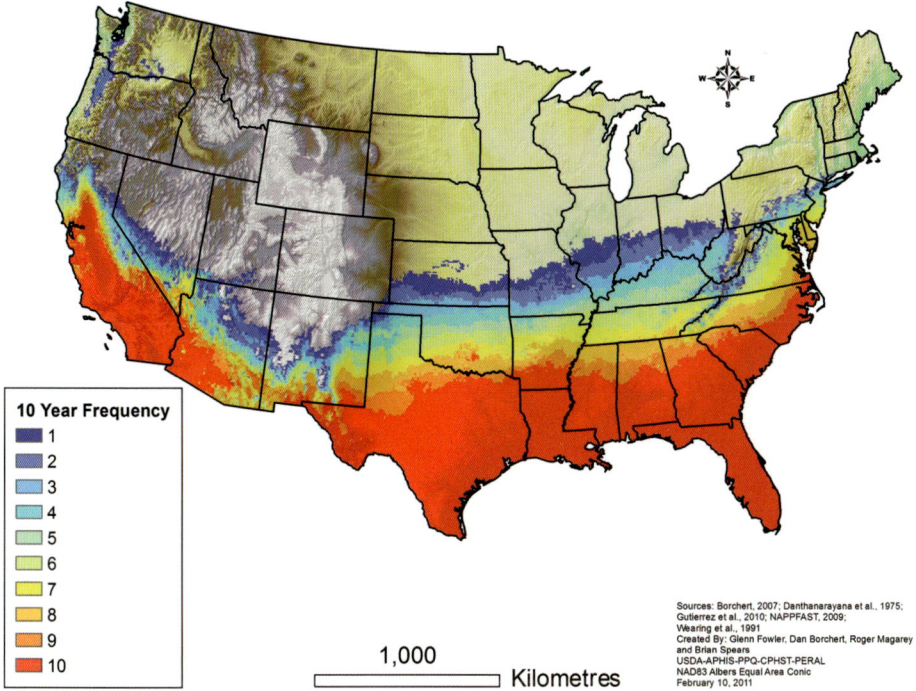

10 Year Frequency
- 1
- 2
- 3
- 4
- 5
- 6
- 7
- 8
- 9
- 10

1,000 Kilometres

Sources: Borchert, 2007; Danthanarayana et al., 1975;
Gutierrez et al., 2010; NAPPFAST, 2009;
Wearing et al., 1991
Created By: Glenn Fowler, Dan Borchert, Roger Magarey
and Brian Spears
USDA-APHIS-PPQ-CPHST-PERAL
NAD83 Albers Equal Area Conic
February 10, 2011

12

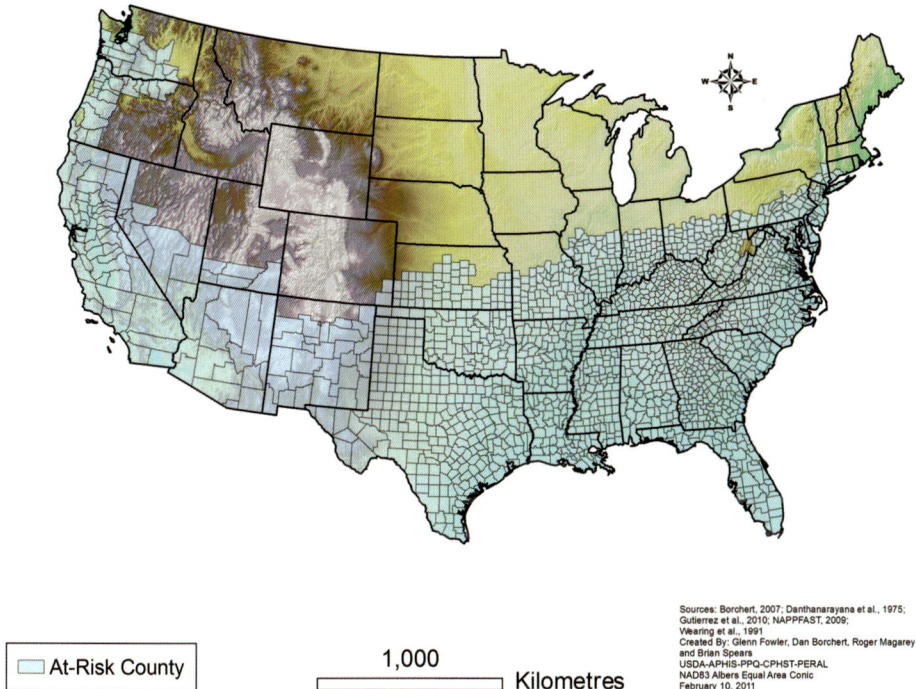

At-Risk County

1,000 Kilometres

Sources: Borchert, 2007; Danthanarayana et al., 1975;
Gutierrez et al., 2010; NAPPFAST, 2009;
Wearing et al., 1991
Created By: Glenn Fowler, Dan Borchert, Roger Magarey
and Brian Spears
USDA-APHIS-PPQ-CPHST-PERAL
NAD83 Albers Equal Area Conic
February 10, 2011

Plate 11. Areas where the light brown apple moth, *Epiphyas postvittana*, could establish as a pest based on the completion of ≥ three generations and ability to survive the winter (Danthanarayana, 1975; Wearing *et al.*, 1991; Borchert, 2007; NAPPFAST, 2009; Gutierrez *et al.*, 2010).

Plate 12. Counties where the light brown apple moth, *Epiphyas postvittana*, could establish as a pest based on the completion of ≥ three generations (Danthanarayana, 1975; Wearing *et al.*, 1991; Borchert, 2007; NAPPFAST, 2009; Gutierrez *et al.*, 2010).

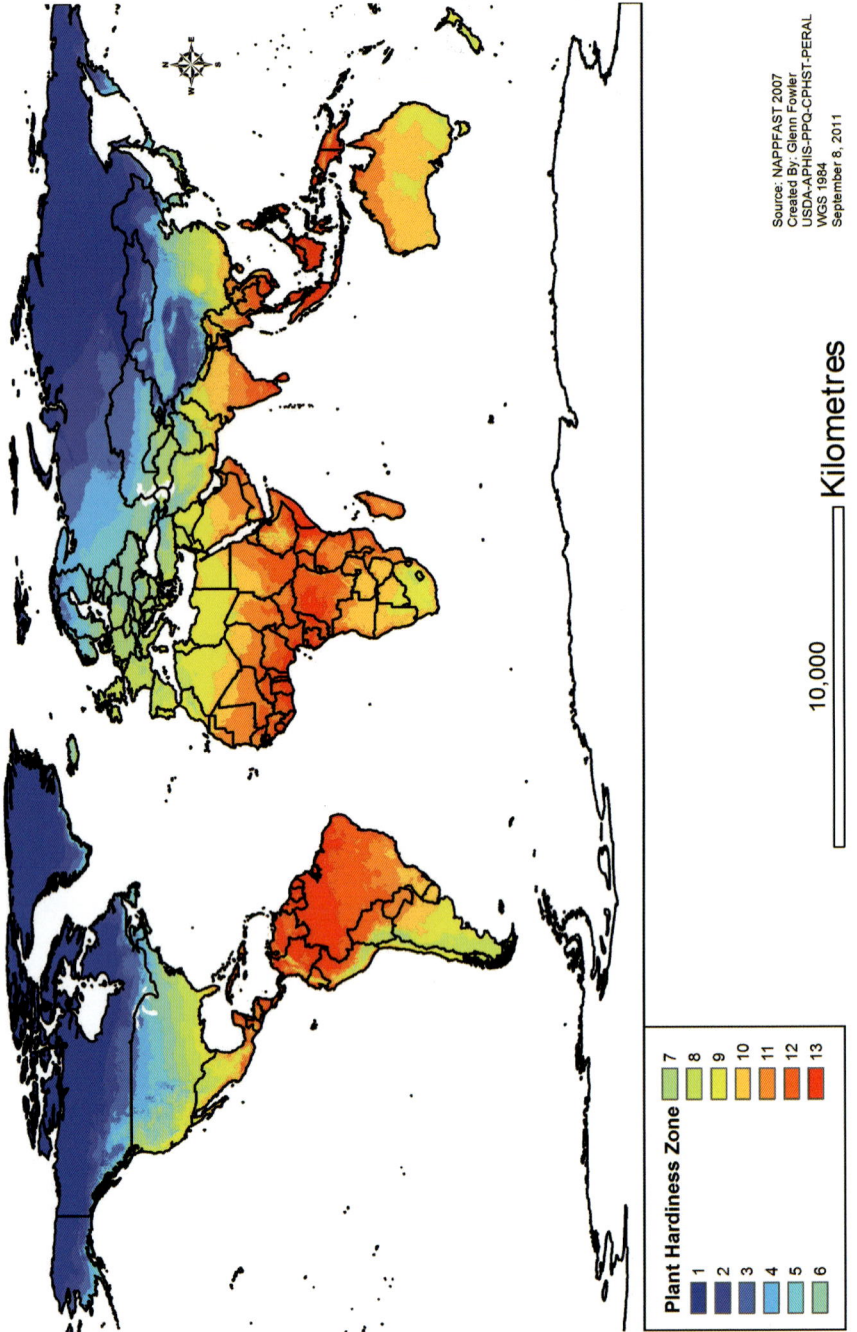

13

Plate 13. Global plant hardiness zones (NAPPFAST, 2007a).

14

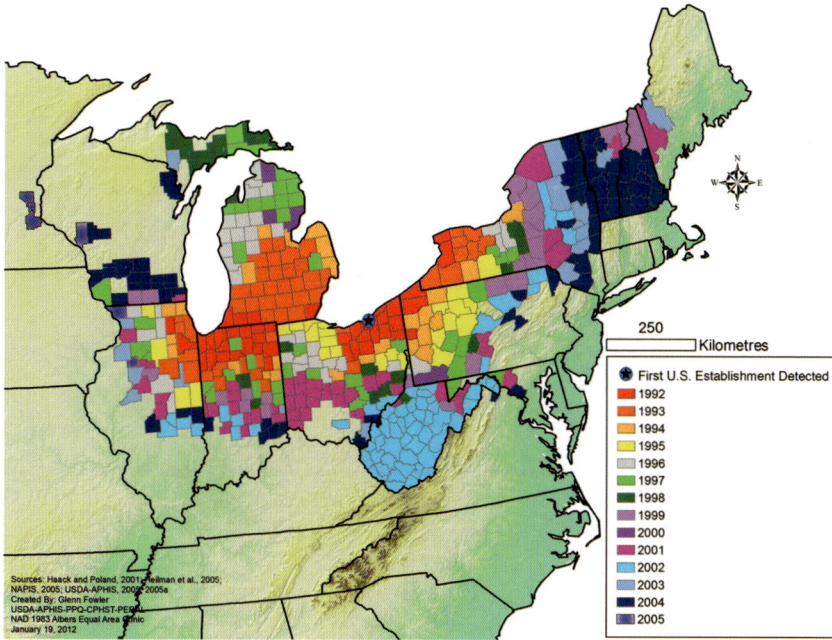

15

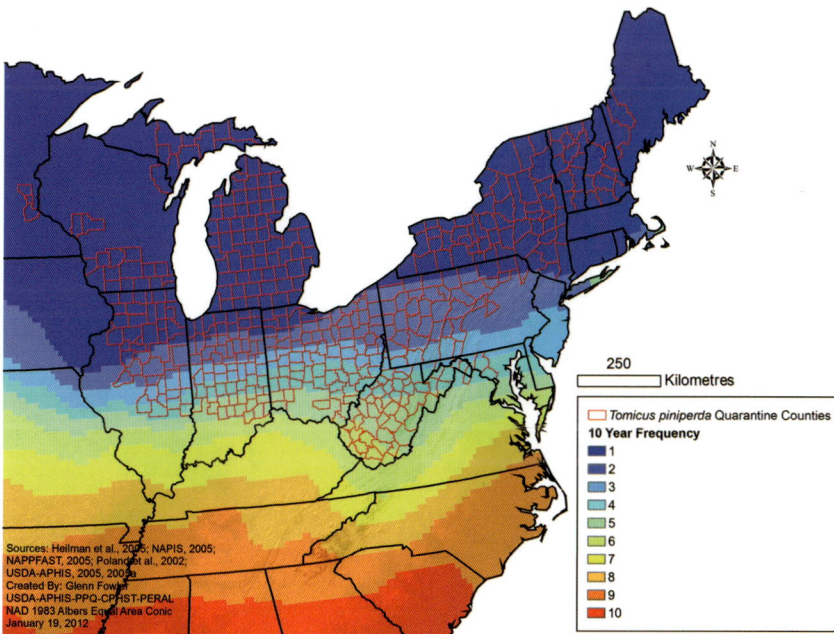

Plate 14. *Tomicus piniperda* quarantined counties by year as of 13 July 2005 (Haack and Poland, 2001; Halman *et al.* 2005; NAPIS, 2005; USDA-APHIS, 2005b).

Plate 15. Ten-year frequency map (1995–2004) visualizing areas where *T. piniperda* emergence is likely between 1 and 7 February because of two days with temperatures greater than or equal to 12°C (Poland *et al.*, 2002).

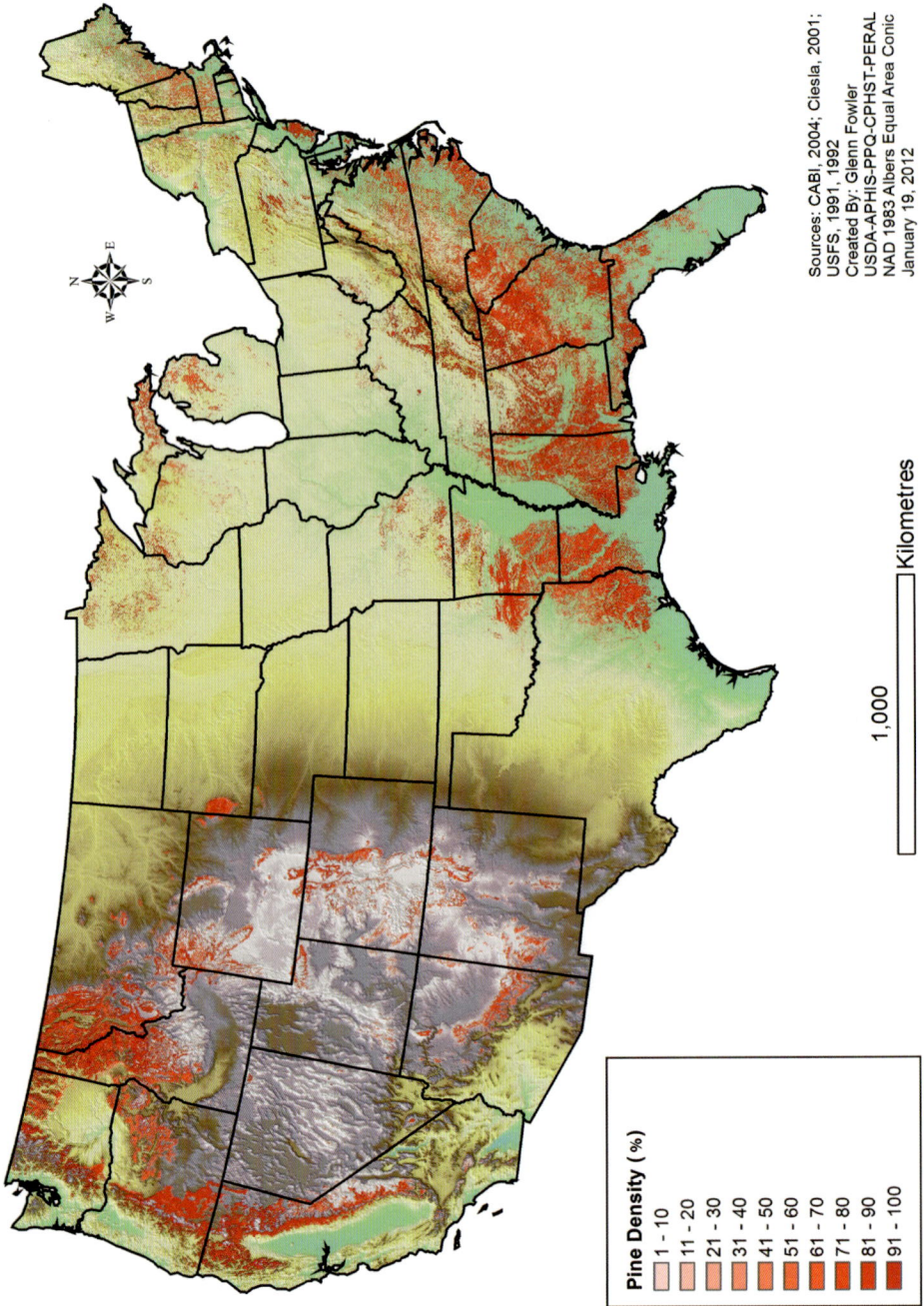

16

Plate 16. US pine host density for *T. piniperda* (USFS, 1991, 1992; Ciesla, 2001; CABI, 2004).

17

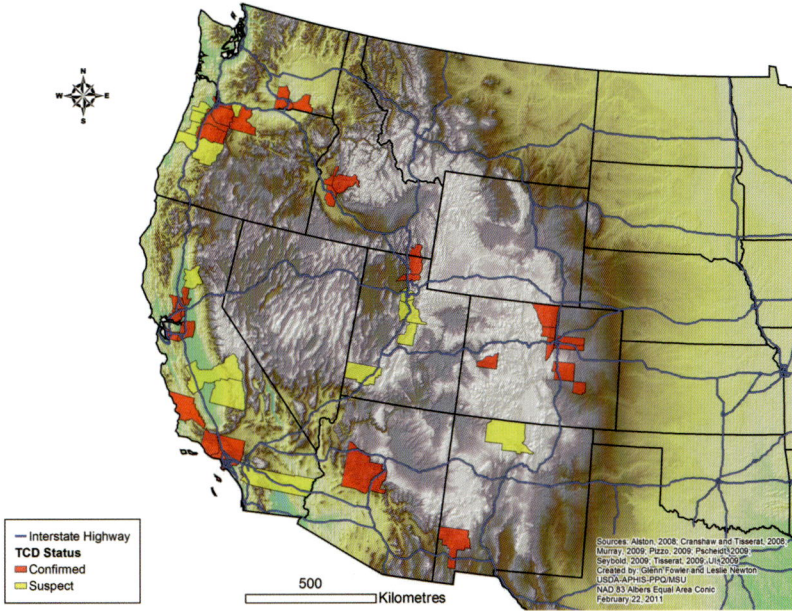

18

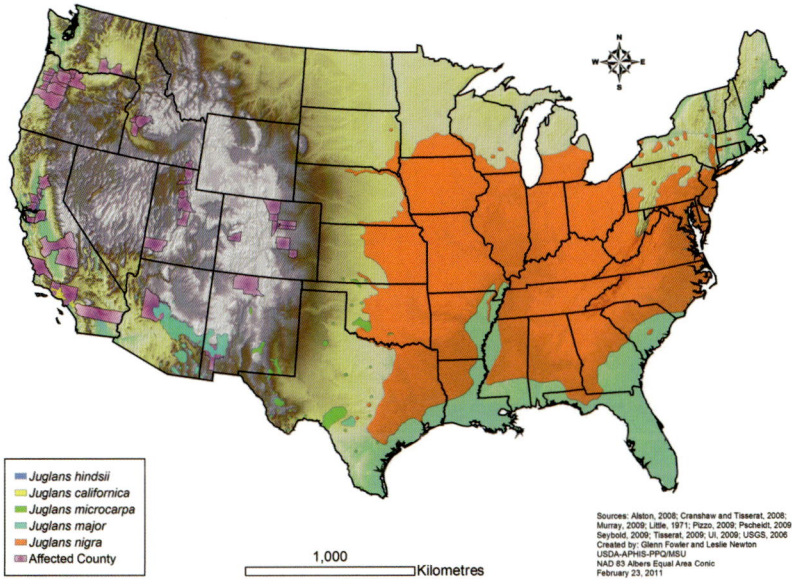

Plate 17. US county distribution of Thousand Cankers Disease (TCD) as of 6 October 2009 (Alston, 2008; Cranshaw and Tisserat, 2008; Murray, 2009; Pizzo, 2009; Pscheidt, 2009; Seybold, 2009; Tisserat, 2009; UI, 2009).

Plate 18. US walnut species distributions and counties affected by TCD (Little, 1971; Vogelmann *et al.*, 2001; USGS, 2006; Alston, 2008; Cranshaw and Tisserat, 2008; Murray, 2009; Pizzo, 2009; Pscheidt, 2009; Seybold, 2009; Tisserat, 2009; UI, 2009).

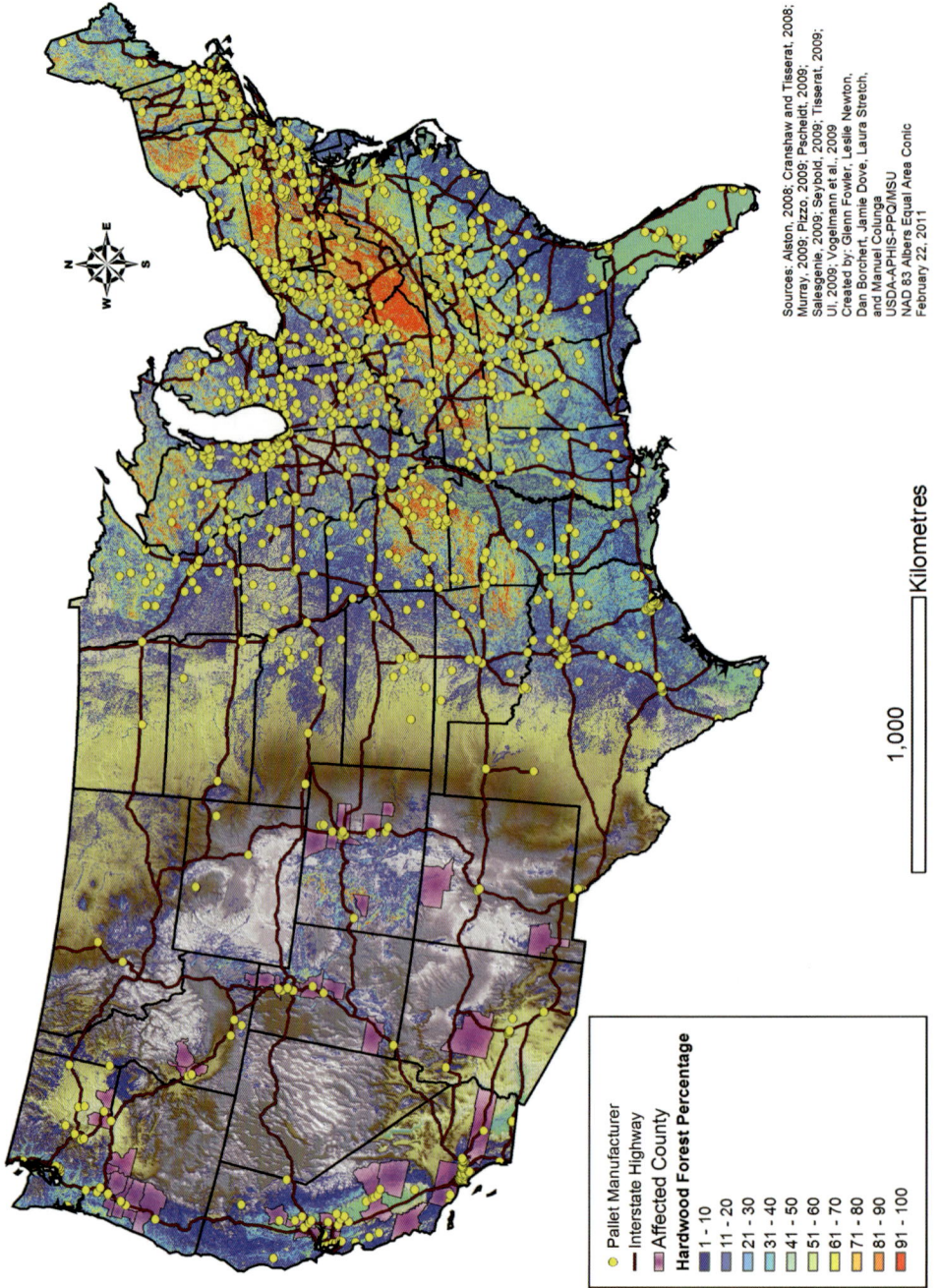

Plate 19. US pallet manufacturers and counties affected by TCD (Vogelmann *et al.*, 2001; Alston, 2008; Cranshaw and Tisserat, 2008; Murray, 2009; Pizzo, 2009; Pscheidt, 2009; Salesgenie, 2009; Seybold, 2009; Tisserat, 2009; UI, 2009).

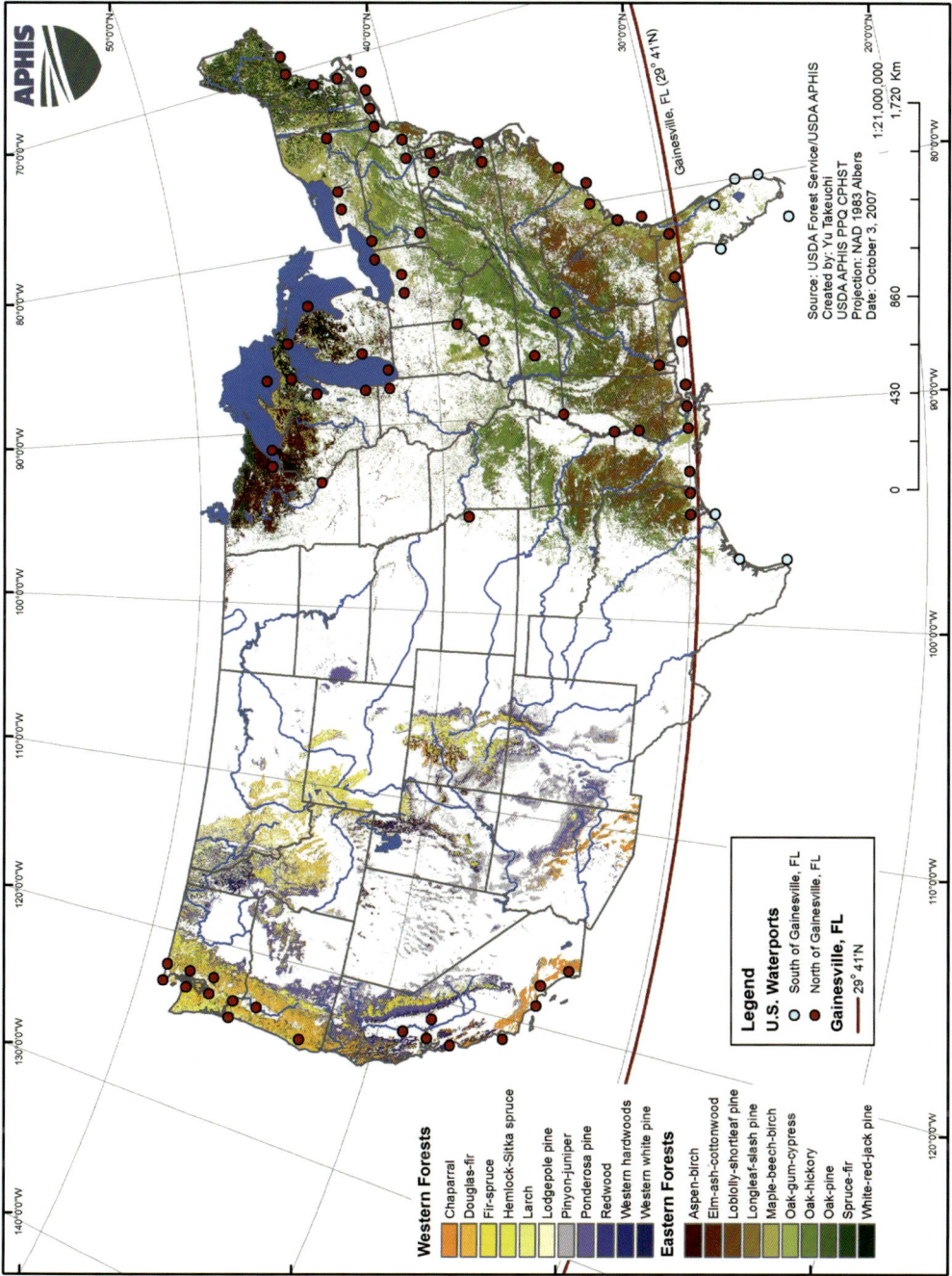

Plate 20. US maritime ports receiving international cargo, and forest types (USDA-Forest Service, 2005; USDA-APHIS, 2007).

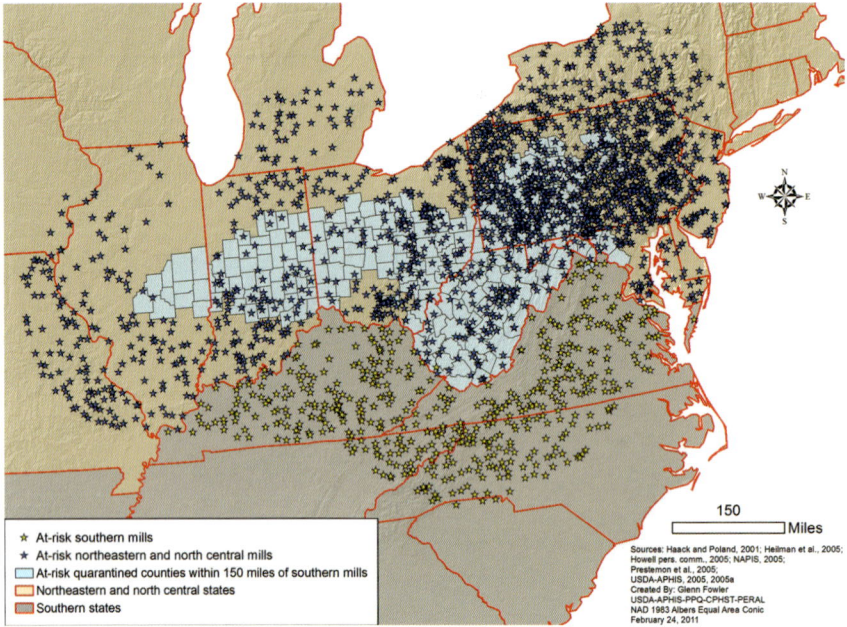

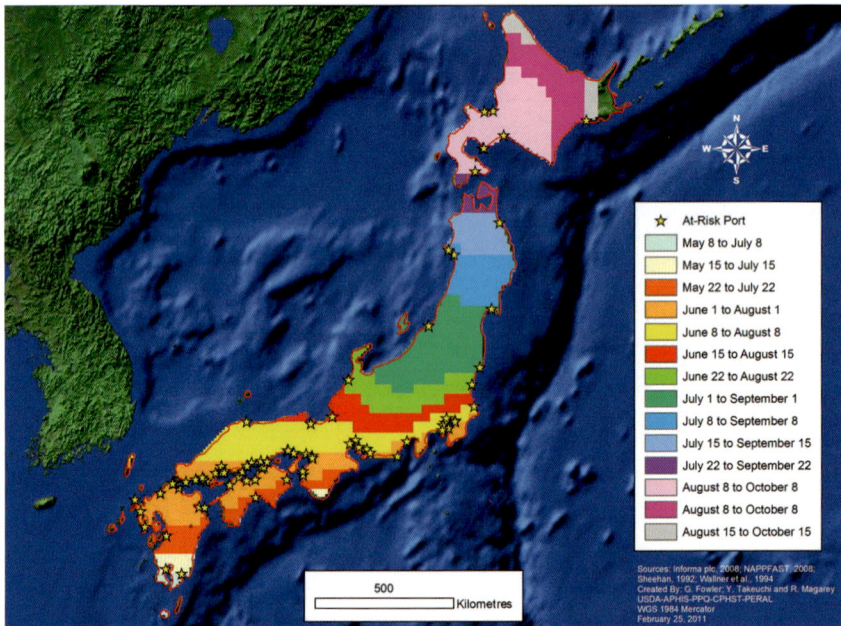

Plate 21. Timber mills considered at risk for *T. piniperda* based on their proximity to the quarantined counties and a timber buying radius of 150 miles (Haack and Poland, 2001; Heilman *et al.*, 2005; Howell, 2005; NAPIS, 2005; Prestemon *et al.*, 2005; USDA-APHIS, 2005a, 2005b).

Plate 22. AGM flight periods at Japanese ports with US-bound shipments (Wallner *et al.*, 1984; Sheehan, 1992; Informa plc, 2008; NAPPFAST, 2008).

12 Mapping, Climate and Geographic Information for Risk Analysis

Glenn Fowler and Yu Takeuchi

12.1. Introduction

Mapping plays an important role in phytosanitary risk analysis. It can be used to visualize where a pest is most likely to become introduced or cause damage, and sets the stage for additional analysis evaluating the risk associated with a given pest. Also, because maps can rapidly convey complex geospatial and/or temporal information in a transparent and easy-to-understand format, they have great utility for a wide audience including: scientists, regulators, industry stakeholders and the general public. These characteristics have made risk maps commonplace in risk analysis and policy making.

In this chapter we will first provide a brief introduction to Geographic Information Systems (GIS), the tool used to generate maps, then explore some mapping methodologies, and finally discuss how mapping can be used in phytosanitary risk analysis.

12.2. Geographic Information Systems

GIS is the technology used to generate risk maps. This powerful tool allows us to generate maps with both spatial and temporal components at a variety of extents from local to global. It also allows us to conduct multiple component geospatial analyses to visualize risk.

GIS utilizes two types of georeferenced data to generate maps, *vector* and *raster*. *Vector data* consists of points, lines and polygons (Wade and Sommer, 2006) (Fig. 12.1). Examples of point data are: latitude/longitude coordinates of pest locations, cities and port locations. Examples of line vector data would be roads and rivers. Polygon data we often utilize include county, state and country boundaries.

Rasters are georeferenced datasets composed of grid cells and their associated attribute values (Wade and Sommer, 2006) (Plate 2). Examples of rasters include satellite imagery and interpolated data sets such as weather and elevation. Rasters come in different resolutions, e.g. 30-m or 10-km grid cells. The extent of the analysis, e.g. field level vs. country level, as well as computing power will determine the resolutions that are most appropriate. For example, it is probably not necessary to generate a global climate map at a 30-m resolution.

Much of the vector and raster data we use comes with the GIS software or is available from other online and electronic sources. However, we also use GIS to generate our own vector and raster data. In addition, GIS allows us to convert data from vector to raster and vice versa.

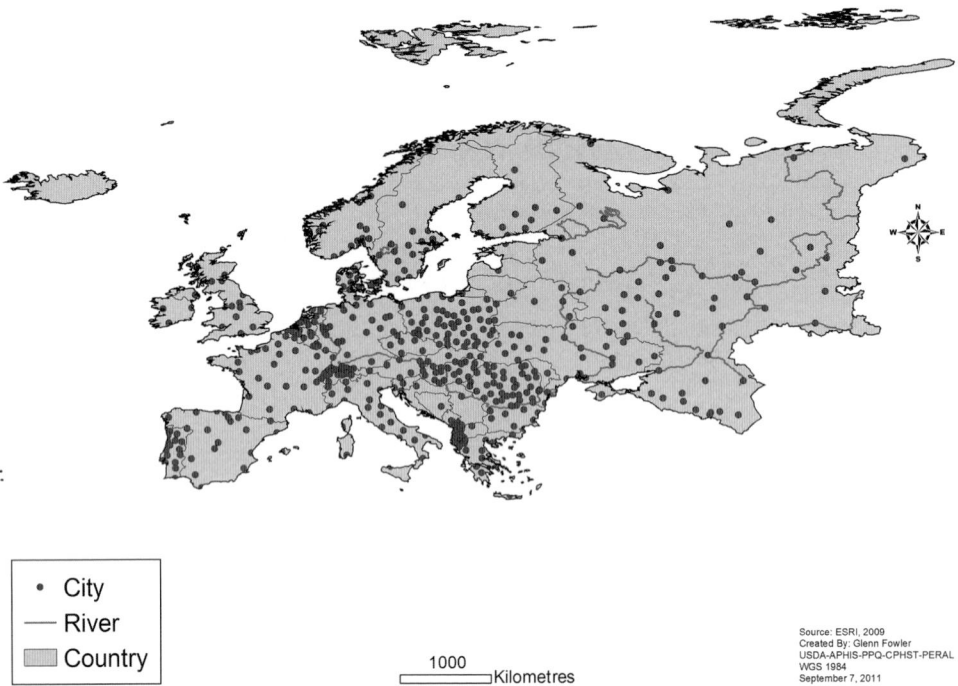

Legend:
- • City
- — River
- ▨ Country

1000 Kilometres

Source: ESRI, 2009
Created By: Glenn Fowler
USDA-APHIS-PPQ-CPHST-PERAL
WGS 1984
September 7, 2011

Fig. 12.1. Examples of vector data using cities (points), rivers (lines) and country outlines (polygons). (From ESRI, 2009.)

12.3. Climate Mapping

We are now ready to explore mapping methodologies that are often used in phytosanitary risk analysis. One of the most important is climate mapping, which allows us to visualize where a pest is likely to become established based on its ability to survive the weather conditions. A good example is fruit flies such as *Ceratitis capitata*, which is a major plant pest but cannot survive freezing temperatures (CABI, 2007). Consequently they could be at risk for establishing in warm areas like California and Florida but would not be a problem in Canada or Sweden.

To generate climate maps we often use weather station data (NCDC, 2000) (Fig. 12.2). This vector point data can be interpolated to produce a raster map (Plate 3).

Climate risk maps are typically based on pest biology and/or distribution, and a variety of methods are available to incorporate climatology into biological models for predicting pest establishment. For a detailed review of commonly used risk mapping approaches and methodologies see Venette *et al.* (2010). Commonly used biological models for invasive plant pests include: degree day, infection, climate-matching and environmental suitability.

In this chapter we will briefly discuss two modelling systems that are used in phytosanitary risk analysis: NAPPFAST and CLIMEX. We have familiarity with these systems and our choice should not be interpreted as product endorsement.

12.3.1. NAPPFAST

NAPPFAST is an acronym for the North Carolina State University Animal and Plant Health Inspection Service Plant Pest Forecast System (Borchert *et al.*, 2007). NAPPFAST was the brain child of the late Dr Jack Bailey and resulted from a cooperative agreement between North Carolina State University and USDA-APHIS.

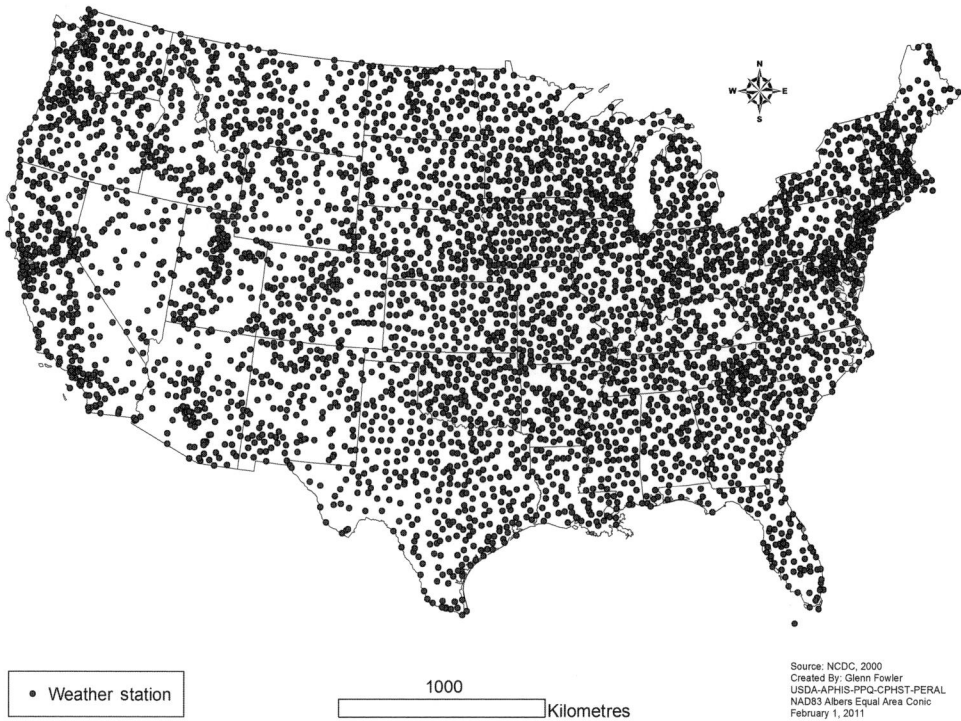

Source: NCDC, 2000
Created By: Glenn Fowler
USDA-APHIS-PPQ-CPHST-PERAL
NAD83 Albers Equal Area Conic
February 1, 2011

Fig. 12.2. US weather stations. (From NCDC, 2000.)

NAPPFAST is structured so that a user can access the system via the internet. Once logged in the user can generate and run predictive models at the global extent. Model requests are submitted to a computer network that queries a robust climate database and outputs a predictive map that can be exported as a pdf or a GIS-compatible geotiff. The resolution of the NAPPFAST output depends on the map extent. The system can generate 10-km maps for North American models and 32-km maps for global models (Borchert et al., 2007; Magarey et al., 2011). NAPPFAST allows the user to choose from three model templates: degree day, infection and generic climate queries.

The degree day template is useful for modelling organisms such as arthropods and weeds whose physiological development rate is temperature-dependent (Allen, 1976; UCIPM, 2003). Degree days are a way of measuring the 'physiological time' needed for development to occur as dictated by the amount of heat present (UCIPM,

2003). Important parameters for generating the NAPPFAST degree day model include the base temperature below which development will not occur and the number of degree days above this temperature that are required for the organism to complete a given developmental stage, e.g. a generation (Borchert et al., 2007).

The infection template can be used to model plant diseases and is based on temperature and moisture (Borchert et al., 2007; Magarey et al., 2005, 2007a). The important parameters for the infection template include the critical temperatures for infection and the duration of leaf wetness.

The climate query (generic) template is useful for identifying geographic areas of interest based on suitable and/or unsuitable climate parameters for pest establishment (Borchert et al., 2007; Magarey et al., 2007a). The template allows the user to choose from a variety of generic model formulas, e.g. X < A, and parameters, e.g. minimum temperature and precipitation.

NAPPFAST models are based on historical climatology and the user can choose from 10, 20 or 30 years ending at the most recent completed year, e.g. 2001 to 2010 for a 10-year query done in 2011. The user can also specify the time period during the year, e.g. 1 January–30 June, when the model will be applied.

NAPPFAST outputs can be generated as probability, e.g. 10-year frequency, or average history maps (Plates 4 and 5). NAPPFAST also allows for models to be run on individual years (Plate 6). In these examples we used the degree day template and generated a model for a hypothetical organism that requires a based developmental temperature of 10°C and 1000 degree days to complete a generation.

12.3.2. CLIMEX

CLIMEX is another popular modelling system that has been produced by CSIRO in Australia since 1985 (Sutherst et al., 2007). CLIMEX was originally developed to predict exotic species, plants and animals (mainly insects) survival if they were introduced to Australia. To do this it compared the climate data where the species occurred in other countries with the climate data in Australia and then predicted suitability.

CLIMEX must be installed on individual computers to run the models and utilizes a global meteorological database consisting of approximately 3000 locations. The models run at those weather station locations or on gridded weather data, which is interpolated across weather stations. Users are able to modify weather data through the CLIMEX MetManager interface. Weather data can be updated with the most recent data, and new weather stations can be added into the database by users. Users are also able to incorporate irrigation and climate change scenarios into the models.

CLIMEX enables the user to parameterize species of interest by optimal and threshold temperatures, moisture requirements, stresses caused by cold, heat, dryness and wetness, degree days, diapause requirements and day length. Parameter fitting in CLIMEX requires the most skill, and estimating them based on accurate biological data, distribution maps, abundance and seasonal activities results in more reliable predictions. Users first fit parameters based on known biological information and examine the output (climatologically preferred areas) based on current distribution. Users need to adjust parameters until the output distributions are similar to the known distributions.

CLIMEX has two distinct modelling functions. These are the species-specific simulation model and the climate-matching model.

The species-specific simulation model has two modes: compare locations for one or two species and compare years. The species-specific simulation model constructs a climatological model for predicting a pest's distribution based on its biology and environmental stress indices, and characterizes the suitability with an Ecoclimatic Index (EI) (Fig. 12.3). EI is a function of growth index, stress indices and species-specific requirements, such as diapause and heat requirement. The growth index is measured by temperature and moisture. The stress index is determined by cold, heat, dryness, wetness and a combination of temperature and moisture stresses. Higher EI values indicate more suitable environments for a particular species. An EI value above 10 means the species could survive, while above 30 is very favourable. The EI model can be calibrated so that its predictions give the best agreement with where the pest naturally occurs.

The climate matching model compares meteorological data without reference to a species' biology. It allows a user to choose a source location and then compare its similarity to other locations or regions based on climate parameters, such as temperature (max, min, average), precipitation (amount and pattern) and relative humidity (Fig. 12.4). Climate similarity is measured by Composite Match Index (CMI). A CMI value greater than 0.8 is considered to be a close match, while a CMI value of less than 0.6 indicates that some or all the climatic variables are quite different. One can also choose the time period during the year

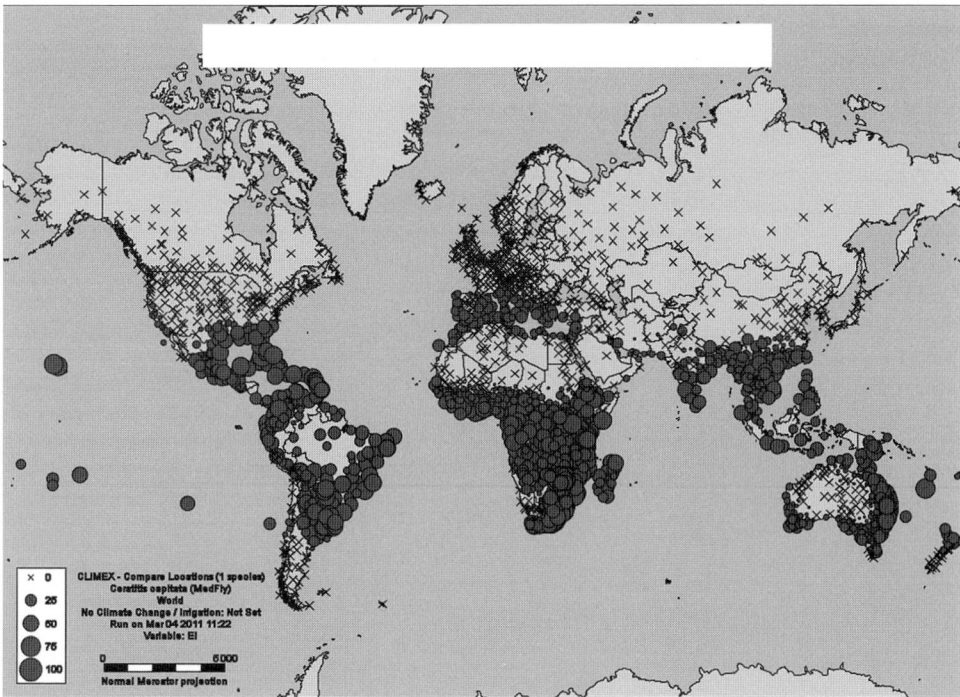

Fig. 12.3. *Ceratitis capitata* suitable areas indicated by CLIMEX EI values. (From CLIMEX, 2011.)

when climate comparisons between locations are made. If home and away locations are in opposite hemispheres, away locations are displaced by 6 months. This function is useful when biological data for an organism are limited because it allows one to predict where it could establish based solely on its geographic distribution.

CLIMEX model outputs (e.g. EI, stress indices, growth index and temperature index) can be visualized as tables, charts or maps. Maps and charts can be exported as pictures (e.g. bmp, jpg, tiff, gif, emf), and tables are exported as comma-separated values (csv) files. CLIMEX does not create geo-referenced outputs; however, all the predicted outputs can be exported with coordinates (altitudes, latitudes and longitudes). CLIMEX outputs are frequently displayed as point data in GIS software (e.g. ArcGIS, GRASS) or as rasters by interpolating output values across the weather stations (Plate 7).

12.4. Host Mapping

Host distribution is another important factor that we use in risk mapping for plant pests. Often, we are interested in mapping forest and agriculture host distributions. Host maps can be used in a variety of risk analysis applications including: establishment potential, economic damage estimates, environmental damage estimates and pathway analysis. Because of the way host data are recorded, we often map host presence based on administrative boundaries like counties (Plate 8). However, more precise data, e.g. satellite imagery and orchard boundaries, are available for some hosts (Plate 9).

12.5. Integrated Risk Mapping

We often need to combine several GIS layers, like host and climate, in order to better visualize the risk posed by a plant pest.

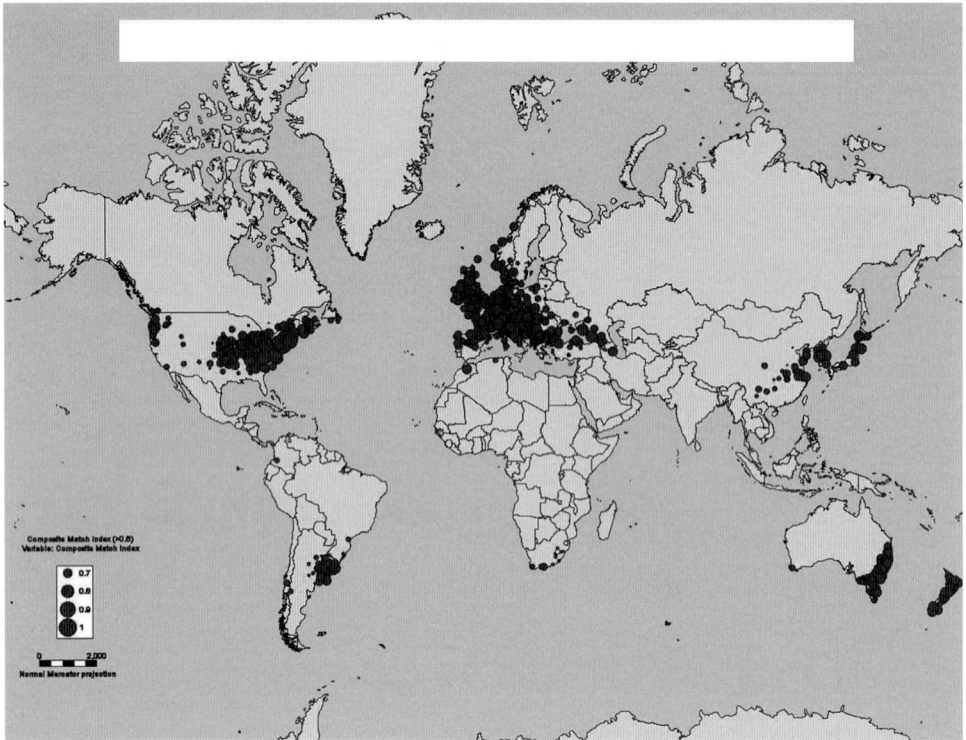

Fig. 12.4. CLIMEX climate match map indicating locations with similar annual climates to New York City. (From CLIMEX, 2011.)

These types of maps can be used in risk assessments, surveys and eradication activities. We will discuss two examples, sudden oak death and light brown apple moth, where multiple GIS layers were combined into a single risk map.

Sudden oak death is a plant disease caused by the pathogen *Phytophthora ramorum* (CABI, 2007). In 1995 it was detected in California and has since killed millions of oaks (Garbelotto *et al.*, 2001; McPherson *et al.*, 2000; USDA-APHIS, 2010c). The USDA, working with UNC-Charlotte and Michigan State University, generated a composite risk map that incorporated understory hosts, overstory hosts and climate suitability for infection to inform surveys and risk analysis for the pathogen (Magarey *et al.*, 2007b) (Plate 10). Their climate infection model was run in NAPPFAST and was based on daily temperature and leaf wetness (Werres *et al.*, 2001; Orlikowski and Szkuta, 2002; Hüberli

et al., 2003; Magarey *et al.*, 2005; Tooley and Kyde, 2005; NAPPFAST, 2007b). The understory hosts were represented by the distribution of seven known hosts for *P. ramorum* in the family Ericaceae (NatureServe, 2002). The overstory hosts were visualized based on hardwood forest percentage and 50% of mixed forest percentage (Vogelmann *et al.*, 2001). In addition to excluding areas where no climate match, understory or overstory hosts occurred, the USDA also excluded areas where the soil temperature reached −25°C as being too cold for *P. ramorum* survival (DEFRA, 2004).

A second example of integrating multiple GIS layers in risk maps is the light brown apple moth (LBAM), *Epiphyas postvittana* (Fowler *et al.*, 2009). This moth was first detected in California in March 2007 and its presence resulted in an APHIS-PPQ domestic programme (USDA-APHIS, 2010b). To inform regulatory policy, the USDA generated a climate risk map based on areas where LBAM could establish as a pest, i.e. complete at least three generations based on its behaviour as a pest in Australia, and survive the winter, i.e. areas where the minimum temperature does not drop below −16°C (Danthanarayana, 1975; Wearing *et al.*, 1991; Borchert, 2007; NAPPFAST, 2009; Gutierrez *et al.*, 2010) (Plate 11). The USDA could then determine which US counties were at risk for LBAM establishment as a pest (Plate 12). To do that the USDA created three generation and lethal cold probability maps in NAPPFAST and subtracted the latter from the former (both were rasters). They then converted the result to vector format. The USDA could then determine the at-risk counties based on the intersection of the US county map and the vector-formatted pest establishment map. The conversion from raster to vector was necessary because the county data was vector.

Once the USDA knew the at-risk counties, they were able to map at-risk apple, grape, orange and pear acreage using county data from the National Agricultural Statistics Service (USDA-NASS, 2009). The USDA then conducted an analysis estimating the amount of economic damage LBAM could do if it established throughout its potential US range.

Risk analysis is a dynamic field and there is other information besides climate and hosts that could also be important in risk mapping. Examples include nursery locations, population centres, ports, transportation networks, elevation and trap locations. For an example of risk mapping that incorporates transportation networks and urban areas see Colunga-Garcia *et al.* (2009).

12.6. Applications in Risk Assessment and Pathway Analysis

Science based analyses are used to inform phytosanitary regulatory policy (WTO, 1994). Generally speaking these analyses usually fall into three categories: commodity risk assessments (CRAs) (also referred to as pathway-initiated risk assessments), organism pest risk assessments and pathway analysis

(IPPC, 2004; USDA-APHIS-PPQ, 2000). Here we will discuss how risk mapping can be used to inform each type.

12.6.1. Commodity risk assessments

CRAs typically determine which pests are likely to follow the pathway for a given agricultural crop (IPPC, 2004; USDA-APHIS-PPQ, 2000). Those pests are then analysed based on criteria that evaluate their potential to become invasive. One commonly analysed criterion is the pest's predicted ability to establish in new areas. This is often based on climate and host presence, and risk mapping can be used to inform this aspect of the risk assessment. This information can serve as the foundation for subsequent analysis of the pests, e.g. economic and environmental effects, since it characterizes the geographic area that could be affected.

For example, in the USDA, the CRA guidelines (version 5.02) use the number of plant hardiness zones that a pest could establish in to characterize the climate–host suitability risk element (USDA-APHIS-PPQ, 2000). These zones are thermal bands representing the average extreme minimum temperatures in 10°F increments (Cathey, 1990; Magarey *et al.*, 2008) (Plate 13). For example, zones 7 and 8 represent areas with temperatures from 0°F to 10°F and 10°F to 20°F. The USDA uses climate and host mapping to estimate how many plant hardiness zones a pest could establish in. If a pest can establish in four or more it is considered high risk, two to three is medium risk and one is low risk for establishment potential (USDA-APHIS-PPQ, 2000).

12.6.2. Organism pest risk assessments

Risk assessments may also be done for specific organisms outside of those conducted in CRAs, often because they represent a high phytosanitary risk (IPPC, 2004). For example, in plant protection and quarantine, the New Pest Advisory Group is responsible for conducting organism pest risk assessments on non-native plant pests that could threaten the USA (USDA-APHIS, 2010a). The EPPO also generates organism pest assessments for pests of concern to its member countries (EPPO, 2010).

Risk mapping can be used to inform these types of assessments. For example risk mapping could identify new areas at risk for establishment as well as provide background information on the pest's distribution.

In 2006 the USDA generated an organism pest risk assessment for the pine shoot beetle, *Tomicus piniperda* (Fowler and Borchert, 2006). This is an introduced bark beetle that is present in the northeastern and north central USA (Haack and Poland, 2001). It has caused negligble direct damage but substantial economic costs to affected industries, e.g. sawmills, due to quarantines. The USDA was tasked with evaluating the risk it posed to the USA as a whole in order to inform domestic regulatory policy.

Using risk mapping, the USDA determined that the northeastern and north central states were at lower risk from *T. piniperda* due to lower pine density and fewer suitable hosts (USFS, 1991, 1992; Price *et al.*, 1998; Ryall and Smith, 2000; CABI, 2004). Conversely, the risk mapping analysis indicated that the southern and western USA were at greater risk because of higher pine density, more susceptible pine species and environmental stress factors like drought, wind and fire (Nilsson, 1976; USFS, 1991, 1992; Price *et al.*, 1998; Swetnam, 2001; CABI, 2004; Eager *et al.*, 2004; NOAA, 2004). Examples of how the USDA used risk mapping in this analysis included: *T. piniperda*'s spatial/temporal US distribution (Plate 14), host distribution and density (USFS, 1991, 1992; Ciesla, 2001; CABI, 2004) (Plate 16) and predicted emergence times based on climate and biology (Poland *et al.*, 2002; Heilman *et al.*, 2005; NAPIS, 2005; NAPPFAST, 2005; USDA-APHIS, 2005a, 2005b) (Plate 15).

12.6.3. Pathway analysis

Of the three analysis categories we've listed, pathway analysis is probably the most dynamic and unique. Commodity and organism risk

assessments often have associated templates or guidelines that risk mapping can be incorporated into. Pathway analyses that are tailored for a given situation are by their nature specific for that problem and consequently can be difficult to assign generic formats to. This characteristic often makes them both more challenging and exciting to generate. Also, in contrast to risk assessments, which provide estimates of introduction likelihood and associated consequences, pathway analyses tend to only characterize the likelihoods associated with pest introduction (IPPC, 2010). The results of pathway analysis are also variable. For example, they may be qualitative, e.g. high risk; semi-quantitative, e.g. high risk with a score of 3; or quantitative, e.g. the probability of a pest surviving transit is 0.88. Because risk maps can be tailored to a variety of situations they are well suited for informing pathway analysis.

An example of a qualitative pathway analysis that utilized risk mapping was conducted by the USDA on Thousand Cankers Disease (TCD) movement from the western USA into the eastern USA (Newton *et al.*, 2009). TCD is caused by the fungus, *Geosmithia morbida*, and is transmitted by the bark beetle, *Pityophthora juglandis* (Kolarik *et al.*, 2011; USDA-NAL-NISIC, 2010). The disease affects walnut trees and there is concern that it could move into the eastern USA where walnut trees are prevalent. The USDA evaluated the approach rate of TCD on the major pathways of concern, e.g. firewood, in order to inform regulatory policy for the disease.

Their methodology was GIS-based and examples of where they used risk maps included: TCD's distribution (Alston, 2008; Cranshaw and Tisserat, 2008; Murray, 2009; Pizzo, 2009; Pscheidt, 2009; Seybold, 2009; Tisserat, 2009; UI, 2009) (Plate 17) and walnut species distribution (Little, 1971; Alston, 2008; Cranshaw and Tisserat, 2008; Murray, 2009; Pizzo, 2009; Pscheidt, 2009; Seybold, 2009; Tisserat, 2009; UI, 2009; USGS, 2006) (Plate 18). They also used composite risk maps, e.g. maps that incorporated TCD distribution, highways, pallet/skid manufacturers and hardwood forest percentage (Alston, 2008; Cranshaw and Tisserat, 2008; Murray, 2009; Pizzo, 2009; Pscheidt, 2009; Salesgenie,

2009; Seybold, 2009; Tisserat, 2009; UI, 2009; Vogelmann *et al.*, 2001) (Plate 19). Incorporating risk mapping, the USDA assigned a qualitative rating, e.g. negligible or low, for each pathway's approach rate and provided a justification for the rating.

An example of semi-quantitative pathway analysis is found in the USDA *Guidelines for Pathway Initiated Pest Risk Assessments* mentioned in Section 12.6.1. (USDA-APHIS-PPQ, 2000). The guidelines use three sections to analyse the risk associated with a pest of concern: consequences of introduction, likelihood of introduction and pest risk potential. All three sections characterize risk in semi-quantitative terms. For example, a pest could receive a qualitative score of 'Low' and an associated numerical score of '1' for economic impact in the consequences of introduction section. The likelihood of introduction section describes the pest movement along the pathway and contains six risk subelements: quantity of commodity imported annually, survive post-harvest treatment, survive shipment, escape detection at the port of entry, moved to an area with a suitable environment for survival and find host material suitable for reproduction. Risk mapping could be of particular relevance to the subelements dealing with suitable environment and host presence since both potentially have geospatial characteristics.

An example of risk mapping being used in semi-quantitative pathway analysis was when the USDA characterized the likelihood of the Asian gypsy moth (AGM) finding host material if imported into the USA (Fowler *et al.*, 2008). In that analysis they determined that 89.8% of the at-risk US ports had forested areas within the reported flight range of AGM (Plate 20). Because the USDA guidelines consider any likelihood greater than 10% to be high risk (USDA-APHIS-PPQ, 2000), AGM was given a semi-quantitative risk rating of 'High (3)' for this risk sub-element.

Quantitative pathway analysis can also use risk mapping. In these cases, probabilistic models are often constructed to simulate the pathway. The models generally consist of fundamental components that represent major transmission points in the pathway, e.g. probability of infestation or shipment volume (Auclair *et al.*, 2005). These components can be quantitatively modelled using point estimates or distributions. In these types of analysis, risk mapping can be used to quantitatively inform the pathway model directly or indirectly.

An example of risk mapping directly informing quantitative pathway analysis was work the USDA conducted estimating the movement of the pine shoot beetle, *Tomicus piniperda*, on logs and lumber with bark from the US quarantined area into the southern USA if deregulated (Fowler *et al.*, 2006).

One of the first things the USDA needed to estimate was the volume of timber that could be shipped from the quarantined areas into the southern states. To do this they assumed a 150-mile timber shipping radius (Howell, 2005). They then used GIS to visualize quarantined counties and southern timber mills within 150 miles of each other (Haack and Poland, 2001; Heilman et al., 2005; Howell, 2005; NAPIS, 2005; Prestemon *et al.*, 2005; USDA-APHIS, 2005a, 2005b) (Plate 21). The USDA could then calculate the volume of timber produced in the at-risk quarantined counties (FIA, 2006).

They then needed to estimate the probability that trees from quarantined counties could be shipped to southern mills. To do this the USDA mapped all timber mills within 150 miles of the quarantined counties in addition to the at-risk southern mills mapped above (Plate 21). This allowed them to tabulate: (i) the number of mills within 150 miles of quarantined counties that could receive infested timber; and (ii) the number of southern mills within 150 miles that could receive infested timber. These two values allowed the USDA to generate a beta distribution for the probability that infested timber could be shipped into the southern United States (Vose, 2000; Palisade, 2002).

This information allowed the USDA to generate a binomial distribution for the timber volume that could be sent south from the quarantined areas. They were then able to continue the analysis, eventually estimating the years until *T. piniperda* colonization in the south via the logs and lumber with bark pathway if deregulated.

An example of risk mapping indirectly informing quantitative pathway analysis was work the USDA conducted estimating the annual number of AGM-infested ships coming from Japan to the USA (Fowler *et al.*, 2008). To do this the USDA first needed to estimate when AGM would be flying at each Japanese port. They used a degree day model and the NAPPFAST system to model the start of AGM flight around each Japanese port and then allowed for a two-month flight period (Wallner *et al.*, 1984; Sheehan, 1992; NAPPFAST, 2008) (Plate 22). This information allowed the USDA to total the number of ships that called at each port during the flight period (Informa plc, 2008). They could then model the number of infested ships coming from Japan each year based on Customs and Border Patrol interception data and using beta and binomial distributions (Vose, 2000; USDA-APHIS, 2008).

12.7. Summary

In this chapter we have tried to provide an overview of risk mapping, associated methodologies and how it can be used in phytosanitary risk analysis. The fields of risk analysis and mapping are extremely diverse and consequently our chapter is only a brief introduction to these topics and how they can be combined. With that caveat, we hope that this chapter has communicated that risk mapping is an extremely powerful tool for informing risk analysis due to its plasticity and ability to convey complex information in a transparent and easily processed format. These characteristics increase the speed of risk communication, hopefully reduce the likelihood of analysis misinterpretation and help provide justification for regulatory decisions.

References

Allen, J.C. (1976) A modified sine wave method for calculating degree days. *Environmental Entomology* 5, 3, 388–396.

Alston, D. (2008) NPAG archives. Personal communication to D. Holzer on 27 May 2008, from Diane Alston (e-mail).

Auclair, A.N.D., Fowler, G., Hennessey, M.K., Hogue, A.T., Keena, M., Lance, D.R., McDowell, R.M., Oryang, D.O. and Sawyer, A.J. (2005) Assessment of the risk of introduction of *Anoplophora glabripennis* (Coleoptera: Cerambycidae) in municipal solid waste from the quarantine area of New York City to landfills outside of the quarantine area: a pathway analysis of the risk of spread and establishment. *Journal of Economic Entomology* 98, 1, 47–60.

Borchert, D. (2007) *Risk Analyst*. USDA-APHIS-PPQ-CPHST-PERAL, Raleigh, North Carolina.

Borchert, D., Magarey, R. and Chanelli, C. (2007) *A Guide to the Use of NAPPFAST*. USDA-APHIS-PPQ-CPHST-PERAL, Raleigh, North Carolina.

CABI (2004) *Crop Protection Compendium*. CAB International, Wallingford.

CABI (2007) *Crop Protection Compendium*. CAB International, Wallingford.

Cathey, H.M. (1990) USDA plant hardiness zone map (web version of USDA. Miscellaneous Publication, 1475).

Ciesla, W.M. (2001) *Tomicus piniperda* pest report. Exotic Forest Pest Information System for North America. Fort Collins, Colorado.

CLIMEX (2011) CLIMEX Version 3. Commonwealth Scientific and Industrial Research Organization (CSIRO) and Hearne Scientific Software, Melbourne.

Colunga-Garcia, M., Haack, R.A. and Adelaja, A.O. (2009) Freight transportation and the potential for invasions of exotic insects in urban and periurban forests of the United States. *Journal of Economic Entomology* 102, 1, 237–246.

Cranshaw, W. and Tisserat, N. (2008) Pest alert: walnut twig beetle and thousand cankers disease of black walnut. Report. Colorado State University, Fort Collins, Colorado.

Danthanarayana, W. (1975) The bionomics, distribution and host range of the light brown apple moth, *Epiphyas postvittana* (Walk.) (Tortricidae). *Australian Journal of Zoology* 23, 2, 419–437.

DEFRA (2004) *Phytophthora ramorum* epidemiology: infection, latency and survival. DEFRA Project Report PH0194, DEFRA, London.

Eager, T.A., Berisford, C.W., Dalusky, M.J., Nielsen, D.G., Brewer, J.W., Hilty, S.J. and Haack, R.A. (2004) Suitability of some southern and western pines as hosts for the pine shoot beetle, *Tomicus piniperda* (Coleoptera: Scolytidae). *Journal of Economic Entomology* 97, 2, 460–467.

EPPO (2010) EPPO activities on plant quarantine. Available online at: http://www.eppo.org/QUARANTINE/quarantine.htm, accessed 5 January 2011.

FIA (2006) Forest Inventory and Analysis Database. (Archived at PERAL).

Fowler, G. and Borchert, D. (2006) *Organism Pest Risk Assessment: Risk to the Continental United States Associated With Pine Shoot Beetle,* Tomicus piniperda *(Linnaeus), (Coleoptera: Scolytidae).* USDA-APHIS-PPQ-CPHST-PERAL, Raleigh, North Carolina.

Fowler, G., Caton, B., Jackson, L., Neeley, A., Bunce, L., Borchert, D. and McDowell, R.M. (2006) *Quantitative Pathway Initiated Pest Risk Assessment: Risks to the Southern United States Associated With Pine Shoot Beetle,* Tomicus piniperda *(Linnaeus), (Coleoptera: Scolytidae), on Pine Bark Nuggets, Logs and Lumber With Bark and Stumps from the United States Quarantined Area (Rev. Original).* USDA-APHIS-PPQ-CPHST-PERAL, Raleigh, North Carolina.

Fowler, G., Takeuchi, Y., Sequeira, R., Lougee, G., Fussell, W., Simon, M., Sato, A. and Yan, X. (2008) *Pathway-initiated Pest Risk Assessment: Asian Gypsy Moth (Lepidoptera: Lymantriidae:* Lymantria dispar *(Linnaeus)) from Japan into the United States on Maritime Ships* (Rev. Original 05062008). USDA-APHIS-PPQ-CPHST-PERAL, Raleigh, North Carolina.

Fowler, G., Garrett, L., Neeley, A., Magarey, R., Borchert, D. and Spears, B. (2009) *Economic Analysis: Risk to U.S. Apple, Grape, Orange and Pear Production from the Light Brown Apple Moth,* Epiphyas postvittana *(Walker).* USDA-APHIS-PPQ-CPHST-PERAL, Raleigh, North Carolina.

Garbelotto, M., Svihra, P. and Rizzo, D.M. (2001) Sudden oak death syndrome fells three oak species. *California Agriculture* 55, 1, 9–19.

Gordh, G. and Headrick, D.H. (2001) *A Dictionary of Entomology.* CABI Publishing, Wallingford.

Gutierrez, A.P., Mills, N.J. and Ponti, L. (2010) Limits to the potential distribution of light brown apple moth in Arizona–California based on climate suitability and host plant availability. *Biological Invasions* 12, 3319–3331.

Haack, R.A. and Poland, T.M. (2001) Evolving management strategies for a recently discovered exotic forest pest: the pine shoot beetle, *Tomicus piniperda* (Coleoptera). *Biological Invasions* 3, 307–322.

Heilman, W.E., Haack, B. and Poland, T. (2005) *Pine Shoot Beetle Outbreaks.* United States Forest Service North Central Research Station, St Paul, Minnesota.

Howell, M. (2005) Forester: Forest Inventory and Analysis (personal communication). USDA-USFS, Knoxville, Tennessee.

Hüberli, D., Van Sant, W., Swain, S., Davidson, J. and Garbelotto, M. (2003) Susceptibility of *Umbellularia californica* to *Phytophthora ramorum.* Eighth International Congress of Plant Pathology, Christchurch.

Informa plc. (2008) Lloyds maritime intelligence unit. (Archived at PERAL).

IPPC (2004) International Standards for Phytosanitary Measures, Publication No. 11: *Pest Risk Analysis for Quarantine Pests Including Analysis of Environmental Risks and Living Modified Organisms.* Secretariat of the International Plant Protection Convention (IPPC), Food and Agriculture Organization of the United Nations, Rome.

IPPC (2007) International Standards for Phytosanitary Measures, Publication No. 2: *Framework for Pest Risk Analysis.* Secretariat of the International Plant Protection Convention (IPPC), Food and Agriculture Organization of the United Nations, Rome.

IPPC (2010) International Standards for Phytosanitary Measures, Publication No. 5: *Glossary of Phytosanitary Terms.* Secretariat of the International Plant Protection Convention (IPPC), Food and Agriculture Organization of the United Nations, Rome.

Kolarik, M., Freeland, E., Utley, C. and Tisserat, N. (2010) *Geosmithia morbida* sp. nov. a new phytopathogenic species living in symbiosis with the walnut twig beetle (*Pityophthorus juglandis*) on *Juglans* in the USA. *Mycologia*, 103, 2, 325–332.

Little, E. L., Jr (1971) *Atlas of United States Trees, Volume 1, Conifers and Important Hardwoods* (Miscellaneous Publication 1146). US Department of Agriculture, Washington, DC.

Magarey, R.D., Sutton, T.B. and Thayer, C.L. (2005) A simple generic infection model for foliar fungal plant pathogens, *Phytopathology* 95, 92–100.

Magarey, R., Fowler, G., Borchert, D., Sutton, T.B., Colunga-Garcia, M. and Simpson, J.A. (2007a) NAPPFAST: an internet system for the weather-based mapping of plant pathogens. *Plant Disease* 91, 4, 336–345.

Magarey, R., Fowler, G., Colunga, M., Smith, B. and Meentemeyer, R. (2007b) Climate-host mapping of *Phytophthora ramorum*, causal agent of sudden oak death. In: Frankel, S.J., Kliejunas, J.T. and Palmieri, K.M. (eds)

Proceedings of the Sudden Oak Death Third Science Symposium. USDA Forest Service, Pacific Southwest Research Station; California Oak Mortality Task Force, Santa Rosa, California, pp. 269–275.

Magarey, R.D., Schlegel, J.W. and Borchert, D.M. (2008) Global plant hardiness zones for phytosanitary risk analysis. *Scientia Agricola* 65, 54–59.

Magarey, R.D., Borchert, D.M., Engle, J.S., Colunga-Garcia, M., Koch, F.H. and Yemshanov, D. (2011) Risk maps for targeting exotic plant pest detection programs in the United States. *OEPP/EPPO Bulletin* 41, 46–56.

McPherson, B.A., Wood, D.L., Storer, A.J., Svirha, P., Rizzo, D.M., Kelly, N.M. and Standiford, R.B. (2000) *Oak Mortality Syndrome: Sudden Death of Oaks and Tanoaks* (Tree Note No. 26). California Department of Forestry and Fire Protection, Berkeley, California.

Murray, M. (2009) Personal communication to L. Newton on 20 August 2009, from Marion Murray.

NAPIS (2005) Tomicus piniperda: *Pine Shoot Beetle*. National Agricultural Pest Information System, West Lafayette, Indiana.

NAPPFAST (2005) North Carolina State University Animal and Plant Health Inspection Service Plant Pest Forecasting System. North Carolina State University and the United States Department of Agriculture. Available online at: http://www.nappfast.org/, accessed 17 May 2012 (archived at PERAL).

NAPPFAST (2007a) NAPPFAST Global Plant Hardiness Zones. North Carolina State University APHIS Plant Pest Forecasting System. Available online at: http://www.nappfast.org, accessed 17 May 2012 (archived at PERAL).

NAPPFAST (2007b) North Carolina State University Animal and Plant Health Inspection Service Plant Pest Forecasting System. North Carolina State University and the United States Department of Agriculture. Available online at: http://www.nappfast.org/, accessed 17 May 2012 (archived at PERAL).

NAPPFAST (2008) North Carolina State University Animal and Plant Health Inspection Service Plant Pest Forecasting System. North Carolina State University and the United States Department of Agriculture. Available online at: http://www.nappfast.org/, accessed 17 May 2012 (archived at PERAL).

NAPPFAST (2009) North Carolina State University Animal and Plant Health Inspection Service Plant Pest Forecasting System. North Carolina State University and the United States Department of Agriculture. Available online at: http://www.nappfast.org/, accessed 17 May 2012 (archived at PERAL).

NatureServe (2002) http://natureserve.org/index.jsp, accessed 19 December 2011.

NCDC (2000) *Climate Atlas of the United States*. National Climatic Data Center, Asheville, North Carolina.

Newton, L., Fowler, G., Neeley, A.D., Schall, R.A. and Takeuchi, Y. (2009) *Pathway Assessment:* Geosmithia *sp. and* Pityophthorus juglandis *Blackman Movement from the Western into the Eastern United States* (Rev. 1: 10.19.2009). USDA-APHIS-PPQ-CPHST-PERAL. Raleigh, North Carolina.

Nilsson, S. (1976) Rationalization of forest operations gives rise to insect attack and increment losses. *Ambio* 5, 1, 17–22.

NOAA (2004) Hurricane Ivan (NOAA 16 AVHRR 1km). National Oceanographic and Atmospheric Administration, Washington, DC.

Orlikowski, L.B. and Szkuta, G. (2002) First record of *Phytophthora ramorum* in Poland. *Phytopathologia Polonica* 25, 69–79.

Palisade (2002) @Risk 4.52 professional, Newfield, New York.

Pizzo, C. (2009) Personal communication to L. Newton on 20 August 2009, from Carolyn Pizzo.

Poland, T.M., Haack, R.A. and Petrice, T.R. (2002) *Tomicus piniperda* (Coleoptera: Scolytidae) initial flight and shoot departure along a north-south gradient. *Journal of Economic Entomology* 95, 6, 1195–1204.

Prestemon, J., Pye, J., Barbour, J., Smith, G.R., Ince, P., Steppleton, C. and Xu, W. (2005) *United States Wood-using Mill Locations*. USDA-Forest Service Western Research Station, Washington, DC.

Price, R.A., Liston, A. and Strauss, S.H. (1998) Phylogeny and systematics of *Pinus*. In: Richardson, D.M. (ed.) *Ecology and Biogeography of Pinus*. Cambridge University Press, Cambridge, pp. 49–68.

Pscheidt, J.W. (2009) Personal communication to L. Newton on 24 August 2009, from Jay Pscheidt, Oregon State University.

Ramankutty, N. and Foley, J.A. (1999) Estimating historical changes in land cover: North American croplands from 1850 to 1992. *Global Ecology and Biogeography* 8, 381–396.

Ryall, K.S. and Smith, S.M. (2000) Reproductive success of the introduced pine shoot beetle, *Tomicus piniperda* (L.) (Coleoptera, Scolytidae) on selected North American and European conifers. *Proceedings of the Entomological Society of Ontario* 131, 113–121.

Salesgenie (2009) http://www.salesgenie.com/, accessed 17 May 2012 (archived at PERAL).

Seybold, S. (2009) Personal communication to L. Newton on 9 September 2009, from Steve Seybold, USFS, UC-Davis.

Sheehan, K.A. (1992) *User's Guide for GMPHEN: Gypsy Moth Phenology Model* (General Technical Report NE-158). United States Department of Agriculture, Forest Service, Northeastern Forest Experiment Station, Washington, DC.

Sutherst, R.W., Maywald, G.F. and Kriticos, D. (2007) *Climex User's Guide Version 3*. CSIRO, Melbourne.

Swetnam, T.W. (2001) Spatial and temporal coherence of forest fire and drought patterns in the western United States. *Fourth Symposium on Fire and Forest Meteorology*, 13–15 November, Reno, Nevada.

Tisserat, N. (2009) Personal communication to L. Newton on 28 August 2009, from Ned Tisserat, Colorado State University (e-mail).

Tooley, P. and Kyde, K. (2005) The effect of temperature and moisture period on infection of Rhododendron 'Cunningham's White' by *Phytophthora ramorum*. *Phytopathology* 95, S104.

UCIPM (2003) *How to Manage Pests: Degree-days*. Statewide IPM Program, Agriculture and Natural Resources, University of California, Davis, California..

UI (2009) *1000 Canker Disease on Walnut. Bugs and Cruds – 2009 Update*. University of Idaho, Extension Forestry, Moscow, Idaho.

USDA-APHIS (2005a) *Pest Detection and Management Programs: Pine Shoot Beetle*. USDA Animal and Plant Health Inspection Service, Riverdale, Maryland.

USDA-APHIS (2005b) *SPRO Notifications*. USDA Animal and Plant Health Inspection Service, Riverdale, Maryland.

USDA APHIS (2007) *Use of HAACP-type Methodology for Mitigating Risk of AGM Introductions*. USDA APHIS PPQ CPHST, Raleigh, North Carolina.

USDA-APHIS (2008) Customs and Border Protection AGM Ship Inspection Data. Data courtesy of Michael Simon, Senior Staff Officer, USDA-APHIS-PPQ-QPAS, Riverdale, Maryland.

USDA-APHIS (2010a) New Pest Advisory Group (NPAG). http://www.aphis.usda.gov/plant_health/cphst/npag/index.shtml, accessed 5 January 2011.

USDA-APHIS (2010b) Plant health: light brown apple moth. Available online at: http://www.aphis.usda.gov/plant_health/plant_pest_info/lba_moth/index.shtml, accessed 12 January 2011.

USDA-APHIS (2010c) Plant health: *Phytophthora ramorum*/sudden oak death. Available online at: http://www.aphis.usda.gov/plant_health/plant_pest_info/pram/, accessed 12 January 2010.

USDA-APHIS-PPQ (2000) *Guidelines for Pathway-initiated Pest Risk Assessments* (Version 5.02). United States Department of Agriculture, Animal and Plant Health Inspection Service, Plant Protection and Quarantine (PPQ), Riverdale, Maryland.

USDA-NAL-NISIC (2010) \thousand cankers black walnut disease. United States Department of Agriculture National Agricultural Library National Invasive Species Information Center. Available online at: http://www.invasivespeciesinfo.gov/microbes/thousandcankers.shtml, accessed 7 January 2011.

USDA-NASS (2009) *2007 Census of Agriculture*. United States Department of Agriculture National Agricultural and Statistics Service. Available online at: http://www.agcensus.usda.gov/Publications/2007/Online_Highlights/Desktop_Application/index.asp, accessed 17 May 2012 (archived at PERAL).

USFS (1991) *Forest Density*. United States Department of Agriculture United States Forest Service, Southern Forest Experiment Station, Forest Inventory and Analysis Research Unit, Washngton, DC.

USFS (1992) *Forest Type Group Data*. United States Department of Agriculture United States Forest Service, Washington, DC.

USGS (2006) Digital representations of tree species range maps from 'Atlas of United States Trees' by Elbert L. Little, Jr. (and other publications). US Department of the Interior, US Geological Survey. Available online at: http://esp.cr.usgs.gov/data/atlas/little/, accessed 17 May 2012 (archived at PERAL).

Venette, R.C., Kriticos, D.J., Magarey, R.D., Koch, F.H., Baker, R.H.A., Worner, S.P., Gomez Raboteaux, N.N., McKenney, D.W., Dobesberger, E.J., Yemshanov, D., De Barro, P.J., Hutchison, W.D., Fowler, G., Kalaris, T.M. and Pedlar, J. (2010) Pest risk maps for invasive alien species: a roadmap for improvement. *BioScience* 60, 5, 349–362.

WTO (1994) *Agreement on the Application of Sanitary and Phytosanitary Measures*. World Trade Organization, Geneva.

Part IV

Pest Risk Management, Risk Communication and Uncertainty

———————————

13 Pest Risk Management Theory and Background

Robert Griffin and Alison Neeley

13.1. Introduction

Risk management, like risk assessment, is distinctly analytical in nature, relying strongly on scientific principles and evidence mixed with expert judgement. In a broader sense however, the concept of risk management is also associated with other procedures and activities needed for the establishment, application and evaluation of phytosanitary measures. Some of these may be considered to be closely related enough to the analytical process to be part of pest risk analysis, while other procedures are clearly not analytical.

Depending on the description of risk management you subscribe to, the concept of risk management may extend to decision making and the implementation or operational aspects of implementing measures. Sometimes risk management also includes the evaluation of measures that are in place to determine their appropriateness and any adjustments that may need to be made as a result of analysis. This monitoring or feedback process, whether or not it is technically part of the analytical process of risk management in pest risk analysis, is crucial for connecting the analytical process to the real world for validating and improving pest risk analyses.

13.2. Risk Management – Analysis and Practice

For the purposes of the discussion here, we will distinguish *risk management analysis* from *risk management practice*, and exclude decision making. This is because decision making represents the point where the analytical processes begin to incorporate political, social and other factors or information that are not directly related to pest risk. Likewise, this discussion will steer clear of risk management practice except to acknowledge that many of the analytical tools and information used for risk management analysis are also useful in an operational context, and much of what we know about the efficacy and feasibility of measures comes from operational feedback.

Risk management analysis involves identifying and evaluating options for reducing, avoiding or eliminating pest risk after we have decided that the risk is unacceptably high and it may be possible to mitigate. Information is required on possible mitigations, their efficacy, feasibility and impacts. Our analysis will be disciplined by a number of key principles and requires a clear understanding of several related concepts and terms. The results of our analysis will typically be conclusions and recommendations. Fig. 13.1 (seen earlier in

Pest risk analysis process overview

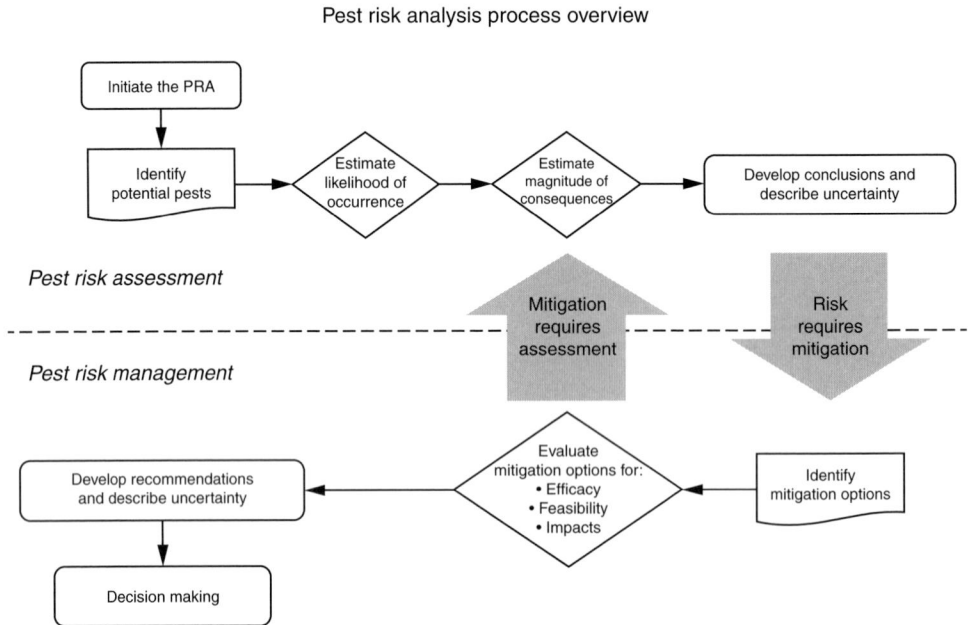

Fig. 13.1. Pest risk analysis overview.

Chapter 11) provides a graphic representation of risk management analysis, where we identify mitigation options for efficacy, feasibility and impacts – and that identification of these options requires analysis.

It is important to recall that the evaluation of risk management strategies for efficacy means that we must revisit risk assessment to determine the degree to which mitigation options change the risk. A baseline level of risk is established from risk assessment for *the unmitigated condition* or some other starting point we describe during hazard identification in the initiation process. In risk management, the baseline risk is compared with the mitigated situations in order to measure the efficacy of specific options. Other elements of pest risk analysis also do not fit sequentially in the process. These include uncertainty, transparency (documentation) and communication. Each of these has significant roles throughout the process.

13.3. Principles and Concepts

The two principles emphasized most in the risk assessment process were transparency and consistency. These continue to be important throughout pest risk analysis, including risk management.

Transparency	Consistency
Transparency is documenting information sources, describing methodologies, identifying assumptions and uncertainties, summarizing results and providing an analysis that is appropriate for the intended audience.	*Consistency* is using the same criteria for judgements made on similar situations to avoid the perception of being arbitrary or biased. Criteria should consistent within an analysis, and between analyses.

The principle of transparency is applied throughout the risk management process, and is crucial for communicating the results of risk management analysis to relevant parties. The principle of consistency is applied alongside other principles in risk management in the analysis, selection and implementation of measures. For example, we may determine that a particular treatment is an appropriate measure even though efficacy data for the treatment

---

Example 1. Application of equivalence.

Imagine that Country A is exporting apples to Country B with a fumigation treatment. Now think about Country C proposing a systems approach for exporting apples with the same pest risk to Country B, assuming that the required level of efficacy must be equivalent to the efficacy of the treatment being used by Country A. In this case, country C is likely to be trying to achieve a higher level of efficacy (stronger measure) than necessary if the treatment requirement was agreed bilaterally with the understanding that the treatment is probably more rigorous than necessary for the pest in question.

The preferred scenario could be for Country C to design and propose a systems approach *which is consistent with the risk* rather than trying to match the existing treatment with equivalent measures. Establishing measures consistent with the risk then provides the opportunity for Country A to use the principle of equivalence as the basis for reducing requirements on its exports to be consistent with Country C. In other words, the principle of equivalence works best when it is linked to the strength of measures.

---

may be for a different but similar pest. In this case, the principle of consistency would be applied in parallel with the principle of equivalence to argue that the treatment may also be extrapolated to other pests that are similar.

Equivalence is a principle we did not consider directly in risk assessment but which has a critical role in risk management. The concept as described by both the SPS and the IPPC is designed to avoid unjustified prescriptive measures (see also Chapter 4). Quite simply, equivalence means that mitigation options that have equivalent or better efficacy and are also feasible should be considered.

One common problem with applying the principle of equivalence in practice is assuming that existing measures represent the appropriate *strength of measures*. In fact, many of the requirements countries have in place have been established prior to the coming into force of the SPS Agreement or are based on bilateral agreements that did not directly link the strength of the measures to a defined level of pest risk.

This brings us to a series of principles (WTO, 1994; FAO, 1997) which are most intimately associated with risk management analysis, including:

- Necessity;
- Managed risk;
- Minimal impact;

- Non-discrimination;
- Technical justification;
- Modification.

Together these principles form the core disciplines applied to both risk management analysis and risk management practice.

The logic of the relationship between these principles begins with *necessity*; recognizing that countries have the sovereign right to protect themselves but they must first demonstrate that a potential hazard exists that justifies the need for protection. At the same time, they must realize that zero risk is not a realistic objective, so a policy of *managed risk* must be adopted based on an appropriate level of protection (or acceptable level of risk). Measures resulting from these policies should aim to have *minimal impact*; representing the least restrictive measures available and resulting in the minimum impediment to trade.

The measures should *not discriminate* between trading partners or between trading partners and domestic producers with similar risks. A proper pest risk analysis, based on scientific principles and evidence, and consistent with international standards, provides the *technical justification* for measures if the measures are not based on international standards or deviate from international standards. Pest risk analyses should be *modified* and measures adjusted without undue delay when new or better information indicates the need. Measures should not be changed without a technical justification and appropriate notifications.

There is one additional 'principle' best described as *rational relationship*. It is not described as such in either the IPPC or the SPS Agreement, but has developed as a central issue in jurisprudence and has a prominent role in risk management (WTO, 2010). Rational relationship is composed of two elements. The first is that the measure in question actually has an effect on mitigating risk. For instance, a fumigation treatment for insects is not appropriate for weed seed and would therefore have no rational relationship to the risk if required as a treatment for weed seed.

The second element of rational relationship follows the idea that the *strength of measures* is proportional to the risk (see also Chapter 19). The concept here is that the magnitude of the risk and the strength of measures applied to mitigate risk are on sliding scales. Higher risks correspond with stronger measures and vice versa. Measures do not have a rational relationship with the risk when they are misaligned based on other effective options that may be available. A simple example: a treatment designed for internally feeding arthropods may be overly rigorous for external feeders and contaminating pests. Table 13.1 demonstrates the conceptual balance of concepts in the IPPC-SPS framework of principles that are most important for risk management.

The IPPC has produced several standards that provide general guidance on pest risk management, as well as some standards that provide specific guidance on particular aspects of pest risk management – for instance, standards that address pest free areas, specific pests like fruit flies or specific types of phytosanitary treatments. The ISPMs particularly relevant to pest risk management, and which may be consulted in identifying options, are shown in Table 13.2.

Table 13.2. ISPMs with particular relevance to pest risk management.

ISPM No. 1 *Phytosanitary Principles for the Protection of Plants and the Application of Phytosanitary Measures In International Trade* (IPPC, 2010a)

ISPM No. 4 *Requirements for the Establishment of Pest Free Areas* (IPPC, 1996)

ISPM No. 10 *Requirements for the establishment of Pest Free Places of production and Pest Free Production Sites* (IPPC, 1999)

ISPM No. 12 *Phytosanitary Certificates* (IPPC, 2011)

ISPM No. 14 *The Use of Integrated Measures in a Systems Approach for Pest Risk Management* (IPPC, 2002)

ISPM No. 18 *Guidelines for the Use of Irradiation as a Phytosanitary Measure* (IPPC, 2003)

ISPM No. 22 *Requirements for the Establishment of Areas of Low Pest Prevalence* (IPPC, 2005a)

ISPM No. 23 *Guidelines for Inspection* (IPPC, 2005b)

ISPM No. 24 *Guidelines for the Determination and Recognition of Equivalence of Phytosanitary Measures* (IPPC, 2005c)

ISPM No. 26 *Establishment of Pest Free Areas for Fruit Flies (Tephritidae)* (IPPC, 2006)

ISPM No. 28 *Phytosanitary Treatments for Regulated Pests* (IPPC, 2009b)

ISPM No. 29 *Recognition of Pest Free Areas and Areas of Low Pest Prevalence* (IPPC, 2007)

ISPM No. 30 *Establishment of Areas of Low Pest Prevalence for Fruit Flies (Tephritidae)* (IPPC, 2008)

ISPM No. 32 *Categorization of Commodities According to Their Pest Risk* (IPPC, 2009c)

Table 13.1. Relationship of key principles associated with risk management.

Sovereign right to protect	→ Demonstrable hazard
No zero risk	→ Manage for minimal impact
Consistency (fairness)	→ No discrimination
Technical justification	→ Scientific principles and evidence (pest risk analysis)
Modification	→ Justification and notification
Rational relationship	→ Strength of measures

13.4. Types of Measures

The discussion above reminds us that pest risk analysis is not strictly a scientific process but exists within an international regulatory framework that directs, defines and disciplines the process. The influence of this framework on risk management is particularly important because both the IPPC and the SPS Agreement are ultimately about risk management, i.e. the measures that are applied in trade. The level of analysis used to justify measures depends on the type of measure, which can be categorized as follows:

- Established measures: Measures which have been put into place as static requirements;
- Provisional measures: Measures taken when there is insufficient scientific evidence to permit a final decision on the safety of a product or process (may or may not be an emergency);
- Emergency measures: Measures taken when an emergency or new or unexpected situation arises (may or may not be provisional).

Established measures should be based on international standards or a pest risk analysis. Established measures define the appropriate level of protection by virtue of the range of risks and the strength of measures they represent. They also offer reference points for equivalency where the measures are linked to the acceptable level of the risk.

Both the IPPC and SPS Agreement also include concepts and terms for provisional and emergency measures, but they are not well aligned with each other and are therefore frequently confused. Direct reference to *provisional measures* is found in Article 5 of the SPS Agreement. Such measures are designed to facilitate trade by making it possible to put in place overly restrictive requirements that will be adjusted later when new or better information or experience is available.

An important point to note regarding provisional measures is that the country imposing a provisional measure must actively pursue the information required for a more objective assessment of the risk and review of the measure within a reasonable period of time. This is one of the few situations under the SPS Agreement where the burden of proof is completely one-sided. In nearly all other circumstances, both the importing and exporting country share the responsibility for providing information necessary to evaluate and agree on appropriate measures.

Although not explicit in the SPS Agreement, *emergency measures* extend from Annex B (urgent problems) and the resulting Emergency Notification format adopted by the SPS Committee (G/SPS/7 Rev 1) (WTO, 1994). The IPPC (the Convention) is explicit about *emergency action* based on the detection of a pest, indicating that such action will be evaluated as soon as possible to ensure that it is justified (Article VII.6). ISPM No. 1 (IPPC, 2010a) (Section 2.11) refers to emergency actions for new or unexpected phytosanitary situations based on a preliminary pest risk analysis and indicating that such measures are temporary and should be the subject of a detailed pest risk analysis as soon as possible.

Provisional measures need not be emergency measures, i.e. not necessarily in response to an immediate threat. Likewise, emergency measures need not be provisional, i.e. require additional information and reconsideration (WTO, 1994; FAO, 1997; IPPC, 2010a).

Other international agreements and organizations may refer to 'precautionary measures', which are variously understood and generally linked to the application of the 'precautionary approach' (see also Chapter 19). The term 'precautionary measures' is not explicitly used or described in either the IPPC or the SPS Agreement. It may be argued however that phytosanitary measures are by their nature more or less precautionary depending on the influence of uncertainty in the judgement regarding acceptable risk. Questions in this regard have surfaced in SPS disputes associated with the interpretation and application of the concept of provisional measures.

Deliberations in this context have resulted in statements from the WTO Dispute Settlement Body that clarify that provisional measures have a precautionary aspect to them but are not intended to be precautionary measures (WTO, 1998). The concept of precaution as a function of uncertainty is therefore implicit in the risk analysis process and does not need to be distinguished as either a principle or a specific type of measure in the WTO-IPPC framework.

It is also important to be clear about the question to be addressed by the pest risk analysis so that risk management can be focused on the correct objective. For instance, risk management strategies to prevent entry will probably be different from those used to prevent spread. Likewise, the risk management analysis done to determine if a pest is an imminent threat will be different than one done to develop recommendations for a pest that is newly established.

13.5. Relationship of Risk Management to Hazard Identification

The risk *assessment* process began with hazards identified in the initiation stage. The risk *management* process begins by identifying possible measures that can be used to mitigate the risk we found associated with the hazards. Before examining this process more closely, we need to briefly revisit the hazard identification process to understand the relationship of risk management to hazard identification and highlight the importance of careful hazard identification.

In broad terms, pest risk analyses are done for *pests, pathways* or *policies* (IPPC, 2004, 2007). The hazards in each of these categories have unique characteristics that will affect aspects of both risk assessment and risk management.

13.5.1. Pests

In the case of *pests*, the key point to remember in hazard identification is to be as specific as possible about the identity of the organism and the question to be addressed. The issues associated with the identity of the organism include the taxonomic level (usually species), the correct scientific name (and synonyms) and the life stage(s) of concern. The identity of the organism has an especially critical relationship to risk management where data on efficacy (such as for a treatment) may be available or required.

13.5.2 Policies

The hazard identification process for *policies* is highly variable. As a general rule, policies are either being created or being changed. In the case of the latter, the analysis may be to determine whether the policy should be changed or to determine whether there is sufficient justification for a policy decision that has already occurred. Policies range from the highest levels of legislation with very formal documentation to the lowest form of local decision making, which may have little or no documentation but legitimately falls under general operational authority. The key point for hazard identification as it relates to risk management in policy-directed pest risk analysis is the scope of the mitigation objective.

For example, consider the challenge associated with defining the scope of a pest risk analysis to justify new regulations on wood packaging. It is probably not feasible to address the range of possible pests and risks individually, so the strategies for risk management will necessarily be generic or categorical and linked to some average risk level which needs to be defined in advance by policy makers or determined as a result of risk assessment. The other end of the hazard spectrum would be a local decision to increase the inspection level for wood packaging and reduce the inspection level for finished lumber. In this case, the relationship of the hazard (pest entry[1]) to risk management (inspection) is sharply focused on the questions of justification and the magnitude of change required.

> **Example 2.** Importance of defining the commodity or pathway – Case 1.
>
> Consider the difference between importing raw logs and finished lumber. The former is likely to have many more pests associated with it than the latter. We make some assumptions about finished lumber that we would not make about raw logs, including that it is debarked, dried, cut and shaped, sanded and packaged. If the import request we have received is only for 'wood', then the safest assumption would be to expect raw wood and the pest risk analysis would include pests and mitigations that may not be necessary if the request were more specific.

13.5.3 Pathways

The importance of careful hazard identification in pest risk analyses carried out for *pathways* is arguably the most critical for risk management. The most common type of pest risk analysis is for commodity imports, a form of pathway analysis that is focused on a particular commodity from a specific origin as the potential means for introducing regulated pests. For these types of pest risk analyses, the description of the commodity for which the pest risk analysis is done will be extremely important for the hazard identification (pest listing) aspect of risk assessment and also for the identification of possible mitigation measures for risk management.

There is a difference between processes that define a commodity and 'normal industry practices'. A further distinction can be made between 'practices that may occur' and 'practices that are required' to produce the product in question. For example, banana fruit are typically grown under bags wherever they are commercially produced around the world. If not, bananas are damaged by myriad pests including birds, and suffer from reduced quality due to physical damage like sunburn, uneven ripening and discoloration.

The assumption therefore is that imported bananas for retail sale will be commercially produced, which includes growing under bags as a normal industry practice. This falls into the category of industry practices that are normally performed. Sometimes however, the bags used to cover the bananas in the field are treated with pesticide(s); other times the bags are not treated. This aspect of the commodity description is an industry practice that may occur. Both the bagging and the pesticide treatment should be confirmed before initiating the pest risk analysis and a decision taken on whether these practices define the commodity or will be ignored in the risk assessment but considered later in risk management.

Citrus fruit, on the other hand, are not defined by a particular process. It may be assumed in many cases that fruit are at least culled and washed, but fruit for juicing or other forms of processing typically receive much less care and may not be inspected or cleaned as carefully as fruit for direct retail sale. For this reason, and because of the variability in processes associated with harvesting, packing and shipping citrus fruit, it is difficult to make assumptions about the commodity for pest risk analysis purposes. In all cases, but especially where erroneous assumptions may be made, it is therefore important for the exporting country to be as forthcoming and detailed as possible when providing the prerequisite information used to initiate a pest risk analysis.

Fewer assumptions will be required for a pest risk analysis on the commodity with detailed information. Such detail does not however mean that the pest risk analysis will have a more or less restrictive result. Based on the description of processes, we would assume that the fruit is very clean and high quality. On the other hand, we notice that 1 cm of the stem (peduncle) is included (a common practice with lemons and tangerines because the fruit are often cut rather than pulled to avoid damage to the rind). Since stems and leaves are often affected by different pests from fruit, there could be a

---

Example 3. Importance of defining the commodity or pathway – Case 2.

Imagine beginning a pest risk analysis for 'lemons' versus a pest risk analysis for 'fresh, ripe, Meyer lemon with 1 cm of stem attached, commercially produced by registered growers, field selected for size and quality, culled at NPPO registered packing houses, washed in soapy water with 2% active chlorine, brushed, waxed, individually wrapped, packed in cardboard boxes of 48 with no vents, inspected, palletized, stored in sealed refrigerated warehouses, transported in sealed containers, shipped by sea within 1 week of harvest and refrigerated in transit at 5°C'.

---

difference in the types of pests we are concerned with and the risk management strategies that will be needed. Failure to include this information could have led to the assumption that fruit were free of stems and the programme would be likely to experience difficulties once it was discovered that the fruit are shipped with stems, especially if pests were found associated with the stems.

Once the commodity or pathway is clearly described, including any processes and actions that are considered relevant to the pest risk analysis, it is necessary to decide the precise starting point for the risk analysis. This may be the *unmitigated* risk. Using the previous example, this would be any lemon from the origin with no assumptions about cultivation or handling. If however, we are looking instead at changes in an existing programme, the starting point may well be the current product, including any processes (e.g. culling and washing) and mitigation measures (e.g. cold treatment) that are already performed. The key here is to establish the baseline for risk according to the description of the product and the objective of the analysis. This means that anything done in risk management to change the risk will be new or additional and focused on the risk associated with the specified commodity. The precise description of the article and any assumptions and uncertainties need to be well documented.

13.6. Efficiency and Economic Analysis in Pest Risk Management

Efficiency is a measure of how the total benefits associated with a particular management option are related to its total costs. Efficiency is achieved when those benefits outweigh the costs. The lower the cost–benefit ratio is, the more efficient the programme is.

The meaning of 'pest risk analysis' and the degree to which economic, social, political and other consequences of pest introduction are a part of 'pest risk assessment' is frequently a source of confusion. Developing a common understanding of the concepts and agreement on the appropriate terms is essential to harmonization in this important area of plant protection.

Historically, a pest risk assessment done for phytosanitary purposes has focused on the biological and technical aspects of pest entry and establishment, primarily relating to the likelihood of introduction. This is a normal extension of the technical and scientific expertise that exists in plant protection organizations and the orientation toward pest exclusion as the primary form of protection. The underlying assumption is that the consequences of pest introduction are generally unacceptable – sometimes even assuming that all consequences are equally unacceptable.

The provisions of Article 5 in the SPS Agreement indicate that the consideration of economic factors such as losses, costs, cost-effectiveness, etc. is clearly within the purview of risk assessment (= risk analysis in IPPC terms) (WTO, 1994; FAO, 1997). It is also a widely held principle in the broad discipline of risk analysis that risk is defined by both the probability of an adverse event occurring and the magnitude of the consequences of that event.

These points would argue for strengthening the role of economics and increasing the proportion of attention given to analyses of 'consequences' in pest risk assessment. It also points to the need for contemporary pest risk assessments to recognize that all consequences are not equally unacceptable. Upon accepting this, it becomes clear that biology and economics have strongly complementary roles in the analysis of pest risk, and that there is a need to handle less tangible factors such as social impacts. What is not always clear is how such a relationship can be structured.

Clarification of this issue begins by recognizing the difference between the consequences directly related to *the introduction and establishment* of the pest organism and the impacts more indirectly related to *making a decision concerning the mitigation* of the pest organism, including the decision to do nothing. Likewise, it is important to distinguish between a risk assessment for the *unmitigated risk* and a risk assessment for the *mitigated risk*.

The primary level of risk assessment addresses the *unmitigated* pest risk and considers impacts *directly* related to the introduction of a pest organism. This establishes the baseline from which risk is measured. The result of this type of assessment is a *conclusion*. Direct consequences *of the introduction and establishment of the pest* are integral to this level of analysis, primarily for the purpose of making a judgement concerning whether the consequences of introduction are unacceptable and therefore worthy of mitigation.

At another level, risk assessment is linked with *risk management* (used here in an analytical context rather than an operational context). It is at this point that options for mitigation are identified and a *risk assessment* must be done for the *mitigated risk*. The result is an iterative and often intuitive process designed to evaluate the efficacy of mitigation options.

The results of this become *conclusions with recommendations* concerning efficacy. These results are compared to the mitigated risk and weighed against criteria, policies and values used to define *safety* or the *acceptable level of risk*.

However, the evaluation of risk management options also creates a situation where the risk associated with the introduction of the pest is changed as a result of some mitigating factor that has a cost (even the option of 'no action' has a cost) as well as other consequences or impacts. Therefore, each mitigation option should also be evaluated for feasibility and considered in economic or other terms for the 'consequences' of implementation (better called 'impacts' to avoid confusion). This involves consideration of the direct economic impacts, as well as the political, social and other indirect economic impacts *of choosing a recommendation*, thereby raising the need for additional, broader or more detailed analysis.

13.6.1. Cost–benefit analysis

Cost–benefit analysis (CBA) – or benefit–cost analysis – is the most common method for measuring economic efficiency. It is a systematic procedure for identifying, calculating and comparing the costs and benefits associated with various policy options. CBA has many applications including feasibility analysis, ranking and prioritizing, maximizing utility, evaluating the value of on-going projects, etc. In terms of risk management analysis, it can be a very powerful tool for evaluating potential phytosanitary measures.

The process for conducting CBA is fairly straightforward – at least in theory. It simply involves comparing the total expected costs of each option (over a specified period of time) against the total expected benefits of each option.

Boardman (2006) outlines the basic steps of conducting a CBA as follows:

1. Identify the projects or programs (e.g. phytosanitary management options) to be compared in the analysis
2. Select and define the measurement criteria you will be using to compare each option against each other. This includes defining your time horizon.
3. Collect all cost and benefits information (according to your defined criteria).
4. Predict the future costs and benefits of each option over the duration of the project. (often collected by year).

5. Monetize all costs and benefits (in other words, describe them in terms of money).

6. Convert the future expected benefits and costs into a present value by applying a discount rate (the discount rate is just a way of weighing the value of future costs and benefits against their value in today's currency. It is generally the interest rate you are currently earning on the money you would invest in the project).

7. Calculate net present value of each project option by subtracting the present value of total cost from the present value of the total benefit.

8. Conduct sensitivity analysis when you come across a great deal of uncertainty.

9. Make a recommendation.

Although the steps above may seem simple, in reality it can be very difficult to collect the data you need for calculating *every* cost and benefit. Technically, benefits must include all indirect effects that would be generated by each option. Costs must include the opportunity cost (the benefits that you would have received had you picked your next best option).

Examples of the kinds of factors that could be included in CBA include:

- A monetary estimate of the impact each proposed option will likely have on existing regulations.
- All direct and indirect commercial, environmental, and social impacts that would result from each proposed option.
- The time it will likely require to implement each strategy.
- The efficacy of each measure (and hence potential additional benefits) against other quarantine pests.

Although CBA is a powerful tool, there are several criticisms of the process. The first, and perhaps most important is that CBA does not take into account distributional impacts and issues of equity. As long as there is a net benefit to society, the option will be efficient. However it is certainly possible that a particular policy will result in great benefits to one sector of society, while causing harm to another. For example, let's

say that you have just completed CBA that compared two possible risk management strategies, requiring the commodity be fumigated at the port of departure, vs. only requiring inspection. The results of your CBA indicate that the second option – inspection – has the greatest net benefits.

The sector of your society that will benefit most from adopting this option is your domestic consumers. However, they are not the same group that is likely to be negatively impacted by a potential pest outbreak. CBA cannot account for the 'unfairness' that domestic consumers receive all the benefits of the policy, whereas your domestic producers are the ones who bear the risk. To counter issues of equity, ideally, there should be a system in place whereby the winners (those who will benefit from the adoption of the policy or programme) compensate those who are harmed by it. In reality this is very difficult to implement and is rarely done.

Additional criticisms of CBA include the fact that the outcome of analyses can be fairly easy to manipulate by changing a few assumptions or parameters (for example, the discount rate). Some important costs – such as social and environmental impacts – are very difficult and controversial to monetize.

It may be argued that considering the feasibility and impact of risk management options falls outside the scope of risk analysis because these elements of the analysis deal more with the 'risk' of making a decision rather than the risk presented by the primary hazard, i.e. the possibility of pest introduction. This view suggests that the 'technical analyses' extend only to the point where efficacy is considered and the degree to which risk is reduced is the only criterion for prioritizing options. Thereafter, the analysis is passed to economists and/or policy analysts and ceases to be a 'pest' risk analysis.

This argument is largely academic because, in the final analysis, the risk manager or decision maker will need as complete a picture as possible. Decision-making is the point where both direct consequences of the pest introduction and indirect impacts of the pest and pest risk management options should be as well characterized as possible.

The full impacts *of deciding upon one or another option* should be included in the analysis at this stage.

Whatever the view taken, the products of risk management are recommendations and should include discussions of associated uncertainties (see Chapter 16) that need to be considered by those who will ultimately make the decisions.

13.7. Summary

The core elements of pest risk analysis are risk assessment and risk management. Risk management is the 'evaluation and selection of options to reduce the risk of introduction and spread of a pest'. Measures should be effective, feasible and technically justified. There are three general aspects to risk management that should be considered: policy, analytical and operational – this text has focused on the analytical aspects of risk management. Several key principles of the SPS and IPPC apply to the policy aspects of risk management, including in particular the:

- Application of the ALOP;
- Principle of least trade restrictive measures (minimal impact);
- Principle of non-discrimination (including national treatment);
- Principle of managed risk;
- Principle of equivalence.

Since zero-risk is not a reasonable option, risk management should focus on reduction of risk that can be justified and is feasible within the limits of available options and resources. In any case, the overarching policy of the NPPO, together with any country/commodity specific policies, is usually the main driver in the ultimate decision making stages for risk management. The analytical aspect of risk management is the component that considers and weighs various options for mitigating risk. Options are usually analysed for efficacy and feasibility, and to some extent costs associated with various options. The analysis of options in pest risk management necessitates a close linkage with the pest risk assessment part of a pest risk analysis. Risk management is therefore dependent on the risk assessment to inform of what mitigation options may be useful, and how various mitigation options will affect overall risk.

Note

[1] The ultimate hazard is pest introduction (entry and establishment), but from a local port perspective, the focus of risk management is on preventing entry. The 'hazard' from a local standpoint is therefore pest entry assuming that the pest is able to establish. Although a valid operational assumption, the same should not be assumed in an analysis done to establish the policy that allowed entry of the commodity which is the pathway.

References

Boardman, N.E. (2006) *Cost–Benefit Analysis, Concepts and Practice.* (third edn). Prentice Hall, Upper Saddle River, New Jersey.

FAO (1997) *New Revised Text of the International Plant Protection Convention.* Food and Agriculture Organization of the United Nations, Rome.

IPPC (1996) International Standards for Phytosanitary Measures, Publication No. 4: *Requirements for the Establishment of Pest Free Areas.* Secretariat of the International Plant Protection Convention (IPPC), Food and Agriculture Organization of the United Nations, Rome.

IPPC (1999) International Standards for Phytosanitary Measures, Publication No. 10: *Requirements for the Establishment of Pest Free Places of Production And Pest Free Production Sites.* Secretariat of the International Plant Protection Convention (IPPC), Food and Agriculture Organization of the United Nations, Rome.

IPPC (2003) International Standards for Phytosanitary Measures, Publication No. 18: *Guidelines for the Use of Irradiation as a Phytosanitary Measure*. Secretariat of the International Plant Protection Convention (IPPC), Food and Agriculture Organization of the United Nations, Rome.

IPPC (2004) International Standards for Phytosanitary Measures, Publication No. 21: *Pest Risk Analysis for Regulated Non Quarantine Pests*. Secretariat of the International Plant Protection Convention (IPPC), Food and Agriculture Organization of the United Nations, Rome.

IPPC (2005a) International Standards for Phytosanitary Measures, Publication No. 22: *Requirements for the Establishment of Areas of Low Pest Prevalence*. Secretariat of the International Plant Protection Convention (IPPC), Food and Agriculture Organization of the United Nations, Rome.

IPPC (2005b) International Standards for Phytosanitary Measures, Publication No. 23: *Guidelines for Inspection*. Secretariat of the International Plant Protection Convention (IPPC), Food and Agriculture Organization of the United Nations, Rome.

IPPC (2005c) International Standards for Phytosanitary Measures, Publication No. 24: *Guidelines for the Determination and Recognition of Equivalence of Phytosanitary Measures*. Secretariat of the International Plant Protection Convention (IPPC), Food and Agriculture Organization of the United Nations, Rome.

IPPC (2006) International Standards for Phytosanitary Measures, Publication No. 26: *Establishment of Pest Free Areas For Fruit Flies (Tephritidae)*. Secretariat of the International Plant Protection Convention (IPPC), Food and Agriculture Organization of the United Nations, Rome.

IPPC (2007) International Standards for Phytosanitary Measures, Publication No. 29: *Recognition of Pest Free Areas and Areas of Low Pest Prevalence*. Secretariat of the International Plant Protection Convention (IPPC), Food and Agriculture Organization of the United Nations, Rome.

IPPC (2008) International Standards for Phytosanitary Measures, Publication No. 30: *Establishment of Areas of Low Pest Prevalence For Fruit Flies (Tephritidae)*. Secretariat of the International Plant Protection Convention (IPPC), Food and Agriculture Organization of the United Nations, Rome.

IPPC (2009a) International Standards for Phytosanitary Measures, Publication No. 15: *Regulation of Wood Packaging Material In International Trade*. Secretariat of the International Plant Protection Convention (IPPC), Food and Agriculture Organization of the United Nations, Rome.

IPPC (2009b) International Standards for Phytosanitary Measures, Publication No. 28: *Phytosanitary Treatments For Regulated Pests*. Secretariat of the International Plant Protection Convention (IPPC), Food and Agriculture Organization of the United Nations, Rome.

IPPC (2009c) International Standards for Phytosanitary Measures, Publication No. 32: *Categorization of Commodities According To Their Pest Risk*. Secretariat of the International Plant Protection Convention (IPPC), Food and Agriculture Organization of the United Nations, Rome.

IPPC (2010a) International Standards for Phytosanitary Measures, Publication No. 1: *Phytosanitary Principles for the Protection of Plants and the Application of Phytosanitary Measures in International Trade*. Secretariat of the International Plant Protection Convention (IPPC), Food and Agriculture Organization of the United Nations, Rome.

IPPC (2010b) International Standards for Phytosanitary Measures, Publication No. 5: *Glossary of Phytosanitary Terms*. Secretariat of the International Plant Protection Convention (IPPC), Food and Agriculture Organization of the United Nations, Rome.

IPPC (2011) International Standards for Phytosanitary Measures, Publication No. 12: *Phytosanitary Certificates*. Secretariat of the International Plant Protection Convention (IPPC), Food and Agriculture Organization of the United Nations, Rome.

WTO (1994) *The WTO Agreement on the Application of Sanitary and Phytosanitary Measures*. World Trade Organization, Geneva.

WTO (1998) *Japan – Measures Affecting Agricultural Products*. Panel Report. WT/DS76/R.

WTO (2010) *Australia – Measure Affecting the Importation of Apples from New Zealand*. Panel Report. WT/DS367/R.

14 Pest Risk Management Applications and Practice

Robert Griffin

14.1. Introduction

The first step in risk management analysis is identifying potential mitigation options that may be used alone or in combination to create risk management strategies. In the case of a pest risk analysis done for a pest, this will be largely dependent on the pathway. In the case of a pest risk analysis done for policy purposes, this will depend on the risk management objective. Using the wood packaging example from Chapter 13 (see 13.5.3, Example 2), we would be looking for measures that can be applied to wood packaging as a pathway. In fact, risk management is typically focused on where measures can be applied to change the risk at one or more points in the series of events that form a pathway (see Fig. 11.2 in Chapter 11). It is therefore most instructional to examine risk management from a pathway pest risk analysis perspective based on common import scenarios.

14.2. Identifying Mitigation Options

The exporting country should be in the best position to understand the pests and mitigations for their export commodities. In an ideal situation, the exporting country's request for market access includes detailed information describing:

- The commodity (specific identification, source, season, volume);
- Pests associated with the commodity in cultivation;
- Pests associated with the harvested commodity;
- Harvesting, handling, processing, packaging and shipping procedures;
- Phytosanitary measures (if any) applied and their efficacy.

It is advantageous for the exporting country to provide as much detail as possible to avoid worst case assumptions that are likely to be made in the pest risk analysis process, but also to promote a sound risk management strategy. In the absence of full disclosure, a risk management programme may be implemented that is either too rigorous or not rigorous enough, making the programme more susceptible to failures. Programme failures are not only undesirable because of the increased risk but failures also reflect poorly on the risk management competencies of both the importing and exporting countries.

Mitigation options cover a range of possibilities and different levels of efficacy. Some possibilities include:

- Inspection and certification;
- Treatment;
- Systems approaches;
- Pest free and low prevalence areas or places of production;
- Shipping windows (designated dates, entry points and distribution areas).

As indicated above, a number of field and industry procedures may also have mitigating effects. The extent to which these are considered to be either mitigation measures or characteristics of the regulated article depends on the decisions taken in the initiation stage of the pest risk analysis to define the article for which the pest risk analysis is being done. For example, if the lemon is defined as being washed, brushed and waxed, then the mitigation effects of these processes need not be considered in risk management but rather affect the risk assessment.

If the processes are not considered to define the article, then the risk assessment needs to address the risk as though none of these procedures occur and the pest mitigation effect of the procedures is evaluated in the risk management analysis. In any case, industry practices that are also determined to be mitigation measures should be identified as such in risk management conclusions.

14.3. Categories of Measures

After clarifying the role and effect of industry practices and identifying mitigations that may be used to reduce risk, the risk management analysis proceeds to evaluate the efficacy of measures. Before beginning this process, it is useful to distinguish the role of different measures in risk management. In the SPS context, *measures* is a broad term that includes a range of possibilities from legislation to operations. For the purposes of risk management analysis, we are primarily concerned with three categories of measures:

1. *Mitigations*: measures that have a direct effect on pests to reduce their prevalence or survivability (e.g. treatment).

2. *Safeguards*: measures that promote phytosanitary security by reducing the probability of pest escape or contamination, entry or establishment (e.g. pest-proof packages, shipping season asynchrony, prohibition).
3. *Procedures*: measures that are not mitigations or safeguards but support or enhance the effectiveness of risk management (e.g. inspection, certification).

Historically, the three most common measures used for risk management are *prohibition, inspection* and *treatment*. A closer look at these measures provides useful insight into the primary points we need to understand for risk management analysis.

14.3.1. Prohibition – a type of safeguard

There is a critical distinction made in the SPS Agreement between 'prohibited', which is a phytosanitary measure and 'not authorized', which invokes a process. Because prohibition is a specific measure, it must be based on an international standard or a pest risk analysis according to the SPS, but the Agreement also recognizes the need for administrative approval processes with provisions found in Annex C. This provides the basis for the 'not authorized' category, which is for those articles that need to be evaluated for their measures to be decided; in other words, for a pest risk analysis to be completed. Although the end result is the same (the article cannot be imported), the rationale and authority behind the condition is extremely important for trading partners to determine the actual status.

A common misunderstanding associated with prohibition is that it is a highly effective measure. Prohibition is generally assumed to close a pathway, but there are at least two situations where this is not the case. One situation is where natural spread is a viable pathway. Any regulatory strategy should be weighed against the likelihood of the hazard occurring naturally. For instance, what would be the value of prohibiting the movement of plant material from affected areas to adjacent areas if the pest of concern is a fungus that spreads by releasing spores into the wind?

The other situation when prohibition is not an effective measure is where there is a strong motivation for the pathway to exist. In such cases, prohibition can actually increase the risk. Consider for instance a prohibition on the importation of a popular fruit or vegetable or a valuable new plant variety. The high demand for these commodities in the face of prohibition will increase the probability for smuggling. A better strategy may be to accept a somewhat less comfortable level of risk but authorize the articles or activity with other measures to reduce the risk and the motivation for smuggling, and increase the ability to monitor imports.

The key point therefore when considering prohibition as a measure is the background risk that may exist whether or not prohibition is implemented and how much more effective prohibition will be in reducing the risk. It is important to realize that prohibition will not result in absolute phytosanitary security in any case, and in some cases, it may actually contribute to increasing the risk.

14.3.2. Inspection – a type of procedure

Inspection concepts

Inspection is by far the most widely applied measure and has historically served as a substantial element of nearly all risk management strategies. Hundreds of decisions are made by phytosanitary officials each day based on inspection, but the meaningfulness of this procedure as a risk management measure is frequently misunderstood.

We should begin by recalling that inspection is a risk management *procedure*. Inspection itself does nothing to change the pest status. It is the actions taken as a result of inspection that will ultimately determine how risk is changed. At an operational level, these decisions will usually be acceptance (no action), rejection or the application of other measures (e.g. treatment). In risk management analysis, we need to first

understand which of these actions is most appropriate and what level of detection should trigger action based on the characteristics of the pest(s) of concern. We aim to link inspection to the level of pest risk deemed to be acceptable and to the operational feasibility of using inspection to manage risk.

Inspection may be broadly interpreted to include a wide range of activities, processes and methods employed for various reasons. For instance, the verification of documentation is an activity commonly associated with a phytosanitary inspection. Likewise, the examination of a site or facility for compliance or suitability under phytosanitary requirements may fall within the broad interpretation of inspection. Inspection may also be used to gather information or to monitor or audit phytosanitary programmes. For the purposes of the discussions here, inspection is concerned mainly with its role as a risk management strategy.

The primary assumption behind the use of inspection is that the pests of concern are detectable. The organism or its signs/symptoms must be visually discernable and distinct enough to minimize the potential for confusion with non-pest organisms or other conditions.

Inspection *may be a good option* for risk management when pests:

- Are large, external, and easily recognized;
- Cause visible damage or have distinct signs/symptoms;
- Have limited mobility or lack mobility;
- Are highly unlikely to be associated with the article in question;
- Are generally eliminated by harvesting and packing procedures.

Inspection should *not* be considered the sole basis for risk management if:

- The pests of concern or their signs/symptoms are not detectable;
- The potentially infested article is difficult or unsafe to inspect;
- The pest of concern is high risk and establishes easily with a few individuals.

In order to understand inspection, and its uses and limitations, it is first useful to highlight some key terms, definitions and concepts:

Term	Concept
1. Prevalence	The level of pests occurring in the population
2. Acceptance sampling	Also known as 'discovery sampling' in which decisions are made to accept or reject articles based on sampling a finite population without replacement
3. Tolerance	The allowable level of prevalence in the *population* being sampled; the prevalence is related to the level of risk that is being accepted
4. Acceptance level	The allowable level of prevalence in the *sample* being sampled
5. Confidence	The relative degree of probability that the sample represents the population
6. Risk-based inspection	Inspecting to find a defined level of possible pest prevalence and a specific level of desired confidence
7. Percentage-based inspection	Inspecting a pre-defined percentage of a population (e.g. 2% sample rate)

Note here that the concept of tolerance applies to the entire population (i.e. the whole consignment), not only the sample. The level of presence in a sample is properly known as the *acceptance level*.

A key assumption associated with inspection is that a certain amount of risk and uncertainty must be accepted. Under normal circumstances, an inspection is not done on 100% of regulated articles, and inspection is not 100% efficient. Since inspection is usually based on a sample and always involves a degree of uncertainty and variability, there will always be some probability that pests will escape detection. Associated with this is a certain degree of confidence in the level of detection achieved using a prescribed level of inspection. The level of possible pest prevalence that is unlikely to be detected may be described as a detection level, threshold prevalence, allowable prevalence or, more often, a tolerance.

Therefore, accepting a tolerance and variability is inherent in the adoption of inspection as a phytosanitary procedure. For this reason, it is not appropriate to use inspection as the basis for phytosanitary decision making if the objective is pest freedom. Further, it must be recognized that inspection cannot be properly used for pest risk management without having an understanding of the level of tolerance and variability that is associated with the procedure.

Acceptance sampling (also called *discovery sampling*) is the discipline that is most critical to understanding the correct application of risk-based inspection. The application of this statistical concept in risk management analysis allows us to determine whether inspection is the most appropriate phytosanitary procedure to use for managing pest risk and the characteristics of a proper inspection design, recognizing the concepts of *tolerance* associated with the probability of detection and consideration of the limitations of *confidence* in acceptance sampling.

For example, finding that two boxes of fruit from a total of ten are free of pests does not provide absolute assurance that all ten boxes are free of pests. There is *some probability that pests occur in the remaining boxes and there is a degree of uncertainty* (both variability and error) associated with the two boxes that were inspected. The issues that must be considered are the levels of tolerance and confidence, which are considered acceptable, and the level of consistency (or the range of variability) in inspection.

Risk-based inspection

A *risk-based inspection* is one that has as its objective a defined level of unacceptable pest prevalence and a specific level of desired confidence. This is in contrast to an inspection that is based on non-transparent criteria (arbitrary or intuitive), or one that is designed only for operational convenience (e.g. percentage based sampling).

Historically, it has been common practice to specify that an inspection sample should be some fixed percentage of a lot; for instance, 2% of a consignment. This specification is based on the mistaken idea that the detection level is constant if the ratio of sample size to lot size is constant. However, the laws of probability argue differently. It is important to understand this mathematical relationship in order to identify the most statistically sound design for risk-based inspection. The mathematical relationship between sample size and allowable prevalence is such that managing for a fixed sample size (such as a 2% sample) results in a detectable prevalence level that fluctuates as the lot size changes. Managing for a fixed prevalence (a defined detection level) results in larger or smaller sample sizes depending on the lot size. This is a fundamental point to understand for risk-based inspection.

Discovery sampling is a type of acceptance sampling where samples are taken from a finite population without replacement and sampled lots are rejected if they are found to contain one or more 'defects' (the characteristic that is to be detected). In the case of phytosanitary inspection, acceptance pertains to regulated pests. A risk-based inspection design will aim to manage for a constant level of prevalence. This means that the *maximum allowable prevalence* would be a fixed value associated with a fixed confidence. The result is a sampling design where the sample size varies according to the lot size. For example, the inspection may be designed to have 95% confidence in the detection of a 10% contamination or infestation rate. In other words, an infestation or contamination rate of 10% or greater would be detected 95% of the time. Proper mathematical modelling of this relationship is based on several assumptions:

- Sampling is done without replacement;
- Sampling is random;[1]
- The population (lot) is finite.

An acceptance level implies some tolerance in the lot according to the statistical relationship between the lot size, sample size, allowable prevalence and confidence

> **Key Concept**
>
> Tolerance refers to the possible prevalence in the entire lot (all similar units of the commodity consignment). Acceptance refers only to the sample. The acceptance level for most phytosanitary inspections will be zero. This means that lots are rejected after a single regulated pest is found. Because a zero acceptance level in the sample does not correspond to zero prevalence in the consignment, some probability of infestation or contamination exists even if the sample is found to be pest free.

level. We typically have no control over the lot size, and the conventional value for confidence is 95%. This leaves only the sample size and allowable prevalence to be managed under most circumstances.

It is well known that sample size increases in a non-linear fashion as population increases. The result is that small lots must be sampled at proportionally higher rates than large lots. The hypergeometric distribution provides an appropriate model for calculating and demonstrating this relationship under the circumstances and assumptions described.

The development and adoption of risk-based inspection programmes enhances the ability of officials to establish priorities for their inspection resources and to design inspection programmes that are transparent for trading partners and the private sector. By establishing reference points (risk-based inspection objectives) and a means to measure the results, it becomes possible to identify, in an analytical context and transparent manner, the areas where inspection resources are most needed and the level of resources required. These determinations then correspond with the acceptable level of risk and the strength of measures to be applied.

14.3.3. Treatment – a type of mitigation

Phytosanitary treatments are mitigation measures that aim to reduce the prevalence or viability of pests by exposing them to conditions and agents that have a detrimental effect (IPPC, 2009). Whether the treatment is a physical or a chemical treatment, the primary aim is to ensure a specific effect on pests while minimizing harmful effects on the articles being treated. The best known and perhaps most widely used post-harvest phytosanitary treatment is fumigation with methyl bromide, but a wide range of other post-harvest treatment options exist, including but not limited to:

- Cold treatment;
- Forced hot air;
- Hot water dip;
- Controlled atmosphere;
- Irradiation.

The historical model for treatment is a single, high-mortality treatment prior to export or immediately upon entry. The possibilities for greater flexibility and creativity have increased substantially as countries continue to explore alternatives to methyl bromide and translate SPS principles into practice. Combination treatments, low-dose treatments, non-mortality treatments and treatments as part of systems approaches are becoming more common, dramatically increasing the options that can be considered for mitigation.

The process of identifying, evaluating and selecting options must consider both efficacy and feasibility. Efficacy is characterized by the *required response*, which is composed of two distinct elements:

1. A precise description of required response (mortality, sterility, etc.);
2. The statistical level of response (metric and methodology).

Measuring treatment efficacy and probit 9

It is not sufficient to only specify a response without also describing how it is measured. Consider 'probit 9 mortality' as an example where mortality describes the required response and probit 9 is used to identify the level of response and the type of statistical analysis that will be applied.

The choice of a required response is considered against the results of the risk assessment, taking into account the principle of minimal impact. Some treatments

(e.g. irradiation) offer the possibility for a range of responses beyond mortality, which, for arthropods, include:

- Complete sterility;
- Limited fertility of only one sex;
- Egg laying and/or hatching without further development;
- Altered behaviour;
- Sterility of the F_1 generation.

Mortality will be the required response for most treatments, but it is useful to explore an example where alternative levels of rigour are possible in order to demonstrate the relationship of international principles, technical requirements, and operational and regulatory realities.

An irradiation treatment for fruit flies that achieves a high level of mortality would probably damage the commodity and be unnecessary in any case since sterility is sufficient to provide quarantine security (prevent establishment). The principle of minimal impact would argue that sterility is the most appropriate objective if the commodity is able to tolerate the lower dose. The problem with this is that sterile flies escaping from imported fruit, although not a risk for establishment, may be found in traps and trigger regulatory responses. As a result, increasing the rigour of the treatment response from sterility to preventing adult emergence is necessary and technically justified.

The key point here is that the selection of a treatment and specification of a required response begins with a clear understanding of the risk that is being addressed, the biology of the organism, the tolerance of the commodity being treated and the operational realities of implementation, which brings us to the questions of feasibility.

Factors to consider regarding the feasibility of treatments include:

- Phytotoxicity;
- Cost;
- Availability of facilities/equipment;
- Use labelling (pesticides).

Probit analysis of treatment efficacy

Probit analysis is one of several regression-type statistical methods that can be used to calculate a dose–response relationship and is commonly used to derive the appropriate dose for a specific degree of response. In most quarantine treatments, probit analysis is used to predict the level of treatment (dose) to achieve a desired degree of mortality (response). It assumes that each organism has an innate tolerance for a certain stimulus to a specific threshold, and when the threshold is exceeded the organisms will respond with an expected and measureable result.

Because an organism's tolerance is typically difficult to measure, the mean and variance from the distribution of tolerances is estimated. The response to the stimulus is dependent on the treatment and may include mortality, sterility or cessation of maturation. A regression of the probit transformation of the proportion for dose–response will be linear if the tolerance of the organism is distributed normally. If the distribution is lognormal, the probit on log (dose) will appear linear.

In a 1939 circular released by the USDA, Baker suggested that probit 9 level treatments would meet the phytosanitary security requirements of the USA for fruit flies (Baker, 1939). Baker's recommendation of probit 9 level treatments was based on the assumption that a lognormal transformation would relate mortality and length of exposure in a linear fashion (Baker, 1939; Liquido et al., 1995). Following Baker's recommendation, probit 9 level treatments became the de facto standard for quarantine security for the USA and many other countries. In the meantime, however, many national and international research publications have questioned the legitimacy of a probit 9 standard, identifying flaws and proposing alternatives (e.g. Landolt et al., 1984; Baker et al., 1990; Vail et al., 1993; Liquido et al., 1995; Mangan et al., 1997). Despite this body of work, little has changed in the last seven decades regarding US requirements for quarantine treatment efficacy.

In 1986, Couey and Chew published formulae for estimating the number of pests that must be treated for a given number of survivors at a given confidence level and treatment efficacy. From this, a probit 9 study with 95% confidence level and no survivors would require that 93,613 insects

be tested. If control mortality is considered, this number would be adjusted to be higher (Follett and Nevin, 2006).

Criteria for requiring probit 9 efficacy

Probit 9 level mortality treatments are typically applied to cases that meet the following criteria:

- The host is suitable;
- The infestation rate is high;
- The host is easily infested;
- Distribution of the pest is highly clumped;
- The pest is internal or difficult to detect (Liquido et al., 1995, 1996).

In other words, probit 9 level treatments would seem to be synonymous with high-risk situations. This level of treatment in most instances is employed without the need for detailed or time-consuming data collection on the prevalence of the pest, the level of infestation or the likelihood of establishment. Probit 9-level treatments are an attractive option for pest risk management because they provide a conservative level of quarantine security, although they may not always provide the desired level of consistency in all situations (Liquido et al., 1995, 1996).

Probit 9 level treatments are convenient mainly due to the relative speed with which they can be developed, tested and implemented for tolerant hosts (Liquido et al., 1996). As a result, a number of precedent-setting treatments have been established that come to be viewed as risk management standards when they are actually based on bilateral agreement for a measure that is believed to result in substantial 'overkill' in order to facilitate trade. Such treatments are difficult to link to consistent policies for a threshold level of risk or arguments of equivalency, and generally violate the principle of 'least restrictive measure'. The legitimacy of such treatments comes from the fact that they are bilaterally agreed, not that they are technically justified.

From a risk standpoint, it is not mortality that is important but rather survivors. Linking the efficacy of the treatment to some level of risk, even if it is only a threshold, requires an estimate of the infestation level in order to predict survivorship. Beyond this, there are biological and other important variables to consider for estimating the likelihood of establishment if the treatment is to have a rationale relationship to the risk. From this standpoint, the technical criteria listed above should be considered in light of a pest risk analysis.

The use of probit 9-level treatments remains an effective tool in quarantine treatments, however, the application of probit 9-level treatments should be justified through treatment models, and not exercised in all quarantine situations. Customizing treatments to specific commodities and pest complexes will demonstrate that all quarantine treatments are justified and based on scientifically sound data. The establishment of such a process would justify and harmonize safeguarding measures with US trading partners and provide compliance with the SPS Agreement, while allowing the flexibility desired for specific pests and pathways.

Statistical considerations

The use of probit 9 level treatments as a *de facto* standard for phytosanitary security does not appear to be based on any specific scientific data. It has been argued that probit 9-level treatments actually provide an insufficient level of security in many instances (Landolt *et al.*, 1984; Chew, 1994; Robertson *et al.*, 1994; Liquido *et al.*, 1995, 1996). With fluctuations in distribution and population levels, probit 9 values may often be inaccurate. Based on normal probit 9 conditions, with a 95% confidence limit, the surviving individuals may range from 29 to 136 individuals per 1,000,000 organisms treated per shipment (Paull and Armstrong, 1994). The fluctuation of surviving organisms under normal conditions, confounded with other variables that may change infestation rate, will drastically alter the risks associated with introduction and establishment of surviving individuals (Landolt, 1984). In other words the efficacy of the probit 9 treatment will vary with infestation rate and distribution, leaving many commodities over- or under-treated. In addition, probit 9

treatments ignore several risk factors that may increase the probability of introduction, such as:

*   The actual infestation rate of the volume of treated fruit;
*   The potential for culling out infested fruit (changing the pest distribution pattern);
*   The infestation rate of good hosts vs. poor hosts;
*   The probability of a mating pair arriving at the same time at the same market (Paull and Armstrong, 1994).

In quarantine treatments, the estimation of a true survivor proportion of pests in a specific treatment is crucial to ensure quarantine security, but is often difficult to estimate. Concentrating on the upper distribution limit will aid in determining the true survival proportion (Chew, 1994). However, assessing treatment efficacy on the narrow criteria of 99.9968% mortality may be too limited to evaluate properly. The upper extreme of the probability distribution is difficult to estimate with any degree of certainty, making it difficult to accurately estimate probit 9 level mortality (Robertson *et al.*, 1994a; Liquido *et al.*, 1995, 1996). A study by Couey and Chew (1986) examining the relationship between confidence intervals and sample size in quarantine treatments showed that sample size drastically affects the level of security achieved. They demonstrated that approximate probit 9-level (or 1/35,000 or 29 per million) treatment with a sample size of 35,000 with one surviving individual with a confidence interval of 95% would result in 0.7 to 159 survivors per million. The inability to accurately predict survival following a probit 9 treatment in shipments of different sample sizes greatly lowers the level of security achieved with a probit 9 treatment.

Accurately estimating quarantine security increases in difficulty due to several factors beyond sample size. Under abnormal or fluctuating conditions the degree of security achieved may fluctuate greatly. Several factors have been identified that may compromise the level of phytosanitary security achieved with probit 9 treatments. These factors include, but are not limited to:

- Pre-shipment cultural practices;
- The size of the shipment;
- Survival and reproductive capacity of the organism;
- Packaging and shipping conditions;
- Seasonality of shipments;
- Distribution of the commodity (Chew, 1996).

The above factors may drastically alter the distribution pattern of the pests within the treated shipment. Probit analysis assumes normal distribution, but with changing ecological and physiological factors, the distribution patterns may in fact be non-random, skewed or clustered (Mangan et al., 1997). If the distribution is not normal, the data may be a poor fit for this statistical model (Robertson et al., 1994b). Thus, a probit analysis may not be the appropriate statistical test for all situations.

14.4. Systems Approaches

A systems approach is the combination of distinctly different pest mitigation measures and phytosanitary safeguards that cumulatively achieve the desired level of phytosanitary security. Although often complex and more difficult to implement than single-mitigation approaches, systems approaches hugely increase risk management possibilities and provide greater flexibility to adjust the strength of measures by combining measures with defined efficacy in a specific way based on risk management analysis.

'Systems approaches' are defined as 'the integration of different risk management measures, at least two of which act independently, and which cumulatively achieve the appropriate level of protection against regulated pests' (IPPC, 2009). Systems approaches may be applied in cases where a single measure, such as a phytosanitary treatment, is either not available or is not likely to achieve the appropriate level of phytosanitary protection, or in cases where the only other alternative is prohibition. The cumulative effect of combining independent measures[2] can provide the necessary level of phytosanitary protection where no other alternatives are available.

A great advantage of systems approaches is that risk mitigation can occur anywhere from the field to consumer – in the growing area, at the packing house, or during shipment and distribution of the commodity (see Table 14.1). Cultural practices, field treatments, post-harvest disinfestation, inspection and other procedures commonly used as single measures can be components of a systems approach. Processes such as pest survey, trapping and sampling can also be components of a systems approach. Safeguards such as maintaining the integrity of lots, designated harvest or shipping periods, restrictions on maturity, the use of resistant hosts and limited distribution at the destination can be key elements of a systems approach. Combination treatments, where two disinfestation treatments are used to achieve the required level of efficacy based on pest mortality, are not usually considered to fall within the scope of systems approaches, but the use of treatment combined with inspection may be considered a systems approach because this mix of procedures involves two distinct types of measures.

Systems approaches are not an entirely new concept. Both the practice and theory of systems approaches predate the use of the term itself. The authorization by the USA for the import of Unshu oranges from Japan is a well-known example of a systems approach dating back over 20 years.[3] The term is simply a descriptive label that has evolved to recognize the similarities in authorizations based on a combination of different phytosanitary measures. In fact, most existing systems approaches have never been so labelled, but were developed through a systematic process of evaluation and negotiation that resulted in an authorization which has many characteristics in common with other authorizations.

Numerous authorizations based on systems approaches now exist – many where

Table 14.1. Examples of measures that can be applied to manage pest risk in a systems approach.[a] (From IPPC, 2002; Jang and Moffit, 1994.)

Pre-harvest	Harvest	Post-harvest handling	Shipping	Distribution	End use
+++++	+++	*LEVEL OF RISK* ++		+	—
Pest free areas or areas of low pest prevalence	Harvesting at specific times or specific stages of ripeness	Post-harvest treatments (e.g. chemical, heat, waxing, washing, brushing, etc.)	Treatment in transit (e.g. cold treatment)	Restrictions on ports of entry	Restrictions on end-use
Resistant cultivars	Culling infested products	Testing	Speed and type of transport	Restrictions on time of year	Post entry processing
Healthy planting material	Field sanitation	Culling	Pre-shipment inspection	Post-entry quarantine	Packaging[b]
Pest mating or development disruption	Harvest technique	Packing house Inspection	Testing[b] Sanitation[b]	Post-entry Inspection	
Sanitation and cultural controls	In-field chemical treatments	Processing (degree and type)	Type of packaging[b]	Post-entry treatment	
Certification schemes	Field surveillance	Method of packing[b]		Packaging[b]	
Testing	Tarping[b]	Screening[b]			
Protected conditions[b]	Sanitation[b]	Sanitation[b]			

[a]This table is not inclusive of *all* potential measures; it lists the most common used on imported commodities.
[b]Indicates a safeguarding measure.

treatments were lost or found unfeasible. The desire to allow the movement of prohibited products or the need to find an approach that does not damage the product has created increased interest in systems approaches. Accelerating trade in fresh and semi-processed agricultural products and a growing recognition of the principle of equivalence, combined with the limitations in the use, availability and acceptance of chemical treatments, have also served to increase the attention devoted to systems approaches as an option for pest risk management.

In general, systems approaches are more difficult to evaluate and manage than point mitigation such as disinfestation treatments because they are usually more complex and require greater effort to develop and implement. In particular, systems approaches require a relatively high level of knowledge and confidence concerning the pest–host relationship and the ability to manage diverse elements of cultivation, harvest, packing and distribution systems. In many cases, the primary limiting factor to the use of a systems approach is the practicality of implementation under specific conditions.

Conditions that favour the success of a systems approach include:

- The pest and pest–host relationship is well known;
- Practical systems exist for pest detection in the field and in consignments;
- Growing, harvesting, packing, transportation and distribution practices are well known and standardized;
- Pests of concern are typically absent or rare;
- The volume/value of the commodity offsets increased programme costs;
- Pest mitigation and safeguard measures can be identified, monitored and corrected;

- Phytosanitary security is apparent through either qualitative or quantitative assessment.

Systems approaches are used by many countries, for a range of pests and commodities. Tropical Tephritid fruit flies (e.g. *Ceratitis capitata, Anastrepha* spp., *Bactrocera* spp., etc.) are a frequent target of systems approaches on commodities such as papaya, mango, avocado, citrus and other tropical and sub-tropical fruits (Podleckis, 2007) due to their importance as quarantine pests. A few examples of other pests or commodities managed through systems approaches include species of Lepidoptera (Follett and Neven, 2006; Grove *et al.*, 2010), and a range of 'hitchhiking' or contaminating pests (MAF, 2008) on different commodities.

The key advantage of a systems approach is that it allows for the consideration of a full range of measures – both existing and prescribed – that add to phytosanitary security. Any procedure or event that can be identified as a mitigation or safeguard which supports pest attrition has the potential to be a component in a systems approach (see Fig. 14.1). This opens many opportunities for growers, packers and others to work with phytosanitary officials to identify the points and controls needed to develop and support a feasible programme. Likewise, it provides substantial flexibility and extensive new possibilities for achieving equivalency where alternatives are sought.

The range of possibilities makes it difficult, indeed disadvantageous, to identify a single structure for systems approaches. However, there are some distinct similarities in existing and proposed applications of the concept that allow for general categories to be described. Three natural groupings are discussed below: mitigation systems, quality systems and control point systems.

14.4.1. Mitigation systems

These involve a combination of official phytosanitary procedures. Such systems do not consider factors other than the prescribed measures in achieving phytosanitary security.

Example:[4] papaya/fruit flies

This may require the cultivation of a variety known to be a poor host for the flies of concern, surveillance in and around the growing area to verify low population pressure and the harvest of only quarter-ripe fruit. The bulk of phytosanitary security is provided by the harvest of only quarter-ripe fruit. Additional measures provide greater security and reduce the likelihood of problems resulting from a failure of the primary measure. The host status of the fruit is determined through research. The variety of fruit

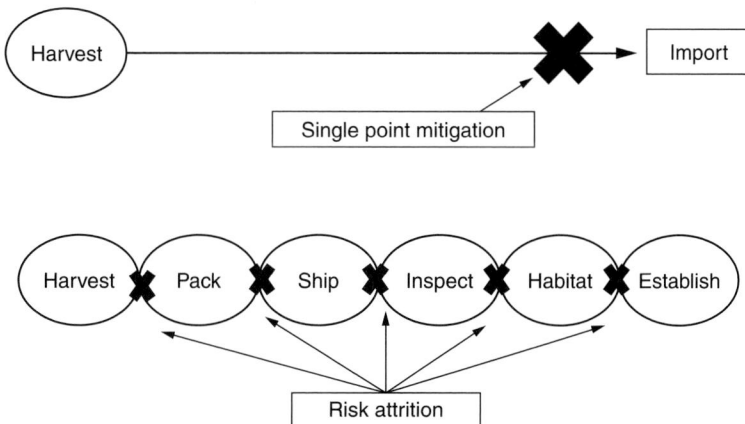

Fig. 14.1. Single point mitigation and risk attrition.

and ripeness at harvest are verifiable through inspection. Population pressure is monitored through routine survey. Factors such as cultivation practices or culling in the packing house may also contribute to phytosanitary security, but are not prescribed components. These measures are either assumed to be static or the degree of their variability is not considered to significantly affect phytosanitary security under the circumstances.

14.4.2. Quality systems

These involve a mix of phytosanitary procedures as well as other procedures. Quality systems typically include a range of processes that are designed to ensure the quality of commodities but also contribute to phytosanitary security.

*Example: cut flowers/leaf miners
and other pests*

Growers may subscribe to a programme that calls for a regime of specific cultivation practices, pest surveillance, and field or glasshouse treatments. Harvest and packing operations have strict culling and sanitation guidelines, including washing and cooling. Statistically designed inspection ensures a high degree of confidence prior to export. Grower accreditation and conformity as well as packing operations may be monitored by phytosanitary officials. Inspection is either performed or overseen by phytosanitary officials. All pests of concern are visually detectable.

In this example, inspection may be considered the only officially prescribed phytosanitary procedure. Other mitigation effects result from the processes that are inherent to maintaining suitable quality, but may be overseen or monitored by phytosanitary officials as necessary. The commodity demands a high level of quality to be marketable, so the commercially necessary processes that reduce the pest load can also be considered part of the phytosanitary programme without being officially

prescribed. The authorization is therefore based on a high degree of confidence that certain commercial processes will occur that have significant mitigation effects. It is the combination of commercial and official procedures that is deemed to achieve phytosanitary security.

14.4.3. Control point systems

Control point systems are the conceptual equivalent of Hazard Analysis Critical Control Point (HACCP) that is widely known and practised in food safety. This type of systems approach involves defined independent events or processes that are measured, monitored and controlled. Control point systems for phytosanitary purposes follow a similar conceptual model:

- Determine the hazards and the objectives for measures within a defined system;
- Identify independent processes or actions that can be monitored and controlled;
- Establish criteria for acceptance/failure to assure control;
- Monitor control points;
- Take corrective action when monitoring results indicate that criteria are not met;
- Review or test to validate system efficacy and confidence.

A control point is a practice, procedure, process or location at or by which a measure can be exercised that will contribute to phytosanitary security (see also Box 14.1). Acceptance criteria at any control point require the systematic observation of measurable characteristics and the ability to exercise action when criteria are not met.

Example:[5] bulbs/soil

In this case, three distinct, independent processes are identified for the certification of freedom from pests and contamination by soil. All these process occur within the closed system of a facility responsible for sorting, culling and packing.

Box 14.1. The role of control points in risk management and systems approaches.

For the purposes of this discussion, it is useful to make a distinction between 'control points' and 'critical control points'. *Control points* are points where controls *can* be applied in the production chain, and the controls are expected to have some effect on the system (i.e. a reduction in risk). We may or may not be able to measure the exact effect of the control, but we would expect that the control would have some (perhaps undefined) level of efficacy (Clay, 2004).

In the phytosanitary context, an example of a control point where efficacy may be difficult to measure would be field sanitation and the removal of fallen fruit from orchards to manage risks associated with fruit flies. We know that sanitation would reduce risk, but the exact level of efficacy would be difficult to measure. In addition, certain measures or conditions exist or are included to compensate for uncertainty. These may not be monitored, verified and corrected as independent procedures (e.g. packing house sorting), or may be monitored but not controlled (e.g. host preference/susceptibility).

In other cases, the control points may be very well defined, and the level of efficacy of that control can be measured, quantified, monitored and verified. In addition, such points in the system may be points where controls *must* be applied in order to reduce risk sufficiently. These would be regarded as *critical control points*. For some commodities, most or all of the specific points in the production chain can be well defined, the hazards and mitigations can be measured, each point can be controlled and finally the efficacy of each mitigation step can be verified, documented and corrected. In these cases, a 'critical control point' system can be applied – this being the most rigid type of systems approach used for phytosanitary risk management (IPPC, 2002).

The use of a control point system for phytosanitary purposes does not imply or prescribe that application of controls is necessary to all control points. These are addressed by risk management procedures whose contribution to the efficacy of the system can be measured and controlled. Even if a critical control point system is not used in risk management, the analysis of control points and critical control points may be useful to identify and analyse hazards, as well as the points in a pathway where risks can be reduced and monitored and adjustments made where necessary.

The first measure involves a cleaning machine whose output is routinely sampled to ensure 95% efficacy. The second measure is a combination of machine sorting and hand-culling to eliminate misshapen, miscoloured and otherwise unacceptable bulbs. The outputs from this process are also continuously sampled to verify 95% efficacy. The last measure is a mechanical packing process which includes the application of a light coating of preservative that has weak pesticidal properties in that it has been demonstrated that it neutralizes most microorganisms associated with small bits of soil. This process is also continuously monitored for efficacy. Lots enter the system and are not released until all sampling and monitoring results indicate normal efficacy. Sampling is statistically designed to ensure 95% confidence for each process.

Clearly, a control point approach is the most rigorous, requiring the greatest amount of knowledge and control. Typically, control point systems also involve the highest technology inputs. Most control point systems are characterized by high-volume, high-value commodities combined with concerns for high-risk pests. The application of a control point system requires a defined system such as a packing facility where measures and controls can be practically managed to achieve the required level of rigour and confidence.

14.4.4. Evaluating systems approaches

Phytosanitary treatments, such as methyl bromide fumigation at a rate sufficient to kill 99.9968% of treated individuals (i.e. probit 9 mortality), provide a familiar framework for determining phytosanitary security. However, when a systems approach that does not include a phytosanitary treatment is used as a phytosanitary measure, this framework is lost and alternative methods for evaluating the measures are needed. Systems approaches involve two or more diverse components, which can present problems when trying to express in similar terms the degree to which these various components mitigate the pest risk and the sum of their combined actions.

The efficacy of the individual mitigation measures can seldom be expressed in common terms, and seldom in the same terms as a treatment. For example, a disinfestation treatment provides a demonstrated level of mortality, whereas pesticide treatments in the growing area reduce the incidence of pests as evidenced by reduced trap captures or fruit cutting. Culling at the packing facility may remove almost 100% of the pests that are visually detectable but only some of the pests that are not detectable. Further, some components of a systems approach are not mitigation measures at all, but instead serve as a means to monitor and verify the prevalence of pests (e.g. trapping surveys, biometrically based inspections). In sum, it is clear that systems approaches require more complex methods for evaluating phytosanitary security.

In order to express the overall efficacy of a systems approach, a 'common currency' (i.e. term of expression) is needed. The chosen endpoint affects how phytosanitary security will be expressed. Examples of endpoints include:

- Prevalence of pests in a consignment or proportion of pests removed;
- Frequency of entry (i.e. number of pests entering per unit time);
- Probability of entry (e.g. probability of pest entry per unit of commodity imported);
- Frequency or probability of establishment;
- Frequency or probability of pest outbreaks.

Regardless of endpoint, the evaluation of measures should be done in the context of a pest risk analysis and should always be directed towards consistency with the endpoint. This is done by breaking the system into its individual components and evaluating the contribution of each component

to achieving phytosanitary security. Qualitative, quantitative or mixed analyses may be used to develop estimates and identify strengths, weaknesses and redundancy.

The key to the acceptance of systems approaches is an objective, analytical view towards pest risk management. As noted in earlier discussions, traditional assumptions and benchmarks for measuring efficacy are not always the most appropriate. Likewise, the principle of equivalence asks to define the level of phytosanitary security required and to accept those that are feasible and demonstrated to achieve this level. As this principle gains recognition in practice, the possibilities for non-traditional alternatives will increase. Foremost among these alternatives will be systems approaches.

Figure 14.2 provides a hypothetical example of a simple systems approach for a single pest. We begin by identifying events in the pathway (under the column 'Events' in Fig. 14.2), for which we can implement measures. Assume we have three measures A, B and C with corresponding levels of efficacy and failure (shown under the column 'Measures' in Fig. 14.2):

- *Measure A: Poor host status* – this measure is 75% effective (meaning that 25% of the time the measure will fail, or that out of 100% of the potential pest population, 25% will survive this measure and remain with the commodity).
- *Measure B: Field treatment* – this measure is 40% effective (meaning that 60% of the treated pest population will survive the measure).
- *Measure C: Cold treatment in transit* – this measure is 80% effective (meaning that 20% of the treated pest population will survive the measure).

Keep in mind that when we are looking at efficacy (e.g. how many pests are 'treated' by the measure), we should really be concerned with how many pests *survive* the treatment (i.e. what is the remaining pest population after we treat or what is the failure rate?). In addition, we are providing point estimates (single discrete values) for the level of efficacy or failure for each measure in this example. In reality, each value

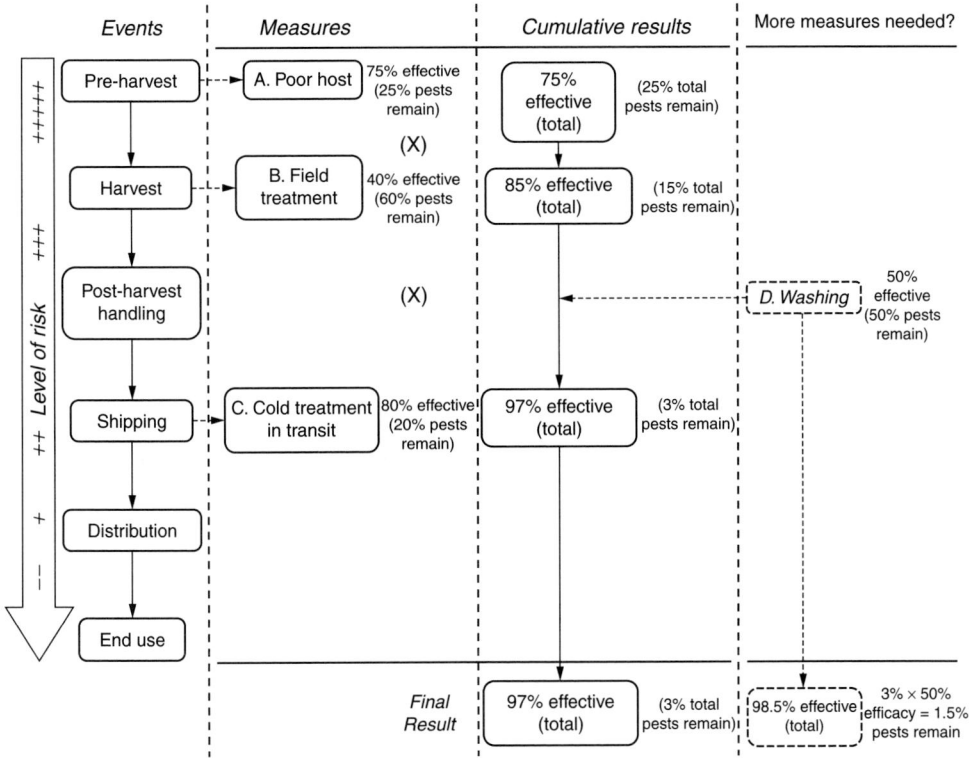

Fig. 14.2. A simple systems approach for a single pest based on specified measures.

would have a corresponding level of variability associated with it. If this were a real systems approach that we were developing, we would want to account for this variability in our calculations.

We can combine measures A, B and C since no one single measure will reach the appropriate level of protection; however, cumulatively, these measures can work together to provide a higher level of protection than any available individual measure.

Now that we have identified possible measures, we can combine them to estimate how effective our systems approach will be. Starting with measure A, we estimate that measure to be 75% effective (or that 25% of the pests will survive and remain with the commodity). If we treat that remaining 25% (remember 75% of our original pest population has been effectively treated so we do not need to treat that part of the pest population any longer) with measure B (field treatment),

we now have only 15% of the pest population that has survived the two treatments combined (and correspondingly, 85% of the pest population has been effectively treated):

> 25% remaining pest population × 40% field treatment = 10% of the surviving pest population treated

> We add 10% (the portion of the pest population from the original surviving 25% successfully treated with the second measure) to the 75% that was successfully treated with our first measure

> = 85% of the original pest population effectively treated

> (and 15% of original pest population surviving)

We can then apply a third measure, measure C, in our systems approach. From the combination of the two previous measures (A × B), we have 15% of the original pest population that has survived the first two combined measures. We apply measure C (cold treatment in transit), which is 80% effective:

> 15% remaining pest population × 80% cold treatment = 12% of the surviving pest population treated

> We add 12% (the portion of the original pest population treated successfully by the third measure) to the 85% we successfully treated with the first two measures

> = 97% of the original pest population effectively treated

> (and 3% of original pest population surviving)

These calculations and results are shown in Fig. 14.2, in the Cumulative Results column.

As a result of combining three measures, we have been 97% effective in treating our pest population (meaning that 3% of the original pest population may still remain with the commodity). At this point, this may be deemed to be sufficient, or we may decide that additional measures are needed to further reduce risk to an acceptable level. In this example, we have added measure D (washing fruit), shown in Fig. 14.2, under 'More measures needed?'. If we add measure D as a fourth measure, we are left with 98.5% of the pests effectively treated, and 1.5% of the pests remaining. If necessary, we can continue adding measures (E, F, G ... X) throughout the pathway (from preharvest through end use), until we reach our appropriate level of protection.

For the purposes of this example, we have addressed a single pest, using a relatively simple combination of measures. In reality, we are often dealing with a few or several different pests. The advantage of a systems approach is that measures applied to one pest often have an effect on other pests as well. This is particularly the case if

we are dealing with species guilds (e.g. aphids, scale insects or other pests that exhibit similar development and feeding habits). However, even if that is not the case, we can still adjust our systems approach (e.g. adding measures E, F, G ... X) to treat for additional pests as necessary.

14.4.5. Redundancy

Somewhere between no measures and the most restrictive measures is the elusive perfect balance of measures with risk. In any risk management process, however, there will be a degree of variability and uncertainty that will generate concern for the need to err on the side of caution by using conservative assumptions and worst-case data to determine the strength of measures. In most instances this will result in risk management strategies designed for the rare instance when measures are failing or operating at the lowest end of their efficacy at the same time pest challenges are maximized. Adding measures or extra strength to measures as a means to compensate for uncertainty is sometimes referred to as redundancy. Redundancy may be used:

- To compensate for uncertainty;
- As a safeguard for lack of experience;
- No less stringent measure is available.

Section 8, paragraph 2 of ISPM No. 14 states:

> A systems approach may include measures that are added or strengthened to compensate for uncertainty due to data gaps, variability, or lack of experience in the application of procedures.

Although most risk management strategies are designed with substantial redundancy built in to the required measures and do not consider other factors that may also reduce pest risk, systems approaches lend themselves most to a holistic approach. This is because in most cases the non-official elements of the system will require a qualitative judgement of their combined contribution to risk mitigation and this is likely to be consistent with the way the official elements are decided.

14.4.6. The nature of uncertainty in systems approaches

The provisions of ISPM No. 14 (IPPC, 2002) refer to uncertainty without explaining the concept or its intended scope. It is assumed (and was indeed intended) that uncertainty be broadly considered to include both *variability* (which cannot be improved) and *error* (which can be improved) as related to measuring the *efficacy* and *consistency* of phytosanitary measures. The implication is that each phytosanitary measure in a systems approach has some specific level of uncertainty associated with its efficacy (in each application) and also with consistency (across applications over time). This also means that the entire systems approach has some level of uncertainty based on the accumulated uncertainties for each prescribed measure.

14.4.7 The purist's view of redundancy in systems approaches

One view of a systems approach only considers those measures that are specified because they have a known pest risk mitigation effect. In this case, the system essentially ignores the effect of any other factor.

Assume that measures A, B and C form the basis for a particular systems approach (see Fig. 14.2). It may be that a certain procedure or condition 'Z' also has a risk mitigating effect, but this additional factor is not 'officially' part of the system. This may be because it is:

- Not desired;
- Not measured or not measurable;
- Beyond the control of system managers;
- A common or natural condition.

According to ISPM No. 14 (IPPC, 2002), the strength of measures A, B and C could be increased, or perhaps a new measure X is added to A, B and C in order to account for uncertainty. This represents the concept of redundancy as it is applied to the specified measures.

ISPM No. 14 gives no guidance on the criteria used to determine the degree to which the strength of measures can or should be increased. Presumably, it is at least justified to increase the strength of measures to meet the appropriate level of protection under conditions of minimum efficacy (i.e. each measure is operating at minimum efficacy and at the upper limit of its uncertainty). Difficulties will arise when the strength of measures is boosted for additional redundancy to cover undefined contingencies and unknown or unanticipated uncertainties for which there is no evidence they actually exist (i.e. possibilities, not probabilities). This additional redundancy will not have a clear technical or operational justification and no basis in the standard. In fact this occurs routinely with all measures, but it is not usually challenged unless the requirements are severely limiting trade.

A problem with systems approaches is the general perception that uncertainty is higher and less easily controlled and therefore greater redundancy is justified. This is partly true because the probability of error and failure generally increases with the complexity of the system. In practice however, phytosanitary systems are not very complex and systems approaches are often more precisely measured and monitored for efficacy so that the uncertainty may actually be significantly less than single-measure strategies.

14.4.8. The holistic view of redundancy in systems approaches

Another way to consider redundancy in a systems approach is to take a holistic view of the entire pathway, beginning with official independent measures that are the core of the systems approach and also including consideration of other conditions or procedures which have a pest mitigation effect.

Using the example described above, this would mean recognizing the effect of 'Z' even though it may not be independent, measured, controlled or prescribed.

Assume for instance that the export of a certain fresh fruit requires: (i) use of a specified non-preferred host; (ii) field surveillance and treatment to ensure a defined level of low prevalence; and (iii) cold storage under specific conditions. Each of these measures is independent and has a known level of efficacy and uncertainty. It is also known that fruit for export will be culled, graded, washed and waxed when packing. It is also known that this packing procedure significantly reduces the prevalence of pests of concern, but the procedure is not officially considered part of the systems approach. There may be other factors that also contribute to reducing pest prevalence or viability which are likewise outside the official systems approach but add to the 'comfort level' associated with the entire system.

In this case, the non-official elements of the system are loosely viewed as adding redundancy, which may partially or fully compensate for the uncertainty associated with the official measures. This possibility is recognized and thought to be legitimate by some experts but it was not included in the standard because there was not consensus on the point. Complications arise from the philosophical difference regarding whether non-official elements are legitimately considered or not. A systems approach designed with a holistic view could not be based on the standard and would therefore require a clear bilateral understanding.

ISPM No. 14 (IPPC, 2002) indirectly makes provision for one aspect of the concept of redundancy without using the term. It focuses on the possibility to increase the strength of officially prescribed measures to account for uncertainty, but the standard does not describe the concept of uncertainty or provide guidance on determining the level of increased strength that can be justified. Some experts consider that redundancy should be limited to incremental increases in officially prescribed

measures, while other experts believe there is value in a holistic view that recognizes non-official elements contributing to overall efficacy. The standard supports the former but not the latter. The less sophisticated types of systems approaches lend themselves more to the possibility of considering non-official risk mitigation factors, but in any case the inclusion of such measures would require bilateral agreement.

14.5. Summary

Traditionally, measures applied to imported commodities have relied on specific high mortality (e.g. probit 9) phytosanitary treatments aimed at reducing the presence of a particular pest (e.g. a species of fruit fly) on a given commodity. However, a wide array of measures may be applied to mitigate risk (either singly or in combination), in addition to traditional phytosanitary treatments. The measures listed below are examples of those that are most commonly applied to traded commodities. The available measures can be classified into broad categories. These include measures (IPPC, 2004):

- Applied to the commodity post-harvest (e.g. brushing, washing, waxing, treatments);
- Applied to prevent or reduce original infestation of the commodity by a pest(s) (e.g. integrated pest management programmes, bagging, safeguarding, sanitation, etc.);
- Applied to ensure the area or place of production of the consignment is free from the hazard (e.g. pest free areas or areas of low pest prevalence);
- Concerning the prohibition of commodities (in the absence of any other feasible options).

Systems approaches may be applied in cases where a single measure, such as a phytosanitary treatment, is either not available or is not likely to achieve the appropriate level of phytosanitary protection, or in cases where the only other alternative is prohibition. The cumulative effect of combining independent measures can provide the necessary level of phytosanitary protection where no other alternatives are available. In many cases, a systems approach can be highly flexible in the number and type of measures applied (as long as at least two measures act independently), even if specific data on efficacy are lacking. The number and types of measures combined in a systems approaches can range from very simple combinations (e.g. two independent measures such as low pest prevalence combined with fumigation) to highly complex, 'control point' systems (IPPC, 2002). Even if a critical control point system is not used in risk management, the analysis of control points and critical control points may be useful to identify and analyse hazards as well as the points in a pathway where risks can be reduced and monitored and adjustments made where necessary.

Notes

[1] True random sampling may not be practical for all phytosanitary inspections. However, maximum randomness should be an objective, recognizing that precision suffers as randomness is reduced.

[2] Note that two or more dependent measures can be combined for risk management (simply a combination of measures, but not a systems approach) *or* dependent measures can also be used in systems approaches (as long as at least two measures in the system act independently).

[3] Citrus from Japan was restricted by the USA due to concerns about citrus canker disease. The measures established for the export of fresh Unshu oranges from Japan to the USA included the use of resistant varieties, a designated growing area surrounded by a buffer zone, routine field survey, surface treatment, inspection and limited distribution at the destination.

[4] Examples for mitigation systems and quality systems are hypothetical but based on one or more actual authorizations.

[5] It is not known if an existing phytosanitary authorization is based specifically on a control point approach, however certain recent authorizations are known to have many of the characteristics of control point systems.

References

Baker, A.C. (1939) The basis for treatment of products where fruit flies are involved as a condition for entry into the United States. *US Department of Agriculture, Circular No. 551*, US Department of Agriculture, Washington, DC.

Baker, R.T., Cowley, J.M., Harte, D.S. and Frampton, E.R. (1990) Development of a maximum pest limit for fruit flies (Diptera: Tephritidae) in produce imported into New Zealand. *Journal of Economic Entomology* 83, 13–17.

Chew, V. (1994) Statistical methods for quarantine treatment data analysis. In: Sharp, J.L. and Hallman, G.J. (eds) *Quarantine Treatments for Pests of Food Plants*. Westview Press, Boulder, Colorado, pp. 33–46.

Clay, H.H. (2004) *Clay's Handbook of Environmental Health*.19th Edition. Bassett, W.H. (ed.) Spon Press, London.

Couey, H.M. and Chew, V. (1986) Confidence limits and sample size in quarantine research. *Journal of Economic Entomology* 79, 887–890.

Follett, P.A. and Neven, L.G. (2006) Current trends in quarantine entomology. *Annual Review of Entomology* 51, 359–385.

Grove, T., De Beer, M.S. and Joubert, P.H. (2010) Developing a systems approach for *Thaumatotibia leucotreta* (Lepidoptera: Tortricidae) on 'Hass' avocado in South Africa. *Journal of Economic Entomology* 103, 4, 1112–1128.

Heather, N. and Hallman, G.J. (2007) Systems approaches to pest risk management. In: Heather, N. and Hallman, G.J. *Pest Management and Phytosanitary Trade Barriers*. CAB International, Wallingford, pp. 47–70.

IPPC (2002) International Standards for Phytosanitary Measures, Publication No. 14: *The Use of Integrated Measures in a Systems Approach for Pest Risk Management*. Secretariat of the International Plant Protection Convention (IPPC), Food and Agriculture Organization of the United Nations, Rome.

IPPC (2004) International Standards for Phytosanitary Measures, Publication No. 11: *Pest Risk Analysis for Quarantine Pests including Analysis of Environmental Risks and Living Modified Organisms*. Secretariat of the International Plant Protection Convention (IPPC), Food and Agriculture Organization of the United Nations, Rome.

IPPC (2009) International Standards for Phytosanitary Measures, Publication No. 28: *Phytosanitary Treatments for Regulated Pests*. Secretariat of the International Plant Protection Convention (IPPC), Food and Agriculture Organization of the United Nations, Rome.

Jang, E.B. and Moffitt, H.R. (1994) The systems approach to achieving quarantine security. In: Sharp, J.L. and Hallman, G.J. (eds) *Quarantine Treatments for Pests of Food Plants*. Westview Press, Boulder, Colorado, pp. 225–237.

Landolt, P.J., Chambers, D.L. and Chew, V. (1984) Alternative to the use of probit-9 mortality as a criterion for quarantine treatments of fruit fly-infested fruit. *Journal of Economic Entomology* 77, 285–287.

Liquido, N.J., Griffin, R.L. and Vick, K.W. (1995) Quarantine security for commodities: current approaches and potential strategies. *USDA Publ. Ser.* 1996–2004, US Department of Agriculture, Washington, DC.

Liquido, N.J., Barr, P.G. and Chew, V. (1996) CQT_STATS: biological statistics for pest risk assessment in developing commodity quarantine treatment. USDA-ARS Publ. Ser. (available at barr@pbarc.ars.usda.gov).

MAF (2008) *Import Risk Analysis: Fresh Coconut* (Cocos nucifera) *from Tuvalu: Draft for Public Consultation*. MAF Biosecurity, Wellington.

Mangan, R.L., Frampton, E.R., Thomas, D.B. and Moreno, D.S. (1997) Application of the maximum pest limit concept to quarantine security standards for the Mexican fruit fly (Diptera: Tephritidae). *Journal of Economic Entomology* 90, 1433–1440.

Paull, R.E. and Armstrong, J.W. (1994) *Insect Pest and Fresh Horticultural Products: Treatments and Responses*. CAB International, Wallingford, pp. 197–200.

Podleckis, E.V. (2007) Systems approach as phytosanitary measures: techniques and case studies. In: Vreysen, M.J.B., Robinson, A.S and Hendrichs, J. (eds) *Area-wide Control of Insect Pests from Research to Field Implementation*. Springer, Dordrecht, pp. 417–423.

Robertson, J.L., Preisler, H.K., Frampton, E.R. and Armstrong, J.W. (1994) Statistical analyses to estimate efficacy of disinfestations treatment. In: Sharp, J.L. and Hallman, G.J. (eds) *Quarantine Treatments for Pests and Food Plants*. Westview Press, Boulder, Colorado.

Vail, P.V., Tebbetts, J.S., Mackey, B.E. and Curtis, C.E. (1993) Quarantine treatments: a biological approach to decision-making for selected hosts of codling moth (Lepidoptera: Tortricidae). *Journal of Economic Entomology* 86, 70–75.

15 Risk Communication in Pest Risk Analysis

Alison Neeley and Christina Devorshak

15.1. Introduction

Throughout this textbook, we have discussed several important themes regarding risk:

- There are inherent benefits and risks associated with all of the activities and technologies we choose.
- Risk management can mitigate these risks, but it cannot eliminate them.
- Every risk management strategy has its own set of risks and benefits.

Making wise choices (such as knowing which risks to focus on and which to ignore, determining an appropriate level protection for a given risk, ascertaining the optimal allocation of resources for managing a variety of risks, etc.) requires an accurate understanding of risks and benefits. Individuals or organizations that are too concerned about relatively low risks or not concerned enough about relatively big ones make poor choices.

Up to this point, we have discussed various methods to understand, identify, describe, estimate and manage the risk of quarantine pests. But as important as each of these activities are, none of them, in and of itself, will necessarily lead to better decision making unless there are effective mechanisms in place for the risk information to be transferred and shared at each stage of the risk analysis process.

In the broadest sense, *risk communication* is simply the process of explaining risk. It is a tool for creating understanding, and as such, is an essential and integral part of a functional pest risk analysis programme. The presence or absence of effective risk communication can have a significant impact on the eventual success of the regulatory process; without effective risk communication, effective risk assessment and risk management are not possible.

This chapter will examine some of the key aspects of the risk communication process for plant protection and how risk communication fits into a pest risk analysis programme.

15.2. Risk Communication and Pest Risk Analysis

Most experienced risk practitioners would agree that successful pest risk analysis requires at least some level of effective risk communication; however, there is surprisingly little consensus on what this actually means. Currently, there are no international standards that explicitly describe how countries should go about incorporating risk communication into their pest risk analysis process and the term *risk communication* has not yet been included in ISPM No. 5: *Glossary of Phytosanitary Terms* (IPPC, 2010).

The most specific guidance on risk communication provided by the IPPC is found in ISPM No. 2 (*Framework for Risk Analysis*). According to the standard, the main purpose of risk communication is to:

> reconcile the views of scientists, stakeholders, politicians in order to develop a common understanding of pest risks, develop credible pest risk management options, develop credible and consistent regulations and policies to deal with pest risks, and to promote awareness of the phytosanitary issues under consideration. (IPPC, 2007)

In the standard, risk communication is described as an 'interactive process' whereby information is exchanged between the NPPO and its stakeholders. According to this framework, unlike risk assessment and risk management, risk communication is not considered to be a distinct step in the pest risk analysis process, but rather something that should be conducted continually throughout the entire pest risk analysis process as information is gathered and documented.

Thus, in an ideal world, the risk analysis process would involve the constant exchange of information and opinion on risk among all stakeholders (including risk analysts, risk managers and policy makers, trading partners, industry and other special interest groups, and all other interested parties) throughout the entire risk analysis process (from initiation, to assessment, to risk management). In reality, NPPOs may face constraints including institutional and cultural barriers that make this level of collaboration very difficult (Lundgren and McMakin, 2009). However, although achieving effective risk communication may be very challenging, there are many benefits that make it worth the effort.

For example, some of the benefits of effective risk communication identified by the IPPC include:

- Ensuring that the outcomes, significance and limitation of pest risk analysis are clearly understood by all the stakeholders.
- Ensuring that all relevant information is included in the pest risk analysis, for example effective risk communication may enable the risk analysis to have access to unpublished data Available online atly from a particular stakeholder.
- Minimizing opposition. Including stakeholders in risk management decisions can help NPPOs minimize opposition, resolve disputes, anticipate concerns, engender trust and increase compliance. Over time, as risk communication channels are opened and trust between stakeholders is established, better decisions result.

15.3. Risk Perception

In order to understand the importance of risk communication, it helpful to know a little about how people process new information and about how they perceive risk.

It is tempting to view risk communication as a process whereby an audience is simply persuaded by experts that the risks associated with a given activity are small, or easily managed and that decision making should be left entirely to those who presumably know best (Gutteling and Wiegman, 1996). However, there can be considerable divergence between expert opinion and risk perception by non-technical laypeople and risk messages that do not account for public concern fail miserably (Gutteling and Wiegman, 1996; Sellnow, 2009).

For example, experts see the risks they study in terms of likelihood and consequences. Non-technical audiences, however, rarely consider the elements of probability and consequence when evaluating whether a particular risk is 'acceptable'. Instead, evaluation of risk is typically based on more emotional factors, as well as on the personal benefits associated with accepting the risk (Lum and Tinker, 1994). Table 15.1 lists some factors that generally lead people to be more or less accepting of a given risk (Lum and Tinker, 1994; Lofstedt, 2011).

Table 15.1. Some factors affecting risk 'acceptability'. (From Lum and Tinker, 1994; Lofstedt, 2011.)

Acceptable	Unacceptable
Voluntary	Imposed
Under individual control	Controlled by others
Beneficial	No clear benefit
Fairly distributed	Unfairly distributed
Natural	Manmade
Statistical	Catastrophic
Familiar	Exotic

People are biased to think of their choices and beliefs as correct, often despite any contrary evidence. When individuals come across new information, it is interpreted according to this bias. Lerner and Keltner refer to this as 'appraisal tendency' or a 'tendency to perceive new events and objects in ways that are consistent with the original cognitive-appraisal dimensions of the emotion' (Lerner and Keltner, 2000). Understanding this principle is key to understanding why different people can interpret the same information in many different ways, and why simply providing information is not enough to change perceptions and behaviours.

This is why most risk practitioners believe that the main purpose of risk communication should be to provide the audience concerned with the information it needs to make its own informed judgements about a particular risk. In a more formal sense, risk communication should be 'an interactive process of exchange of information and opinion among individuals, groups and institutions' (NRC, 1989).

15.4. Types of Risk Communication

As mentioned above, risk communication is an interactive process with information flowing between the concerned parties. Depending on the objective, risk communication can be placed into one of three basic categories (Lundgren and McMakin, 2009). 'Crisis communication' entails communicating about a risk in the midst of a crisis often for the purpose of motivating the audience to take some sort of appropriate action. 'Care communication' involves communicating about a risk for which the assessment is done, and which is well understood and accepted by the audience. It is particularly applicable where the audience may need to modify its behaviour to reduce risk. 'Consensus communication' is communicating about a risk to bring a number of parties to consensus on how to manage the risk taking into account who bears the risk and ensuring a fair distribution of benefits.

15.4.1. Crisis communication

In emergency situations, such as a new pest introduction or an outbreak, NPPOs may have to act quickly to contain or eradicate a pest. Under these circumstances, the NPPO often must rely upon the actions of others, such as homeowners and private businesses, to ensure the success of an eradication or containment programme. The NPPO is then faced with the challenge of communicating the relevant information to its audience, as well as motivating this audience to take some sort of appropriate action to reduce or mitigate risk. It is important to note that there should be a risk communication plan in place, in conjunction with the risk management plan, *before* a crisis occurs in order to ensure the best possible implementation. Namely, the NPPO should consider:

- Why the communication should take place;
- Who will be targeted for communication;
- What information needs to be communicated;
- How that information should be shared (e.g. mass media, public meetings, brochures, etc.);
- What actions need to take place, and who is responsible for carrying out these actions.

Crisis communication can take a variety of forms, depending on the severity of the situation, the urgency of required actions and who is affected by a particular hazard. Mass media, including press releases and web pages, can be used to reach the widest sector of the population very quickly. More targeted campaigns such as posters, brochures and pamphlets can be highly informative and provide the audience with material that they can literally carry away with them.

The disadvantage is that these materials can take more time to produce and for time-sensitive issues, this may not be practical. Public meetings or forums can also be a highly effective means of communicating about a crisis situation, and have the advantage of giving the audience the opportunity to provide direct feedback or ask questions to the communicators. The risk communicators can also use this mechanism to make a call for action, if the audience is directly involved with reducing or mitigating the risks in question. Regardless of how the information is communicated, the crisis communication process will involve:

- Ensuring delivery of relevant information to the appropriate audience;
- Conveying what actions may be required of whom;
- Providing a mechanism for a two-way exchange of information when necessary.

The actions taken after the Asian Longhorned Beetle (ALB) was introduced into the USA from China provide a useful example of crisis communication for plant protection. After the introduction of this pest, APHIS issued several press releases to alert the public to the situation. However, it was also necessary to ensure that some people, such as homeowners, modified their behaviours in certain ways to reduce the risk of ALB spreading to new areas (e.g. not moving infested wood from one area to another). Likewise, visitors to forests were alerted about the beetle through the distribution of educational materials and requested to contact APHIS if they detected signs of ALB. The relevant information that was communicated to the target audiences included the biology of ALB, its host plants, the type of damage it causes, how to recognize the beetle and its damage on host plants, means of spreading and what to do if the beetle or damage was detected (i.e. contact APHIS). Involving the public in the campaign to eradicate infestations of ALB has helped to ensure that the programme will be a success. In addition, sensitizing the public to the damage caused by this pest can help to minimize resistance and hostility when potentially objectionable measures need to be carried out by the agency, including the necessary removal of infested trees from neighbourhoods or the application of pesticides to control the beetle.

15.4.2. Care communication

Like crisis communication, care communication can be used to motivate the audience to modify its behaviour in order to reduce or mitigate risks. However, care communication differs from crisis communication in that it typically involves communicating about risks that are relatively well understood by the experts and laypeople alike. It is especially applicable to situations where the risk is a longstanding or regular issue, rather than a new or unexpected event.

A well-known example in plant protection would be the risks associated with travellers introducing pests if they are carrying certain types of materials from one place to another. The relative advantage of care communication is that the information that needs to be communicated to an audience is not necessarily urgent. This affords the NPPO the opportunity to employ a far-reaching strategy in its communication plan in order to maximize participation and success in a particular programme. As with crisis communication, NPPOs should consider several elements in a typical care communication plan, including:

- Who is the target audience(s)?
- Who should deliver the information (e.g. expert panels, agency spokesperson, etc.)?
- Why does the information need to be communicated?
- What information is relevant?
- How should the information be delivered?
- When should the information be delivered (e.g. is there a seasonal problem?)?

- What sort of actions or behaviour modifications are necessary, and who is responsible?

There are many methods for delivering information for care communication depending on the target audience, and what the desired outcome of the communication might be. Mass media, including the internet, news media, commercials and advertisements, are useful for communicating to large numbers of people. The message can vary in complexity and the amount of information delivered from relatively short, simple messages to in-depth analyses depending on which type of medium is used.

Posters and brochures in opportune places (e.g. airports) may also be useful for communicating simple messages. An excellent example of care communication can be seen with the Australian Agricultural Quarantine and Inspection Service educational fact sheets designed for schools (see Box 15.1: example of an AQIS educational fact sheet 'What is quarantine and why do we have it?'). The fact sheets contain information on the importance of quarantine, including activities for students to reinforce the message of the importance of quarantine.

Clearly, one benefit of these fact sheets is that they are informative on particular issues. Perhaps more importantly, another benefit of these fact sheets is that they create long-lasting awareness and sensitization not only about the particular topic of the fact sheet, but for quarantine issues as a whole.

Notably, these fact sheets are part of an education programme that also includes support for teachers so that education about quarantine issues can become an integral part of a teacher's programme. Over time, one can expect that educating people from an early age about quarantine issues would lead to a population that is sympathetic and supportive towards quarantine activities. It is also interesting to note that this form of communication reaches an audience that is not typically targeted in most risk communication plans – children.

15.4.3. Consensus communication

Situations that call for care or crisis communication will often have well defined hazards, the risks are well described, and the risk management options understood. In both cases, the objective of the communication is frequently to motivate the audience to modify its behaviour or to take some action to mitigate a risk. Consensus communication is the third type of risk communication that NPPOs may undertake, although it differs considerably from both crisis and care communication.

As mentioned above, consensus communication is communicating about a risk to bring a number of parties to consensus on how to manage the risk. The audience may or may not be responsible for taking action or modifying its behaviour. Consensus communication is most useful for addressing particularly contentious, controversial or divisive issues. It is also useful for communicating about risks that are considered exotic, involuntary, catastrophic, not well understood by experts and laypeople, where there is disagreement between experts on the risks of a given activity or where there are considerable levels of uncertainty that need to be addressed.

Because risk analysis, by its very nature, frequently deals with high levels of uncertainty, it is not uncommon to see significantly different interpretations of the same scientific information depending on the expert. When uncertainty is ignored or improperly addressed, especially when there is disagreement amongst experts, it can result in non-technical audiences becoming hostile, confused and losing trust in regulatory agencies. Under these circumstances, it is especially important to develop a risk communication plan where uncertainty is communicated to the audience. Likewise, some activities that are clearly beneficial to some groups can create risks for others groups who may or may not see any benefits arising from the activity. This is yet another important factor to consider in both the communication plan and in the risk management and decision making stages.

Box 15.1. What is quarantine and why do we have it? (Partially excerpted from http://www.daff.gov.au/aqis/about/public-awareness/education/fact-sheets/what-why, AQIS, 2011.)

What's quarantine?

Quarantine is any work designed to keep out living things that do not belong in a particular place. We can create a quarantine zone whenever we decide to draw a real or imaginary boundary around a place, and keep out anything we think should not be there. You could turn your school (or bedroom!) into a quarantine zone, if you decided for some reason to stop certain goods from coming in. In Australia, we have made a quarantine zone of our entire country. That means certain things are not allowed in, and others must be inspected or treated before they can come in.

Why we have quarantine

Australia has quarantine measures to protect the health and wellbeing of the plants, animals, people and other organisms that live here. Quarantine is very important to Australia.

Because it's an island, Australia has evolved in isolation. That's the reason our plants and animals are so unique. Quarantine protects our unique environment by keeping out weeds, animals, insects, diseases and other organisms that could destroy, disrupt or compete with native species.

Quarantine protects the animals and plants we use for food. We enjoy diversity and healthy food in Australia that helps us live well. Food exports to other parts of the world are important for our economy.

Australia has great biodiversity. Biodiversity means a variety of plants, animals and other organisms. Biodiversity comes from Australia's isolation, vast undisturbed wilderness areas, climatic range and other factors. Quarantine helps protect diversity by keeping outside organisms from disrupting native ecosystems.

Who's job? [*sic*]

In Australia, quarantine is the job of government agencies, including the Australian Quarantine and Inspection Service, and health authorities in the case of human diseases. But we all have a role to play – by knowing and following quarantine rules when we travel, and by showing respect for all the living things that make up our environment.

Methods for consensus communication must rely more heavily on effective feedback mechanisms between the communicators and the audience than do care or crisis communication. A common method of consensus communication is conducting a public hearing that addresses a given topic. In this case, any interested party could attend the public hearing and in principle, could provide comments on a given activity. A similar method is stakeholder participation, where representatives of various groups may be invited to participate in a forum to address a problem. A frequent downfall of both of these methods of communication is that in practice, they may be implemented well after the risk assessment process has started, and frequently even after all of the decisions have been made.

Under these circumstances, the purpose of communicating is merely to persuade the audience to agree with those creating the risk in the first place. The audience may then become hostile if it feels its concerns are not taken into account, lose faith in the communication process, and in some cases, the communication may have an effect of creating even more problems. One key, therefore, in consensus communication is to ensure stakeholder or audience participation as early as possible in the process, and to continue including the audience or stakeholders throughout.

As with other forms of risk communication, the NPPO should consider several factors before deciding to implement a consensus communication plan to address a particular problem. These factors include:

- Timing (short and long-term options);
- Who should be targeted (e.g. grower groups, public interest groups, etc.);

- Types of consultative processes (e.g. public hearing, open or closed forum, internet survey);
- Venues for public consultations;
- Who will deliver the information (expert panels, agency communication specialists, etc.);
- Information delivery (need for brochures, pamphlets or more detailed documents);
- Mechanisms for feedback from the audience;
- How to ensure audience response is incorporated into the process.

An example of consensus communication used by an NPPO is the 'Public Forum on Plant Molecular Farming' conducted by the Canadian Food Inspection Agency (CFIA) (see Box 15.2). The risks associated with genetically modified organisms (GMOs), in this case plants, are a highly emotive and contentious problem that many NPPOs are increasingly required to address. Because of the high level of media attention, these risks are frequently in the public eye and as such, there is a great deal of interest from many sectors, including private citizens, environmental groups, consumer advocacy organizations, growers and private businesses.

Each of these groups carries its own beliefs and value systems, and each of these groups may be variously affected (either beneficially or not) by the use of GMOs in agriculture. For instance, some growers may expect to use fewer pesticides than with conventional crop varieties, and therefore stand to benefit. Environmental groups have a range of concerns over GMOs, including gene escape to weedy relatives and negative effects on non-target organisms.

15.5. Best Practices

In his book *Risk Management in Post-Trust Societies* (2005), Ragnar E. Lofstedt states, 'There is no such thing as a formula for risk communication. The same strategy may have different outcomes depending on the audience, the country, and context in which it is used. A strategy for managing

risk in the USA, for example, may be wholly inappropriate in a European context.' There are many different models of risk communication.

In developing and implementing a sound risk communication plan, it is important to identify who the audience could be, and how best to reach that audience. Methods of information delivery will vary depending on who needs to be reached, and what works for one group may not work for others. For instance, it is tempting to rely on using the internet as a means to both deliver and receive information since the internet is in principle widely available to a large sector of the population. However, the use of the internet to deliver and receive information is also dependent on the audience actively taking an interest in and pursuing an issue of its own accord. In many cases, this will not sufficiently reach the target audiences, and the NPPO may need to more actively target certain groups and reach out to them via means other than the internet in order to make sure that the communication process is ultimately successful.

A good example of using appropriate methods would be targeting campers in national parks to prevent the spread of a pest like gypsy moth. In this case, the internet may not be a particularly useful means of transmitting information to the target audience; informative posters and brochures distributed in parks would have a better chance of reaching the target. As discussed in previous sections, other factors that affect the choice of methods of risk communication will largely be dependent on the objectives of the communication, time sensitivity and the type of information that needs to be conveyed.

The purpose of risk communication is to present necessary information so that the audience can make its own informed judgements about a particular risk. The educational background and level of interest amongst audience(s) will vary so it is important for NPPOs to adjust the delivery of information to the needs of the audience as appropriate. Technical and scientific information should not be simplified

---

Box 15.2. Consensus communication: public forum on plant molecular farming. (Excerpted from the CFIA website: http://www.collectionscanada.gc.ca/webarchives/20071123180225/http://www.inspection.gc.ca/english/plaveg/bio/mf/mf_fore.shtml, CFIA, 2001.).

Plant molecular farming uses the science of genetic engineering to produce substances for scientific, medical or industrial use. Some potential products of molecular farming are antigens for vaccines that can be mass produced in plants and used to respond to diseases such as cancer, diabetes, rabies, foot and mouth disease and the common cold.

October 30, 2001, Ottawa – the Canadian Food Inspection Agency (CFIA) held a Public Forum to solicit the public's views on plant molecular farming. The CFIA is responsible for regulating plants with novel traits and recognizes that plant molecular farming poses a number of regulatory challenges. The Agency is conferring with a wide range of people while products of molecular farming are still years away from the marketplace.

The public forum on plant molecular farming was held in the Auditorium of the Museum of Nature in Ottawa. The moderator, Alain Rabeau of Intersol Consulting Associates Ltd., opened the forum at 7 pm by introducing the four panelists. Bart Bilmer, Director of the Office of Biotechnology, was invited to give some opening remarks, after which the panelists gave presentations covering a range of issues related to plant molecular farming:

Dr. Louis Vézina, from Medicago Inc., gave an introduction to plant molecular farming, and an overview of some potential medical applications of the technology. Mr. Harry Richards, of the University of North Carolina at Greensboro, presented some of the potential environmental impacts of plant molecular farming applications.

Dr. Bill Leask, of the Canadian Seed Trade Association, briefly discussed some of the production challenges and considerations applicable to different applications of plant molecular farming.

Finally, Dr. David Castle, of the University of Guelph, discussed the ethical debate surrounding the introduction of new technologies, such as plant molecular farming.

Some 75 to 100 people attended the forum – following the presentations, audience members were invited to present their views and concerns, and to ask questions of the panelists. Among the issues and concerns raised with regards to the application of plant molecular farming were:

How to achieve an equitable distribution of the potential benefits of plant molecular farming.

How to address intellectual property and liability questions.

How to monitor and assess potential long-term health and environmental effects.

A summary of these discussions was presented the following day at a technical consultation meeting, organized by the CFIA's Plant Biosafety Office, and also held in Ottawa (see Multistakeholder Consultation on Plant Molecular Farming). Participants to the multi-stakeholder consultation represented academia, government, industry, provinces and non-governmental organizations interested in health and the environment. The consultation will assist in the development of guidelines specifically related to plant molecular farming.

---

or diluted to reach non-technical audiences; rather, the information should be delivered in a simplified manner to the extent necessary while retaining technical accuracy. Omitting details or leaving out difficult concepts can backfire as the audience may be left with the impression that it has all the facts it needs when it really doesn't. Even worse, the audience may be insulted and suspicious if it feels that experts are trying to patronize them, trick them into complying or cover up some horrible truth by leaving out relevant details.

15.6. Summary

Effective risk communication is an integral and essential part of any risk analysis programme. Whether the problem calls for care, crisis or consensus communication, there are no prescriptive answers; what works in some cases is inappropriate or ineffective in other cases. Risk communication is a broad field that seeks to integrate scientific knowledge with social values, and the approaches, objectives and outcomes of risk communication will be as varied as the risks themselves. Nonetheless, risk analysis experience in other disciplines can provide a useful paradigm for risk communication applied in pest risk analysis (see Box 15.3). In short, these rules state that risk communicators should plan their efforts, be honest, involve the public and work to develop a reputation as being trustworthy.

By including risk communication as part of the normal routine of pest risk analysis programmes, NPPOs can ensure that their risk management and decision making will be fair, more positively received by affected groups and more effectively implemented in the long run.

The identification of particular interest groups and their representatives should comprise a part of an overall risk communication strategy. This risk communication strategy should be discussed and agreed upon between risk assessors and managers early in the process to ensure two-way communication. This strategy should also cover who should present information to the public, and the manner in which it will be done.

Decisions on risk communication, including what, whom and how, should be part of an overall risk communication strategy. Risk communication is most effective if undertaken in a systematic way, and generally starts with the gathering of information on the risk issue of concern. Therefore

---

Box 15.3. Seven cardinal rules of risk communication. (Derived from a pamphlet drafted by Vincent T. Covello and Frederick H. Allen, US Environmental Protection Agency, Washington, DC, April 1988, OPA-87-020.)

1. Accept and involve the public as a legitimate partner.
2. Plan carefully and evaluate your efforts.
3. Listen to the public's specific concerns.
4. Be honest, frank and open.
5. Coordinate and collaborate with other credible sources.
6. Meet the needs of the media.
7. Speak clearly and with compassion.

---

the risk manager and risk assessor must be able to briefly and clearly summarize what this issue encompasses, at an early stage, in order to elicit interest and stakeholder input. Communication must then continue throughout the entire process. Once available information has been used to fully identify the hazards, and decide on and assess the appropriate risks, then the preparation and dissemination of this information is required. This will be followed by further discussion with stakeholders, leading to corrections, amendments and additions as appropriate, resulting in the final risk assessment and risk analysis reports.

References

AQIS (2011) What is quarantine and why do we have it? Australian Quarantine and Inspection Service Fact Sheet. Available online at: http://www.daff.gov.au/aqis/about/public-awareness/education/fact-sheets/what-why, accessed 20 December 2011.

CFIA (2001) Public forum on plant molecular farming. 31 October 31 2001. Ottawa, Canada. Available online at: http://www.collectionscanada.gc.ca/webarchives/20071123180225/http://www.inspection.gc.ca/english/plaveg/bio/mf/mf_fore.shtml, accessed 22 November 2011.

Covello, V.T. and Allen, F.H. (1988) *Seven Cardinal Rules of Risk Communication*. US Environmental Protection Agency, Washington, DC, OPA-87-020.

Gutteling, J.M. and Wiegman, O. (1996) *Exploring Risk Communication*. Kluwer Academic, Dordrecht.

IPPC (2007) International Standards for Phytosanitary Measures, Publication No. 2: *Framework for Pest Risk Analysis*. Secretariat of the International Plant Protection Convention (IPPC), Food and Agriculture Organization of the United Nations, Rome.

IPPC (2010) International Standards for Phytosanitary Measures, Publication No. 5: *Glossary of Phytosanitary Terms*. Secretariat of the International Plant Protection Convention (IPPC), Food and Agriculture Organization of the United Nations, Rome.

Lerner, J.S. and Keltner, D. (2000) Beyond valence: toward a model of emotion-specific influences on judgment and choice. *Cognition and Emotion* 14, 473–493.

Lofstedt, R.E. (2005) *Risk Management in Post Trust Societies*. Palgrave Macmillan, London.

Lofstedt, R.E. (2011) Risk perception and communication: an introduction. In: Harvard School of Public Health (ed.) *Effective Risk Communication: Theory, Tools, and Practical Skills for Communicating About Risk*. Harvard School of Public Health, Cambridge, Massachusetts.

Lum, M.R. and Tinker, T.L. (1994) *A Primer on Health Risk Communication Principles and Practices*. US Department of Health and Human Services, Washington, DC.

Lundgren, R.E. and McMakin, A.H. (2009) *Risk Communication: a Handbook for Communicating Environmental, Safety, and Health Risks*. Wiley, Hoboken, New Jersey.

Morgan, M.G., Fischoff, B., Bostrom, A. and Atman, C.J. (2002) *Risk Communication: A Mental Models Approach*. Cambridge University Press, Cambridge.

NRC (1989) *Improving Risk Communication*. National Acadamy Press, Washington, DC.

Sellnow, T.L. (2009) *Effective Risk Communication: a Message-centered Approach*. Springer, New York.

16 Uncertainty in Pest Risk Analysis

Robert Griffin

16.1. Introduction

The role of uncertainty in risk and its use in risk analyses are topics with a long history of discussion in the sciences and more recently in policy circles (Morgan and Henrion, 1990). The concept of risk is inherently tied to uncertainty. There is no risk without uncertainty. If we know the probability and the consequences of the hazard in question, it becomes a certainty and not a risk.

Although uncertainty is a critical component of risk, it gets far less attention in most risk analysis methodologies than the more tangible aspects of probability and consequences. This is especially true for risk analysis in the WTO-IPPC framework because of the strong emphasis placed on evidence as the basis for justifying measures.

16.2. The Nature of Uncertainty

Uncertainty in pest risk analysis is typically much more extensive than what is encountered with risk analysis in similar disciplines such as food safety and animal health. There are many reasons for this difference, but the two main factors contributing to uncertainty in the phytosanitary context are: (i) fewer data, and (ii) a larger universe of variability. Thousands of potential pests, hundreds of potential host commodities and an unlimited number of possible trade scenarios can sometimes make it seem like the phytosanitary world is a tiny island of evidence in a vast sea of uncertainty.

Despite the high levels of uncertainty, regulatory decisions are necessary and, for WTO members, such decisions are expected to occur in the SPS-IPPC framework with a certain amount of confidence (for stakeholders) and predictability (for trading partners). Building confidence and predictability into pest risk analysis begins by ensuring that the process consistently handles uncertainty – however large – in ways that are predictable, scientifically defensible and responsive to the needs of decision makers (National Research Council, 1994).

The broad concept of uncertainty has many different interpretations. For pest risk analysis purposes, it is convenient to group uncertainties into three categories based on distinctly different characteristics:

- Insufficient information (or lack of knowledge);
- Variability;
- Imprecision (including error).

Insufficient information is one large and common category of uncertainty in pest risk analysis (Yoe, 2012). It is also the easiest to understand, identify and express in the pest risk analysis process. Information gaps may be about the identity, biology, distribution, host range or other aspect of a pest, situation or regulated article. Gaps often exist regarding the economic significance of pests, hosts or measures and the efficacy of processes, procedures or treatments. The most important point to understand about insufficient information is that it may be reduced by further study and research

In contrast, *variability* is unpredictable natural variation in biological and physical processes (Yoe, 2012). Variability can occur because of changes over time, with location or in an organism or group. Biological quantities and processes are inherently variable; hence variability is a ubiquitous source of uncertainty in pest risk analysis. Variability cannot be reduced by gathering additional data, but it can usually be represented more accurately and communicated better with additional data. Mathematical techniques such as the Monte Carlo simulation (see Chapter 10) may be used to represent variability (Cullen and Frey, 1999), but such analyses represent a level of sophistication that is rarely practised in pest risk analysis.

To help understand the relationship of insufficient information and variability, consider the significance of a pest interception in passenger baggage, which represents the first instance of the pest associated with a particular origin. Such an interception may represent important new data about a potential risk that requires a regulatory response. It could also be completely incorrect considering that inference from baggage interceptions is generally fraught with high uncertainty. The result is information that is both insufficient and variable. More interceptions would increase our confidence that the pest is indeed present in a new area, especially if the interceptions came from cargo or another pathway with greater reliability, but it would do nothing to increase our confidence in the quality of information from baggage interceptions generally.

The third large area for uncertainty is the *imprecision* that comes from using incorrect or conflicting data, simplifications and approximations. Imprecision also results from incorrect methodologies, assumptions, concepts and terms. It also includes ambiguity or vagueness that results from misunderstandings or looseness in the interpretation of language (Regan *et al.*, 2002). A classic example of this latter condition is the undefined use of the terms *high, medium*, and *low* as relative indicators in pest risk analyses; a problem that is easily mitigated by creating precise definitions and specifying the context for their application in a pest risk analysis.

A very large and important aspect of imprecision involves *assumptions*. This is one of the most important and also one of the most overlooked problems in pest risk analysis because it is so easy for assumptions to be missed if they are not recorded, and they are often not recorded because they seem either obvious or unnecessary, or they are deliberately hidden. In any case, good analysts will train themselves to identify assumptions in their own work and strive for transparency in describing assumptions in their analyses, while ferreting out assumptions in the analyses of others. A complete accounting of assumptions and a clear understanding of the rationale for their inclusion is essential for true understanding of the uncertainty.

Insufficient information and variability are areas of uncertainty that a risk analyst should both identify and characterize. Imprecision, on the other hand, is usually a consequence of ignorance or sloppiness and is therefore unlikely to be apparent to the analyst unless deliberately intended. Providing guidelines for interpreting information consistently and having analyses reviewed by independent assessors provide the best means to identify this type of uncertainty (Regan *et al.*, 2002). The more experience an analyst has with expert review, the greater the likelihood that imprecision will not be a significant factor.

16.2.1. Data quality and uncertainty

Two other important concepts that mix with uncertainty are the overarching issues of data quality and probability. Judging the quality of information is one of the most common challenges for an analyst concerned with uncertainty. A piece of information may be considered insufficient, variable or imprecise, or any combination of these, but it nonetheless constitutes evidence and therefore requires a critical eye for its relative value in the pest risk analysis based on known policies or a political sense for the acceptable limits of data quality. The acceptability of data quality depends on the situation in which it is to be used and the comfort level demanded by decision makers. Because these factors are subject to change and may not be well understood by analysts, consistency and transparency in the application of data quality criteria are commonly lacking in pest risk analyses, making this one of the main areas where difficulties arise in the technical dialogue between trading partners.

The reliability of different sources of information is a key element of data quality judged by analysts. ISPM No. 8 (IPPC, 1998) provides guidance on judging the reliability of different sources of information for the determination of pest status in an area. As a general rule the hierarchy presented in the Reliability Table of ISPM No. 8 (see also Chapter 6 on information sources) is useful for relative comparisons for most any scientific or technical information used in pest risk analysis, but it should be noted that the examples are not exhaustive and their ranking is not exclusive. ISPM No. 8 also does not identify criteria for completely discarding any evidence. Analysts therefore have considerable latitude for applying their own criteria and establishing internal policies on the criteria to be applied and the circumstances for their application.

Whether done intuitively or explicitly, the analyst considers the probability that information is wrong or the probability that the results from using the information will be wrong. The critical question that arises is what impact the data will have if they are incorrect. In the absence of policies or guidelines on how to handle uncertainty, analysts typically adopt a conservative position, preferring to err on the side of precaution. Unfortunately, this practice is often neither consistent nor transparent, and may

be subject to wide variation across analysts and analyses that may be masked by adopted historical practice or standard procedures. The result is non-transparent and unspecified levels of conservatism, making it difficult for others to understand the process and interpret the results.

Let's say for instance that the only report of a particular pest–host relationship is more than 100 years old but the source is a highly respected author published in a well-known journal. Despite the age, it is possible that the information is perfectly valid. It is also possible that the information is outdated – perhaps incorrect. In the absence of a precedent or policy for consistency, the natural tendency of the analyst is to consider the implications of being wrong by giving the evidence a high or low degree of credibility. Another way to think of it is that the analyst is applying expert judgement to the uncertainty.

16.2.2. Expert judgement

The role of expert judgement in all types of risk analysis is a topic of significant discussion among experts. Expert elicitation techniques and quantitative models for the expert evaluation of information and uncertainty abound. Without specific guidance from the SPS Agreement and associated jurisprudence or the IPPC and associated international standards, any of these techniques may be used if they are able to withstand the scrutiny and possible challenge of other experts and trading partners. The advantage of this situation is a wide field of options that may be explored. The disadvantage is that controversy necessarily precedes a better understanding of appropriate methods.

Expert judgement takes two forms: the first is the informal judgement used by analysts in the process of completing a pest risk analysis; the second involves expert elicitation from others. In most cases, the informal type of expert judgement will be associated with evaluating the quality of evidence as described above. It typically involves one or more analysts who are internal to the organization and usually closely associated with the process.

When large uncertainties result from a combination of lack of information and a lack of conceptual understanding, it is not uncommon to elicit expert judgement from external subject matter experts to fill the gaps or establish default assumptions. The context and objective is normally explained to experts in advance, and background information is provided to ensure that all have the same basic understanding of the situation. Experts may be asked to provide information or opinions, or both. Highly structured expert elicitation exercises may also include an exercise to 'qualify' the experts or training and calibration exercises, which help establish confidence intervals for the estimates.

Formal expert elicitation processes are usually only used for highly significant pest risk analyses and should be facilitated by professionals to ensure the highest degree of credibility associated with the process and results. From an SPS-IPPC standpoint, the main points to keep in mind, as noted above and in SPS jurisprudence, are: (i) clearly distinguishing evidence from uncertainty, and (ii) ensuring that the analysis is focused on probabilities and not only possibilities.

16.3. Analysis and Presentation of Uncertainty in Pest Risk Analysis

Different kinds of uncertainties warrant different kinds of analysis (Regan *et al.*, 2002). As noted above, the type of uncertainty caused by imprecision comes largely from ignorance and is not analysed but rather discovered and should be acknowledged or corrected. Ambiguity and vagueness are to be avoided or clarified as much as possible with particular attention being given to correct terminology based on standards and conventional references, and defining other terms within the context they are used. Assumptions and extrapolations should be identified as such and the rationale for their application explained for transparency and to facilitate the type of review which would identify shortcomings.

16.3.1. Insufficient information

Insufficient information, where encountered, should be identified and its significance explained in terms of its relationship to the results. Information gaps that have no significant effect on the results should be noted, just as information that significantly impacts results must be highlighted. The same information could have different impacts in different situations. For instance, the uncertainty associated with the taxonomy of a particular bark beetle may not be significant if a number of other bark beetles are also being analysed, but this same uncertainty could be very significant if the species in question is the only bark beetle of concern.

Expert judgement may be used to fill information gaps, in which case care needs to be taken to ensure that the process and results are as consistent as possible with the disciplines of the SPS-IPPC framework. Recall also that the uncertainty resulting from insufficient information is useful for identifying and prioritizing research needs. Highlighting these research possibilities and their significance is easily done and can be a valuable addition to a pest risk analysis, especially when research that is feasible also has strong relevance to the results.

16.3.2. Variability

The other type of uncertainty that requires analysis is variability. Variability is typically characterized as a range or distribution. Where variability is relatively low, it may be ignored but should nevertheless be acknowledged. This is a rare situation in pest risk analysis. What is more likely is that variability will be significant and requires careful handling.

In some cases the analysis of variability can be simplified by disaggregating it. For instance, the variability around the effects of a quarantine treatment for a commodity may be separated for individual pests or groups of pests with similar characteristics or results. Disaggregating variability can make it easier to analyse and understand.

16.4. General Approaches to Uncertainty

A common technique for dealing with uncertainty is to rely upon average values. Although perceived to be more objective, this approach is often less comfortable than the alternative, which is to use the maximum or minimum values that represent the worst case estimate; again, erring on the side of precaution based on the analyst's expert judgement. In cases where maximum or minimum values are selected, care needs to be taken to first be transparent about the choice and rationale, and second to avoid combining a number of worst-case judgements, which compounds the conservatism in pest risk analyses making the results highly precautionary to an unknown extent.

This compounding phenomenon can occur with any aspect of the pest risk analysis as analytical elements within a pest risk analysis are linked through chains of reasoning in each element of the analysis and finally to the result. Thus an exaggeration in a particular aspect of either the analysis of the evidence or the uncertainty does not stand alone but rather affects the overall analysis (in the results). Applying the same criteria and reasoning to other elements of the process tends to compound the problem. It is tempting for the analyst to allow this to occur because it results in highly conservative (implicitly safer) results. The danger is that it is also likely to lead to more restrictive measures that are liable to be challenged. As noted earlier, a certain degree of conservatism based on uncertainty may be considered tolerable, but unreasonably exaggerating the risk or the uncertainty invites challenge on a point which SPS jurisprudence has consistently found unjustified (see Chapter 19 on disputes).

16.4.1. Probabilistic methodologies

One useful technique that makes compounding more transparent is the use of probabilistic methodologies. The probability curve representing the evidence and results conveys variability explicitly in its size and shape as well as the type of curve (e.g. normal versus uniform) (see also Chapter 10). Expressing this variability with the same consistency and precision in a qualitative assessment is much different but the results are often more appealing to non-technical readers. For transparency's sake, the best possible situation is to use both methodologies; probability curves to visually represent the variability in data, complemented by a narrative that interprets the analysis and results qualitatively.

Despite the advantage of high transparency, probabilistic methods are used sparingly. This is mainly because they generally require more data to be meaningful, and also because non-technical audiences sometimes have difficulty understanding the methodology and presentation. Alternative techniques use rankings like high, medium and low, or assign numbers, ranges and weights (see Chapter 9 for more information on ratings). These approaches are more subjective and can be seen as arbitrary, requiring careful justification and explanation (World Organization for Animal Health, 2010).

16.4.2. Rankings

Methods that rely on numbers to represent ranks, ranges or weights are sometimes referred to as 'semi-quantitative' and, because they use numbers, may facilitate combining estimates for a summary result. Unfortunately, they may give a misleading impression of objectivity and precision associated with a mathematical relationship that does not exist. An easy way to check this is to look closely at the units associated with numbers and follow the logic of their progression through the process. The units (years, individuals, outbreaks, etc.) quickly become meaningless if no real mathematical relationship exists between the data and the rankings.

16.4.3. Sensitivity analysis

Sensitivity analysis is another way to systematically examine uncertainty in pest risk analysis processes. By adjusting input parameters and noting the change in results, the analysis may be tested for its limits on various parameters. In practice, this technique has been used infrequently, but the simplicity of its application makes it a useful tool for testing and challenging assumptions.

For example, assume that the estimate for the probability of pest establishment is somewhere between high and very high. The judgement of 'very high' comes from an outside guess associated with the range of possibilities. The estimate of 'high' is closer to the medium in the range of possibilities and is therefore a stronger probability based on the information available to the analyst. Adjusting the input estimates up and down by single increments for various elements of the analysis gives the analyst the ability to test the significance of uncertainty in the results. If any of the adjustments result in a significant difference in the results, then the uncertainties associated with that aspect are significant. The analyst may look for additional information to reduce the uncertainty or take the conservative estimate until better information is available. On the other hand, if adjustments in the input parameter make no difference in the result, then the uncertainties may be considered insignificant.

Other mathematical techniques can be used to analyse uncertainty including in particular Bayesian methods (Kaplan, 1993) and the application of techniques associated with systems approaches. The former have gained some favour in food safety and animal health applications but have not seen wide use in the phytosanitary community. The latter however is becoming increasingly more attractive as systems approaches grow in importance. The main difference in this approach is that rather than associating uncertainty with the elements of probability for establishment in a pest risk analysis, uncertainties are associated with the change in pest prevalence across a series of events having measurable effects on pest attrition (see Chapter 14 for more on systems approaches and risk management).

16.5. Practical Options for Expressing Uncertainty in Pest Risk Analyses

There is no 'right' or 'wrong' way to describe uncertainty in a pest risk analysis – so long as uncertainty is incorporated. Just as there are methods for analysing risk that are more or less appropriate for different situations, there are methods for describing uncertainty that are more or less appropriate. At a minimum, uncertainty should be explained within the pest risk analysis – either incorporated throughout the pest risk analysis at relevant points; or summarized at the end. For instance, Fig. 16.1 (provided earlier in Chapter 9) shows that uncertainty is incorporated into the analysis of probability and consequences, and then summarized at the end of the assessment.

In many cases, we may simply provide a verbal description of the uncertainty, particularly if our analysis is a qualitative analysis. For example, in discussing a newly described pathogen's host range we may state that 'the known hosts of the pathogen are X, Y and Z, but there is uncertainty as to whether the pathogen also infects hosts A, B and C'.

Furthermore, for the sake of this example, we may know that hosts X, Y and Z are closely related to hosts A, B and C (e.g. all plants occurring in the same genus). As part of the analysis, and in the absence of other information, the analyst may make assumptions, based on their own judgement and experience, that the pathogen could also infect hosts A, B and C if it readily infects known hosts X, Y and Z. This assumption, however, should be explicitly stated, for example as 'because the pathogen is known to infect hosts X, Y and Z, we assume that it is likely the pathogen will also be able to infect hosts A, B and C since these plants are closely related'.

Yet another option for describing uncertainty is to rank the uncertainty right along with various risk elements. Table 16.1 provides an example of uncertainty ratings. Table 16.2 (seen earlier in Chapter 11) shows how those uncertainty ratings can be incorporated into a hypothetical analysis along with corresponding risk ratings.

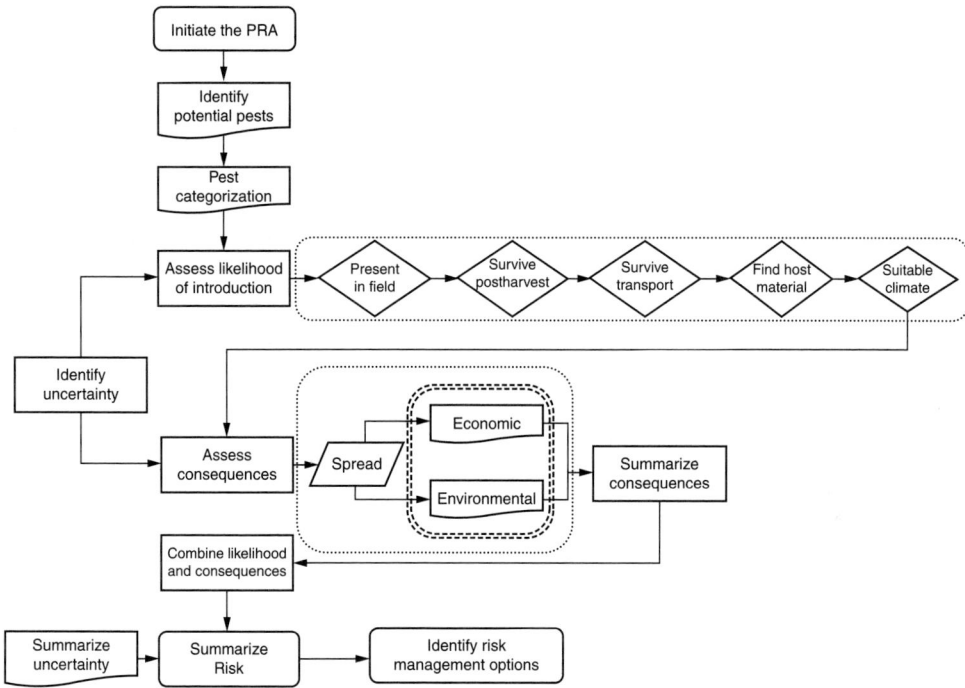

Fig. 16.1. Model for analysing risk of introduction of pests on a commodity, showing where in the analysis uncertainty is considered.

Table 16.1. Verbal descriptions of uncertainty using codes.

Degree of uncertainty	Uncertainty code	Description
Certain	C	Additional or better information is very unlikely to change the rating
Moderately certain	MC	Additional or better information probably will not change rating
Moderately uncertain	MU	Additional or better information may or may not change rating
Uncertain	U	Absence of reliable information

In this case, uncertainty is explicitly linked to each element, and we have communicated a level of uncertainty as well. For instance, in this hypothetical analysis, the risk element 'pest finds suitable host material post-entry' was given a 'moderately uncertain' – we may not know the full host range of the pest, or there may be conflicting reports about the host range. In any case, we can express how certain or uncertain we are, and then provide a further explanation in addition to the rating the uncertainty. The rating is guided by the descriptions provided in Table 16.1 – but the analyst must still make a judgement based on the quality of the information and their experience.

While this method allows us to be explicit about our uncertainty, we have not integrated the uncertainty into our risk rating, nor have we 'quantified' that uncertainty beyond indicating a general level. Other options for expressing uncertainty in pest risk analyses have been developed that use risk ratings (as discussed in Chapter 9), which effectively quantify the risk rating, and then assign a distribution to that risk rating based on the level of uncertainty. This approach was explored in the PRATIQUE project and has been incorporated into the

Table 16.2. Hypothetical example of ratings for a fruit fly on imported fruit showing uncertainty ratings.

Element	Risk rating[a]	Uncertainty[a]	Supporting evidence
Pest is present in field	H (3)	C	*In this section we would provide evidence (e.g. information from journals, databases, compendia or other sources) that supports a particular rating for each element. This may include data or information on a pest's prevalence in the field, host range, geographic range, fecundity, virulence, dispersal ability, or any other biological or technical information relevant to particular elements. We may also provide a verbal explanation of uncertainty here.*
Pest survives post-harvest handling	M (2)	C	
Pest survives transit	M (2)	MC	
Pest finds suitable host material post-entry	M (2)	MU	
Pest finds suitable climate post-entry	M (2)	MU	
Summary/combined ratings	M (11)	*Describe overall uncertainty*	*In this section we would summarize the likelihood of introduction and discuss overall uncertainty*

[a]These ratings are purely hypothetical and for example only.

online pest risk analysis scheme CAPRA, developed by EPPO (Holt *et al.*, 2010).

For instance, let's say we have two risk elements that we are combining: one is a 'low' with a high level of uncertainty and the other is a 'high' with a medium level of uncertainty. If we were assigning discrete values the low would be rated '2' (weighted 100% as that score), and the high would be rated '4' on a 5-point scale from very low to very high. We could combine these ratings in a risk matrix (refer to Chapter 9 for more on combining scores in matrices), shown in Fig. 16.2a to have a single overall risk rating.

Alternatively, we could incorporate uncertainty and assign a distribution to each rating. In this case, the rating for low has some probability of being low (the most likely possibility and therefore weighted more), but some lesser probability (weighted less) of being very low, medium, high or very high (based on a pre-determined distribution, in this case a Beta distribution adapted from the IPCC definitions of uncertainty). Combining the two ratings (high and low) with uncertainty incorporated would then yield the result as shown in

Figure 16.2b. Note that in this hypothetical example, the maximum of scores was used in the final result (implying a more conservative approach to risk).

In Fig. 16.2a, the combined risk score is communicated simply as 'high' as the result of combining a 'low' and a 'high' together and using the maximum of scores approach. In Fig. 16.2b, the combined risk score is communicated as 'high' as well, but there is now a distribution around that score that corresponds to the level of certainty we have for that result. This gives us a degree of transparency in communicating our uncertainty. However, analysts should understand the implications of ratings of uncertainty when using a model such as this, where quantitative methods are built into a qualitative approach (for a semi-quantitative analysis). Although the results appear transparent, the model itself may not be – thus, we still need to communicate information about uncertainty in the analysis itself rather than leaving it to the model and the results.

Another option for quantitatively addressing uncertainty in our analysis is probabilistic modelling, mentioned earlier

(a)

Percent weighted			0%	100%	0%	0%	0%	Sum	
		Rating	VL	L	M	H	VH		
0%		VL	0%	0%	0%	0%	0%	0%	VL
0%		L	0%	0%	0%	0%	0%	0%	L
0%		M	0%	0%	0%	0%	0%	0%	M
100%		H	0%	100%	0%	0%	0%	100%	H
0%		VH	0%	0%	0%	0%	0%	0%	VH

(b)

Percent weighted			21%	35%	28%	14%	2%	Sum	
		Rating	VL	L	M	H	VH		
0%		VL	0%	0%	0%	0%	0%	0%	VL
5%		L	1%	2%	1%	1%	0%	3%	L
28%		M	6%	10%	8%	4%	1%	25%	M
50%		H	11%	18%	14%	7%	1%	54%	H
17%		VH	4%	6%	5%	2%	0%	19%	VH

Fig. 16.2. (a) Maximum matrix (no uncertainty shown) Adapted from Holt et al., 2010. (b) Maximum matrix showing uncertainty integrated into scores. (Adapted from Holt et al., 2010.)

in Section 16.4.1. If we are doing a quantitative analysis, we may use discrete values (e.g. deterministic modelling), or we may use ranges (e.g. stochastic modelling). Ranges more often reflect biological situations – where we have variability in populations, and uncertainty arising from insufficient information. Using methods such as Monte Carlo simulations (discussed in Chapter 10) can allow us to incorporate uncertainty from both insufficient information – where we use estimates and expert judgement – and from variability directly into the analysis. The results are given often in the form of a probability distribution, and the distribution itself communicates the level of certainty in the overall estimate of risk.

Regardless of the technique used to identify and analyse uncertainty, the principle of transparency requires a full accounting of the nature and significance of the uncertainty. Even where uncertainty is quantified (for instance through Monte Carlo methods), analysts should still clearly document major sources of uncertainty. In a relatively complex analysis this is best accomplished by noting uncertainty where it occurs in the process and then summarizing the uncertainty and impacts in a narrative or graphic way (or both) as a conclusion. In relatively simple analyses, a summary alone may be sufficient.

Two key areas of the uncertainty analysis that are frequently overlooked are the summary of research possibilities and priorities, and the accounting of assumptions. Both may be critical components for next steps after a pest risk analysis is completed.

16.6. The WTO-SPS Agreement and Uncertainty

To fully grasp that dealing with uncertainty is not simply an academic exercise, we must also understand its place in the international framework, which defines the application for risk analysis as a fundamental aspect of phytosanitary policy making in the SPS-IPPC framework.

The SPS Agreement (WTO, 1994) is not explicit about uncertainty or the analysis of uncertainty except in Article 5.7 where insufficient scientific evidence is identified as the justification for provisional measures (see also Chapter 4 and Chapter 19 for more on provisional measures). This reference to a direct relationship between a specific type of uncertainty and a specific type of measure is the only explicit treatment of uncertainty in the SPS Agreement. The criteria for arriving at a judgement on insufficient evidence are lacking because the Agreement assumes that such details are better addressed in international standards, guidelines and recommendations from the designated standard setting organizations (the IPPC in the case of plant life and health). Likewise, an explanation of the relationship of these concepts to risk analysis is absent. This leaves open the question as to whether the judgement on insufficient evidence is determined *before* performing a risk analysis or *as a result of* performing a risk analysis. A purist would argue the latter, but based on the types of issues that have been disputed under the SPS Agreement, it is clear that not all risk analysis practitioners in the SPS world agree on this point.

SPS jurisprudence has helped to clarify the situation somewhat by explaining that while provisional measures may have a precautionary element, they are primarily designed to facilitate trade and do not have precaution as their main objective, i.e. provisional measures are not precautionary measures. This tells us that provisional measures are like other measures – requiring a technical justification (risk analysis), and while they may be precautionary in nature, provisional measures are ultimately designed to facilitate trade, not unduly restrict it.

Despite this guidance, Article 5.7 (WTO, 1997) continues to be seen as an area for interpretation often linked to the concept of precaution. In extreme instances, a precautionary approach is interpreted to mean that no risk analysis (and by implication, no authorization for trade) should occur where the judgement is made beforehand that scientific evidence is insufficient. From a risk analysis standpoint, a central objective of the process should be to identify the type and magnitude of all uncertainty, including information gaps, and the degree to which specific information can affect the results. In other words, the analysis of uncertainty is inherent in the risk analysis process. A risk analysis therefore provides a much more transparent argument to justify the conclusion that scientific evidence is insufficient for a final recommendation on measures.

On the other hand, one can rightly argue that all scientific information has more or less uncertainty and the dynamic nature of the biological risks we manage makes it necessary to re-visit our analyses whenever new information has the potential to impact our results. From this standpoint, all SPS measures may be considered more or less provisional, which would further complicate the interpretation of Article 5.7 (WTO, 1994) if we did not consider other disciplines in the Agreement. The distinction that exists between established measures (e.g. prohibition) and approval procedures (not authorized) is particularly critical to understanding this point (see also Section 14.3.1. of Chapter 14 for more on this point).

As discussed above, provisional measures represent a specific type of measure and therefore require a technical justification. By invoking provisional measures, WTO Members create the obligation for themselves to seek the additional information necessary 'for a more objective assessment of risk and review the sanitary or phytosanitary measure accordingly within a reasonable period of time' (SPS Article 5.7; WTO, 1994). This assumes that only additional information is needed for the type of uncertainty associated with provisional measures.

Variability (the type of uncertainty that cannot be addressed with additional information) would not be either an appropriate justification for provisional measures or a useful strategy for an importing country because it would not be possible to satisfy the requirements of Article 5.7 for a reassessment (WTO, 1994). Establishing provisional measures therefore requires a thoughtful commitment by the importing country, recognizing that this type of measure is more one-sided regarding the burden of proof than other types of SPS measures.

In some cases, WTO Members take the argument of insufficient information a step further by deciding in advance that there is insufficient scientific evidence to even perform a risk assessment. Under these circumstances, SPS discipline comes from the obligations in Annex C (Control, Inspection and Approval Procedures) regarding fees, timely handling and non-discrimination. Such an approach is necessarily based on a 'white list' regulatory structure (regulations listing what is authorized rather than what is prohibited) for the label of 'not authorized' to be legitimately applied.

Aside from being inconsistent with the spirit of the SPS Agreement, the approach of 'not authorized because there is insufficient scientific evidence for a risk assessment' has several disadvantages. Most importantly, it implies that there is a threshold for information that will fill the gap. Unless the type and amount of information is specified, the 'not authorized' approach has the potential to create a technical barrier to trade if the trading partner is unable to determine how much and what type of information is sufficient. If the needed information is available, then the importing country has unfairly shifted the burden to the exporting country, which is in the awkward position of finding information to justify the measures that will be placed against it.

From a risk analysis standpoint, the idea of having basic information to initiate the pest risk analysis is a practical matter. To this end, it is advantageous for the exporting country to provide as much information as possible for this purpose. For the importing country to also extend this by establishing an arbitrary threshold for the scientific evidence needed for the analysis is inconsistent with the purpose of the risk analysis process, i.e. to characterize the risk based on the available evidence, which if done via a risk analysis, is complemented by an objective accounting of uncertainty. This point demonstrates why risk analysis practitioners need to understand not only the conceptual aspects of risk analysis generally but also how these concepts fit in the unique framework created by the WTO-IPPC.

16.7. The IPPC and Uncertainty

The IPPC (the 1997 Convention text; FAO, 1997) makes no explicit reference to uncertainty but includes numerous references to the provision of information. Those that are most relevant to risk analysis include:

- Article II.1 (Definition for pest risk analysis)...justified on the basis of

conclusions reached by using an appropriate pest risk analysis or, where applicable, another comparable examination and evaluation of *available scientific information* (emphasis added).
- Article III.2b...reporting the occurrence, outbreak, and spread of pests...
- Article III.3a...distribution of information within the territory...regarding regulated pests and the means of their prevention and control...
- Article VI.2b...publish and transmit phytosanitary requirements, restrictions, and prohibitions...
- Article VI.2f...inform the exporting contracting party...of significant instances of non-compliance with phytosanitary certification...on request, report the result of its investigation...
- Article VI.2i...establish and update lists of regulated pests...and make such lists available to the Secretary, to regional plant protection organizations of which they are members and, on request, to other contracting parties.
- Article VI.2j...develop and maintain adequate information on pest status in order to support categorization of pests...information shall be made available to contracting parties, on request.
- Article VIII.1a...exchange of information on plant pests, particularly the reporting of the occurrence, outbreak, or spread of pests that may be of immediate or potential danger...
- Article VIII.1c...providing technical and biological information necessary for pest risk analysis (FAO, 1997).

The overall aim of these provisions is twofold: first to establish the baseline expectations for information needed by national plant protection organizations to be effective; and second, to encourage the exchange of information necessary for trading partners to engage in a transparent and meaningful dialogue on trade questions – in other words to reduce the uncertainty associated with negotiating market access, expansion and retention issues.

It is useful to note that the IPPC definition of pest risk analysis refers to *available* scientific information in contrast to the SPS Agreement, which refers to scientific principles and evidence. The difference is subtle, but a key distinction intended by the IPPC is to acknowledge that the availability of information and the capability of Members to obtain information will vary, and also that information will be dynamic. The IPPC therefore recognizes that uncertainty due to information gaps is a normal and expected condition that is always subject to change. The IPPC also does not assume that all countries have the same access to information, so uncertainty on any given point is situational. This latter point emphasizes the need for technical dialogue between trading partners.

The most substantial guidance provided by the IPPC on uncertainty is found in ISPM No. 2 (IPPC, 2007) and ISPM No. 11 (IPPC, 2004). Section 3.1 of ISPM No. 2 contains the full discussion of uncertainty as follows:

> Uncertainty is a component of risk and therefore important to recognize and document when performing PRAs. Sources of uncertainty with a particular PRA may include: missing, incomplete, inconsistent or conflicting data; natural variability of biological systems; subjectiveness of analysis; and sampling randomness. Symptoms of uncertain causes and origin and asymptomatic carriers of pests may pose particular challenges.
>
> The nature and degree of uncertainty in the analysis should be documented and communicated, and the use of expert judgement indicated. If adding or strengthening of phytosanitary measures are recommended to compensate for uncertainty, this should be recorded. Documentation of uncertainty contributes to transparency and may also be used for identifying research needs or priorities.
>
> As uncertainty is an inherent part of PRA, it is appropriate to monitor the phytosanitary situation resulting from the regulation based on any particular PRA and to re-evaluate previous decisions.

Three key points emerge from this guidance beyond what has already been discussed in the context of the SPS Agreement. The first is that uncertainty arises from more than just information gaps. Natural variability, subjective analyses and sampling randomness are also identified as sources of uncertainty. The role of variability and the other listed factors was not evident from the SPS Agreement.

The second point to note regards transparency associated with expert judgement and strengthening measures to compensate for uncertainty. These statements imply a relationship between expert judgement and uncertainty. This is indeed the case, and a key point of discussion in most academic treatments of uncertainty, especially as it pertains to policy decision making (Morgan and Henrion, 1990).

ISPM No. 11 (IPPC, 2004) also refers to identifying research needs or priorities that may arise in part due to uncertainty. This point ties back to the discussion above about the importance of performing risk analysis in order to properly characterize uncertainty. A very useful by-product of risk analysis is the ability to identify research needs as a result of properly characterizing uncertainty. Not only are research needs identified, but the relative significance of different pieces of information becomes clear. The analyst may find, for instance, that relatively simple, quick and inexpensive research provides critical information that vastly improves the conclusions or credibility of the analysis. Likewise, the uncertainty may be such that the missing information will be difficult or cost prohibitive to obtain and the likelihood of improving the results is marginal. Such insights are extremely useful for more closely linking the research and regulatory communities to policy making.

ISPM No. 11 provides additional guidance on the treatment of uncertainty in pest risk analysis, but nothing specific regarding the analysis of uncertainty. Section 2.1.2 establishes a direct relationship between the pest categorization process and the question of sufficient information. Presumably, at this stage the analyst will have a good understanding of information gaps and should ensure that this is adequately documented, but also continue with the pest risk analysis process.

Section 2.3.2.3 explains that the choice of analytical techniques for risk assessment may depend on the nature of available data and uncertainties. Section 2.4 speaks to the need for documentation regarding extrapolations and hypothetical situations where expert judgement is involved. The same section also refers to the relationship of uncertainty to identifying and prioritizing research possibilities. These points reinforce similar guidance provided in ISPM No. 2.

Special attention is given to uncertainty associated with information gaps and additional complexity that is likely for unmanaged and uncultivated plants. This appears more as a warning regarding what to expect rather than guidance on how to handle it.

Finally, in Section 3.6.1, the standard refers to the principle of modification and relates this to the availability of new information and the need to update analyses as appropriate. Although ISPM No. 11 doesn't associate this with the earlier point regarding the identification and prioritization of research needs, there is clearly a practical advantage from understanding in advance what information makes a significant difference and would have sufficient impact to justify a revision of the analysis.

16.8. Jurisprudence and Uncertainty

The patchwork of guidance available to analysts in the WTO-IPPC Agreements and standards is useful for emphasizing the importance of uncertainty in risk analysis, in particular as regards meeting the demands of transparency – something natural for science but sometimes difficult in the policy world. Precious little guidance is provided however on unique aspects of handling uncertainty in this framework, and possible methodologies for the analysis of uncertainty are not discussed.

Fortunately, ample information on methods exists in other sources, but unfortunately

the lack of discipline we find in the regulatory framework also leaves open the potential for extreme creativity and abuse. Without an agreed reference point for understanding reasonable limits, the only options for settling differences based on the treatment of uncertainty are bilateral negotiations or formal disputes. It is therefore not surprising that a central aspect of many SPS disputes has been related to questions of uncertainty – not so much from the uncertainty itself, but rather how it is interpreted and applied in the process of characterizing risk.

Two main points associated with uncertainty seem to be repeated in SPS dispute processes. One theme is the distinction between *possible* and *probable*, and the role of expert judgement in this distinction. The other is distinguishing the estimate of risk that is strictly associated with the *evidence* from aspects of the estimate that are mainly attributed to the *uncertainty* (see also Chapter 19).

As noted previously, uncertainty is a critical component of the concept of risk, but it has a unique status in the context of the SPS Agreement which requires that it be treated separately in order for the relationship of the evidence to the risk to be clear. This latter point arises in disputes where uncertainty is used to exaggerate the risk or to justify a precautionary approach. Chapter 19 provides more detailed information on SPS jurisprudence, including aspects related to the treatment of uncertainty and includes a discussion of the precautionary approach and its relationship to uncertainty.

16.9. Summary

The most critical message for pest risk analysis practitioners to understand is the need to be faithful to the separation of evidence and uncertainty. Risk, in the WTO-IPPC context, must be assessed and managed as objectively as possible according to the evidence. Uncertainty is handled in parallel but finds its most critical role in policy judgements following the risk analysis process.

For SPS purposes, risk and uncertainty are two complementary results of risk analysis that are used with other information in the decision making event that follows risk analysis. This separation of basic risk elements is unique to the SPS Agreement but a necessary approach if trading partners are to determine what aspects of the measures applied against them are based on evidence (as required under the Agreement) and what aspects are based on judgements regarding uncertainty. The former is disciplined mainly by science, while the latter is disciplined mainly by consistency.

It is useful to remind ourselves once again that although uncertainty is an essential component of risk, the expression of risk in the SPS-IPPC framework must be linked to the evidence. The role of uncertainty is not to affect the risk but rather to affect the judgements that will be made on (i) the acceptability of the risk; and (ii) the strength of measures required to mitigate the risk. From this standpoint, the analyst must conscientiously distinguish the evidence from the uncertainty throughout the process and in the conclusions so that reviewers and policy makers can clearly understand the difference. This unique aspect of pest risk analysis requires a trained eye and practice to master.

References

Cullen, A.C. and Frey, H.C. (1999) *Probabilistic Techniques in Exposure Assessment: a Handbook for Dealing with Variability and Uncertainty in Models and Inputs*. Plenum Press, London.

FAO (1997) *New Revised Text of the International Plant Protection Convention*. Food and Agriculture Organization of the United Nations, Rome.

Holt, J., Leach, A. and Knight, J. (2010) Risk matrix models. PRATIQUE/EPPO Workshop on Pest Risk Analysis. Hammamet, Tunisia, 23–26 November 2010. EPPO, Paris.

IPPC (1998) International Standards for Phytosanitary Measures, Publication No. 8: *Determination of Pest Status in an Area*. Secretariat of the International Plant Protection Convention (IPPC), Food and Agriculture Organization of the United Nations, Rome.

IPPC (2004) International Standards for Phytosanitary Measures, Publication No. 11: *Pest Risk Analysis for Quarantine Pests Including Analysis of Environmental Risks and Living Modified Organisms*. Secretariat of the International Plant Protection Convention (IPPC), Food and Agriculture Organization of the United Nations, Rome.

IPPC (2007) International Standards for Phytosanitary Measures, Publication No. 2: *Framework for Pest Risk Analysis*. Secretariat of the International Plant Protection Convention (IPPC), Food and Agriculture Organization of the United Nations, Rome.

Kaplan, S. (1993) The General Theory of Quantitative Risk Assessment – Its role in the regulation of agricultural pests. In: NAPPO. *International Approaches to Plant Pest Risk Analysis: Proceedings of the APHIS/NAPPO International Workshop on the Identification, Assessment, and Management of Risks due to Exotic Agricultural Pests*, Alexandria, Virginia, 23–25 October 1991. Ottawa, Canada. Bulletin (North American Plant Protection Organization) No. 11.

Morgan, M.G. and Henrion, M. (1990) *Uncertainty: A Guide to Dealing with Uncertainty in Quantitative Risk and Policy Analysis*. Cambridge University Press, New York.

Regan, H.M., Colyvan, M. and Burgman, M.A. (2002) A taxonomy and treatment of uncertainty for ecology and conservation biology. *Ecological Applications* 12, 618–628.

World Organization for Animal Health (OIE) (2010) *Terrestrial Animal Health Code*. World Organization for Animal Health, Paris.

WTO (1994) *The WTO Agreement on the Application of Sanitary and Phytosanitary Measures*. World Trade Organization, Geneva.

Yoe, C. (2012) *Principles of Risk Analysis: Decision Making Under Uncertainty*. CRC Press, Boca Raton, Florida.

Part V

Special Topics for Pest Risk Analysis

————————

17 Special Applications of Pest Risk Analysis – Beneficial Organisms

Stephanie Bloem and Kenneth Bloem

17.1. Introduction

In this chapter, we bring together international and regional guidance as well as discuss current approaches to assessing the risks of using beneficial organisms. Even though beneficial organisms are directly or indirectly advantageous to plants or plant products, their use in pest management has some inherent risks. For example, biological control agents may themselves harbour undesirable contaminants (such as diseases or hyperparasites), may have negative effects on the environment such as attacking species other than their intended targets or may spread beyond intended release areas. In the case of sterile insects, the sterilization process might fail resulting in the release of large numbers of fertile target pests into the environment. NPPOs should assess the risks posed by beneficial organisms just as they would examine the risks posed by organisms regarded as true pests of plants.

17.2. Risk Assessment for Beneficial Organisms – Guidance from the IPPC

As discussed in Chapter 4, the objective of the IPPC (FAO, 1997) is to secure coordinated, effective action to prevent and control the introduction and spread of pests of plants and plant products. The IPPC includes the protection of wild flora. The Convention considers both direct and indirect damage by pests. While its main focus is on plants and plant products moving in international trade, the convention also includes other articles or actions that can facilitate the spread of plant pests.

The scope of the IPPC includes beneficial organisms. Biological control agents and other beneficial organisms are intended to be beneficial to plants, but can nonetheless present risks to plant health. When performing a pest risk analysis for beneficial organisms, the main concern is identifying and documenting potential non-target effects. Other concerns may include contamination of cultures (of beneficial organisms) with other species, the culture thereby acting as a pathway for pests, and the reliability of containment facilities when such facilities are required.

Three IPPC ISPMs are relevant when examining the potential risks of using beneficial organisms: ISPM No. 2: *Framework for Pest Risk Analysis* (IPPC, 2007), ISPM No. 11: *Pest Risk Analysis for Quarantine Pests, Including Analysis of Environmental Risks and Living Modified Organisms* (IPPC, 2004a) and ISPM No. 3: *Guidelines for the Export, Shipment, Import and Release of*

*Biological Control Agents and Other Bene-
ficial Organisms* (IPPC, 2005). Furthermore,
Section 4.1 of ISPM No. 20 – *Guidelines for
a Phytosanitary Import Regulatory System*
(IPPC, 2004b), names biological control
agents as examples of regulated articles.

General requirements for pest risk analysis include stage 1 – initiation, stage 2 – pest
risk assessment and stage 3 – pest risk management. In the case of beneficial organisms,
stage 1 is initiated by a request to import (for
research, analysis or other purposes) or
release a beneficial organism that may become
a potential (direct or indirect) pest. Stage 2 –
pest risk assessment, broadly divided into
three interrelated steps (pest categorization,
assessment of the probability of introduction
and spread, and assessment of the potential
economic consequences including environmental impacts), considers the biology of the
organism in question. The pest categorization
step examines whether the criteria in the definition of a quarantine pest are met by the
organism in question.

For exotic biological control agents
these criteria (obviously) are satisfied – they
are not present in the pest risk analysis area,
they are 'intended' to establish and spread
in the pest risk analysis area, and they are
expected to have environmental consequences. An assessment of the probability
of entry does not apply in the case of beneficial organisms. Risks, in this case, arise
because there is some probability of unintended environmental impacts. Available
information and expert judgement is used
to assess the potential and extent of such
impacts in the pest risk analysis area. Annex
1 of ISPM No. 11 (IPPC, 2004a) includes
guidance for pest risk assessment in relation
to environmental risks.

ISPM No. 11 notes that greater uncertainty
is involved in the assessment of the probability
and consequences for pests affecting uncultivated and unmanaged plants. This is due to
lack of information, additional complexity
associated with natural ecosystems and greater
variability of hosts or habitats. As such, pest
risk analysis for phytophagous beneficial organisms released into unmanaged ecosystems as
well as for those that can potentially spread
from managed into unmanaged habitats would

have higher uncertainty. ISPM No. 11 also recognizes that official control of pests that
present environmental risks (such as beneficial organisms) may require the involvement
of other agencies besides the NPPO.

In stage 3 – pest risk management –
information on options to reduce risk to an
acceptable level should be gathered and
data on their efficacy and impact should be
sought. For beneficial organisms, ISPM No. 3
(IPPC, 2005) provides guidelines for risk
management related to export, shipment,
import and release of beneficial organisms.
The standard identifies the obligations of the
NPPO and other responsible parties including importers and exporters. It also addresses
importation into containment facilities.

As described in ISPM No. 3, the NPPO
or other responsible authorities should:

- Do pest risk analysis prior to import or
  release of a beneficial organism.
- Ensure, when certifying exports, that
  all phytosanitary import requirements
  are met.
- Obtain, provide and assess documentation relevant to export, shipment, import
  or release of beneficial organisms.
- Ensure that beneficial organisms arrive
  at their intended destination, which can
  be a containment or mass-rearing facility or the designated release area(s).
- Encourage post-release monitoring to
  assess possible non-target effects.

Exporters should:

- Ensure consignments comply with phy-
  tosanitary import requirements.
- Provide appropriate documentation.
- Ensure appropriate packaging.

Importers should:

- Provide appropriate documentation to the
  NPPO and other responsible authorities
  in the importing country.

ISPM No. 11 (IPPC, 2004a) points out
that for environmental risks (such as those
potentially posed by the release of beneficial
organisms), the variety of information
sources will be wider and broader inputs
may be required. Additional sources of

information may include environmental impact assessments. The standard recognizes that such assessments usually do not have the same purpose as a pest risk analysis and cannot substitute for pest risk analysis.

17.3. Definitions

ISPM No. 3 and ISPM No. 5: *Glossary of Phytosanitary Terms* (IPPC, 2010) – provide the following definitions: A beneficial organism is any organism directly or indirectly advantageous to plants or plant products, including biological control agents. Biological control agent is a natural enemy, antagonist or competitor, or other organism, used for pest control. Finally, biological control is a pest control strategy making use of living natural enemies, antagonists, competitors or other biological control agents.

Organisms traditionally considered biological control agents can be entomophagous – defined by NAPPO, RSPM No. 5: *NAPPO Glossary of Phytosanitary Terms* (NAPPO, 2010) and No. 12: *Guidelines for Petition for First Release of Non-indigenous Entomophagous Biological Control Agents* (NAPPO, 2008b) – as organisms that eat insects, or phytophagous – defined in RSPM No. 7: *Guidelines for Petition for First Release of Non-indigenous Phytophagous Biological Control Agents* (NAPPO, 2008a) – as organisms that eat plants.

The definition of beneficial organism in ISPMs Nos 3 and 5 is broad enough to allow the inclusion of sterile insects. Sterile insects are defined as insects that, as a result of a specific treatment (usually gamma radiation), are unable to reproduce. Sterile insect technique (SIT) is defined as a method of pest control using area-wide inundative release of sterile insects to reduce reproduction in a field population of the same species. A relatively recent variation in the application of SIT has been proposed, whereby the arthropod species that is mass-reared, treated with gamma radiation and released is a phytophagous biological control agent (Carpenter *et al.*, 2001; Moeri *et al.*, 2009). In this approach to SIT, the goal is to have effective population suppression of the target pest plant by the actions of the reproductively inactivated phytophagous agent without the risk of establishing a breeding population of the agent in the field. This approach would work well for arthropod species that feed on their target host plants as adults. In the case of phytophagous arthropods that feed only in the larval stages, a modification of the SIT termed inherited sterility could be used to effect population suppression as above (Carpenter *et al.*, 2005).

Other biological control texts, for example Bellows and Fischer (1999), use the following classification for biological control agents:

Enemies of invertebrate pests can be:

• Parasitoids – which typically develop on/in a single host, however, adult parasitoids may use multiple hosts or lay multiple eggs on/in a single host. Parasitoids tend to be host specific. Ectoparasitoids feed externally and endoparasitoids develop internally on host tissue. Typical examples include wasps and flies.
• Predators – which completely consume their prey and are capable of killing large numbers of individuals. Many are not specific in their choice of prey species although there are exceptions. Typical examples include beetles (e.g. lady beetles) and predatory mites.
• Pathogens – which include viruses, bacteria, fungi, protozoa (and nematodes). They cause disease in their hosts and may be developed commercially as biopesticides.

Enemies of pest plants (weeds) can be:

• Phytophagous insects and mites – which can specialize on seeds, foliage, stems or roots, and combinations of agents attacking different parts of the pest plant are often employed together.
• Pathogens – which cause microbial or infectious disease in their hosts.

Natural biological control as defined by Ehler (1990) is that which occurs without the aid of humans, such as the regulation of native phytophagous insects by native natural enemies. On the other hand, applied biological control involves some type of human intervention. Applied biological control (*sensu* Ehler) has four main strategies, depending on how the organisms are used – classical, inundative, augmentative and conservation (Andow *et al.*, 1995; De Clerck-Floate *et al.*, 2006).

The aim of classical or innoculative biological control is the importation and permanent establishment of an exotic, host-specific organism on a target pest species through one or a small number of introductions. Inundative biological control involves the repeated and controlled release of large numbers of a specific biological control agent to immediately reduce the target pest population, similar to the use of biopesticides. Inundative biological control may or may not use exotic natural enemies, is not self-sustaining and, as such, is conducive to commercialization. In the context of this chapter, SIT is considered to be a type of inundative biological control. Augmentative biological control seeks to improve the impact of an established (usually native) natural enemy by releasing additional individuals at critical times of the year (e.g. early in the year to speed population build-up). Augmentative biological control agents are typically purchased from commercial facilities. Finally, conservation biological control pertains to management of habitat or environment such that survival and performance of an established biological control agent are enhanced.

17.4. Potential Risks Associated with the Use of Beneficial Organisms/ Why Do Pest Risk Analysis on a Beneficial Organism?

Not all beneficial organisms have the same potential risks. For example, conservation biological control has low inherent risks, as it primarily involves habitat manipulation to improve conditions for beneficial organisms already established in an area. Examples of this approach include maintaining native

plants in the vicinity of crops to provide beneficial organisms with alternative food sources and refugia, or selective use of insecticides (both in timing and type) to minimize impacts on natural enemies (Barbosa, 1998; Pickett and Bugg, 1998).

Inundative biological control requires the mass production and repeated release of a specific beneficial organism to control a target pest population. Inundative biological control is not self-sustaining. Usually, beneficial organisms used in inundative biological control have been extensively studied and, as such, uncertainty surrounding their potential risks is reduced. Nevertheless, some potential risks can include inaccurate taxonomic identification of the beneficial agent (especially if several congeners are reared in the same commercial facility) and the possibility that the colony of beneficial organisms might be contaminated with other potentially injurious species such as hyperparasites (Mason *et al.*, 2005), diseases or other contaminants. Similar risks are associated with the application of augmentative biological control, even though environmental release of additional beneficial organisms (already present in the area) is done on a smaller spatial and temporal scale.

SIT is a type of biological control (Dyck *et al.*, 2005) whereby the target pest is mass-reared, sterilized and released (repeatedly) to compete with wild conspecifics for mates. If sufficiently large numbers of sterile insects are released and the majority of matings involve a sterilized individual, field populations of the target pest can be suppressed or eradicated (Robinson, 2005). As in the above-mentioned types of biological control, colony contamination is a risk in the application of SIT. However, the most important potential risk for SIT is the failure of the sterilization method, which would result in the release of large numbers of fertile pests into the environment. Depending on the species, the accidental release of fertile insects could include only one (e.g. male, for Mediterranean fruit fly, *Ceratitis capitata* (Wiedemann)) or both genders (e.g. false codling moth, *Thaumatotibia leucotreta* (Meyrick)).

Classical biological control is irreversible, self-perpetuating and self-dispersing. These characteristics are what make classical biological control a successful component of sustainable pest management programmes. However, these attributes are also the ones that have alerted researchers and regulatory officials to the potential negative consequences of such introductions.

Publications questioning the environmental safety of (mostly) classical biological control began appearing in the late 1980s and early 1990s (Howarth, 1983, 1991; Simberloff and Stiling, 1996), and for a few years the views expressed in the scientific literature became highly polarized. Proponents of biological control argued that there was lack of evidence for the negative environmental impacts from biological control. Others suggested that the lack of evidence for negative environmental impacts was a result of limited study rather than the absence of impacts (Delfosse, 2005).

Identifying/predicting the risks of using classical biological control is much more advanced for phytophagous than for entomophagous beneficial organisms (Strong and Pemberton, 2001). A study by Pemberton (2000) reported that only one of 117 established phytophagous agents (including insects, fungi, mites and nematodes) attacked a native plant unrelated to the target pest plant. This strongly suggests that most risks of non-target effects are borne by native plant species that are closely related to the target pest plant. Thus, the elements of protection for native flora lie in the selection of pest plant targets that have few or no native congeners and in the introduction of phytophagous biological control agents with suitably narrow diets (Strong and Pemberton, 2001).

The field of biological control has used host specificity testing as a form of risk assessment for many years (Lonsdale et al., 2001). Host specificity testing still provides the best assurance that the biological control agent will suppress the target species without harmful or unintended side effects. For phytophagous biological control agents, the centrifugal–phylogenetic method developed by Wapshere (1974, 1989) is used to

select test species for host range (host specificity) testing. The method is based on taxonomic affinities of related plant taxa. Recent studies have demonstrated that host range for phytophagous biological control agents is limited to phylogenetically related plant taxa (Bernays, 2000; Pemberton, 2000; Kuhlmann et al., 2006).

Louda et al. (2003) published a retrospective analysis of ten classical biological control programmes – three for pest plants and seven for pest insects – that contained data on non-target effects. The following ten 'risk' patterns emerged from this analysis:

- Relatives of the target pest are most likely to be attacked.
- Host specificity testing defines physiological host range but not ecological range.
- Prediction of ecological consequences requires population data.
- Level of impact can vary often in relation to environmental conditions.
- Information on magnitude of non-target impact is sparse.
- Attack on rare native species can accelerate their decline.
- Non-target effects can be indirect.
- Agents disperse from agro-ecosystems.
- Whole assemblages of species can be perturbed.
- No evidence on adaptation is available.

In the last 15 years, the published literature on the potential risks associated with the use of beneficial organisms has been vast. Below we list additional identified risks of using beneficial organisms and provide relevant citations that give historical context and case studies and, in some cases, methodologies and frameworks to use in the assessment of each identified risk.

- Competition and displacement (Messing et al., 2006);
- Interbreeding between beneficial and target species (Hopper et al., 2006);
- Entomophagous agents causing plant damage (Albajes et al., 2006).

Comparative risk analysis is one of the four components of risk analysis as defined by Davis (1996a,b). Comparative risk analysis has two forms. The first – also known as risk ranking – consists of comparing two types of risk. For biological control, risk ranking might involve comparing the risks of non-target effects by the biological control agent with those expected when other means of control – for example, pesticides or herbicides – are used (Lonsdale *et al.*, 2001). This type of risk ranking analysis is part of the environmental assessment documents prepared as part of the permitting process for phytophagous biological control agents in the USA.

The second type of comparative risk analysis (*sensu* Davis) is broader in scale. It is also known as programmatic analysis and is used for setting government priorities among a large number of risks. In general, this type of analysis is usually reserved for high-impact pests and has had little application in the area of classical biological control. However, in some cases, programmatic analysis has been used when examining the application of the SIT in the USA (e.g. strategies to deal with light brown apple moth, *Epiphyas postvittana* (Walker)).

Safe use of classical biological control can lead to enormous economic and environmental benefits. As such, regulatory entities should take into account the risk of rejecting potentially useful species by making over-cautious decisions. In general, options for using biological control against pests are typically not developed unless clear and serious expectations or evidence exist to suggest that the target pest will negatively impact agriculture or the environment. The question then becomes, is it better to live with the known or expected negative impacts of the pest or the unknown impacts of introducing a potentially beneficial biological control agent to control/suppress the pest? Unfortunately, current processes in the United States only provide legislative authority to assess the risks and do not have a framework to consider whether the potential benefits outweigh the risks of biological control.

17.5. Pest Risk Analysis Approaches for Importation of Beneficial Organisms into countries/regions other than the USA – Australia, New Zealand, Canada and the European Union

Hunt *et al.* (2008) recently reviewed the regulations and requirements for environmental release of invertebrate (phytophagous and entomophagous) biological control agents (IBCAs) in Australia, New Zealand, Canada and the USA. The objective of this review was to make recommendations for the development of a harmonized 'evidence-based' regulatory system for IBCAs for the European Union. The article discusses the following criteria for each country:

- Legislation and administration;
- Application procedures;
- Decision making processes and decision maker(s);
- Data requirements;
- Time frames;
- Availability of information about regulation to aid applicants;
- Public participation;
- Length of permit validity;
- Safe lists for organisms exempt from regulation.

Due to the recency of the review by Hunt *et al.* (2008), the reader is referred to it for information pertaining to importation of IBCAs into Australia, New Zealand and Canada. However, information on regulations and requirements for the USA needed updating. The most current information for the USA is contained in the next section.

Australia and New Zealand also have excellent websites (Australia, http://www. daff.gov.au/ba/reviews/biological_control_ agents; New Zealand, http://www.ermanz. govt.nz/new-organisms/Pages/default.aspx) with the latest information on requirements and processes for importation and release of IBCAs. In Australia, information on IBCA risks is posted alongside that of other risk assessments, for example, those pertaining to the importation of fruits and vegetables. Furthermore, the process followed for risk assessment of IBCAs is essentially the same as it is for other imports. However, since IBCAs intended for release are deliberately introduced, distributed and assisted in their establishment and spread, it is considered inappropriate to evaluate the probability of these events during the risk assessment process. In 2009, Australia enacted new biosecurity guidelines for the introduction of exotic biological control agents for the control of weeds and plant pests. The guidelines provide stakeholders with a clear roadmap to follow including who is responsible for each step of the process. The reader should also consult the *Guide for the Importation and Release of Arthropod Biological Control Agents* (De Clerck-Floate *et al.*, 2006) for additional information concerning the processes for IBCAs in Canada.

Most, if not all, European countries are signatories of the CBD and, as such, have the obligation to prevent the introduction and, as far as possible, to control alien species that threaten indigenous ecosystems and habitats (Bigler *et al.*, 2005). EPPO recently published a standard with guidelines for import, release and required application procedures for non-indigenous biological control agents (EPPO, 2010). This document attempts to harmonize application procedures, documentation and risk assessment processes for all countries in the European Union. However, contrary to the advice provided in Hunt *et al.* (2008), the authority to enforce the suggested guidelines lies with the National Authority of each member country.

17.6. Importation of Beneficial Organisms – a View from the USA

17.6.1. Legislation/regulations supporting importation/permitting of phytophagous and entomophagous beneficial organisms

The Plant Protection Act (PPA) of 2000 provides the US NPPO (USDA-APHIS – Plant Protection and Quarantine) with the authority to regulate organisms that may directly or indirectly harm plants or plant products. Unlike the Federal Plant Pest Act of 1957, the PPA names biological control organisms and defines them as any enemy, antagonist or competitor used to control plant pests or noxious weeds. This provision recognizes that not all organisms in need of permits are plant pests and that some are actually beneficial to US agriculture and the environment.

Non-indigenous, classical biological control agents for weed suppression are direct plant pests according to the PPA. Any organism that is indirectly injurious to plants also is considered to be a pest. Predators and parasites/parasitoids of phytophagous organisms are considered indirect plant pests according to the PPA, and thus, their importation and release is also regulated.

17.6.2. Processes for importation/permitting of phytophagous and entomophagous beneficial organisms

In the USA, under the authority of the PPA, a Plant Protection and Quarantine permit (PPQ 526) is required for importation, interstate movement and release of beneficial organisms. Each action requires the issuance of a separate permit. A permit petition should contain sufficient information to facilitate the evaluation of the potential risks associated with the proposed action (importation into containment, interstate movement or release).

Generally speaking, importation of an exotic organism into a containment facility presents minimal risks and, as such, the information required in a petition is straightforward – scientific name, life stages, purpose of importation, how organisms will be handled. Containment facilities must be inspected and approved by the US NPPO as meeting the standards to receive and maintain shipments of the specific exotic organism in question. The risks associated with movement of beneficial organisms into (other) areas where they are already present are also low. This situation occurs in augmentative or inundative programmes, when commercial production of a biological control agent takes place in a different state, or in classical biocontrol programmes, when a field insectary in one state serves as a source for redistribution of beneficial organisms to other states. As a result, the information required for this type of petition is similarly brief.

Releasing beneficial organisms that are new to the US environment, present significantly greater risks and require a much larger body of information to justify the proposed action and to ensure that potential non-target impacts are minimal/acceptable. Depending on the type of beneficial organism, petitioners are directed to follow the guidance provided by the US NPPO Technical Advisory Group (TAG) – for petitions to release phytophagous biological control agents – or consult NAPPO RSPM No. 12 – for petitions to release entomophagous biological control agents. It should be noted that NAPPO also has RSPM No. 7 – for phytophagous biological control agents that largely parallels the TAG process.

In general terms, a petition for the first environmental release of beneficial organisms should include the following information.

Data on the proposed action including:

- Justification for the action itself;
- Justification for the selection of beneficial organism;
- Where and when the proposed action will take place, including justifications for these decisions;
- Detailed information on proposed rearing and containment facilities for the beneficial organism;
- Detailed information on safeguards (e.g. limited access, disposal, risk mitigation);
- Identification of agencies or individuals responsible for the proposed action.

Data on the beneficial organism including:

- Scientific name, describing authority, common name and full synonymy;
- Useful characters (morphological, molecular) to identify the organism and location of reference specimens;
- Source of the beneficial organism;
- Life history, including dispersal capability;

- Habitat preference and climatic requirements, including expected attainable range in the USA;
- Host range;
- History of use;
- Information on rearing protocols and possible contaminants if appropriate.

Data on the target species including:

- Scientific name, describing authority, common name and full synonymy (in the case of sterile insects the target species and the beneficial organism will be the same species);
- Higher level classification;
- Useful characters to allow for unambiguous recognition;
- Life history, including vulnerable life stages to attack by the beneficial organism;
- Distribution;
- Pest status in the USA;
- Listing of economically and ecologically important species related to the target that are present in the USA (either naturally occurring or introduced);
- Economic impacts.

Data on possible economic and environmental impacts of the proposed action including:

- Direct impacts on target and non-target species (including humans);
- Effects on the physical environment (natural resources – such as water, soil, air);
- Impacts of not taking the proposed action (e.g. increase in pesticide use);
- Indirect effects (for example, competition with other beneficial organisms);
- Effects (either direct or indirect) on threatened and endangered species.

Information on post release monitoring including:

- Data on predicted versus observed performance and behaviour;
- Data on establishment and spread;
- Data on target species densities over time;

- Changes in the target species;
- Changes in community structure and species diversity;
- Data on economic and environmental impacts of the action.

Phytophagous biological control agents

For issues relating to the importation and release of phytophagous biological control agents, the US NPPO makes use of a long-standing Technical Advisory Group, or TAG. The TAG provides guidance to petitioners and recommendations to regulatory agencies concerning the risks posed by the proposed action. The TAG brings together representatives from all major groups that have resources at risk. TAG members:

- Review the scientific data (contained in a petition) to evaluate the risks of the proposed action to agriculture, human health and the environment;
- Identify and consult with subject matter experts as needed;
- Represent their agency's perspective during review;
- Serve as a communication conduit during the review and evaluation process.

The TAG, in cooperation with the US NPPO, developed detailed formats and guidelines to assist petitioners interested in importing and releasing phytophagous biological control agents in the USA. The guidelines are consistent with (and, in fact, formed the basis for) the current regional standards dealing with phytophagous and entomophagous biological control agents. The guidelines can be found on the US NPPO website (http://www.aphis.usda.gov/plant_health/permits/tag/index.shtml).

If the TAG recommends release of the phytophagous agent, the US NPPO conducts an internal review to confirm their concurrence with the TAG recommendation. The US NPPO prepares an Environmental Assessment (EA) in accordance with the National Environmental Policy Act (NEPA). This law requires that all US federal agencies evaluate the potential environmental health and safety risks of their programmes and activities. It should be noted that ISPM No. 11 recognizes that environmental impact assessments usually do not have the same purpose as a pest risk analysis and cannot substitute for pest risk analysis.

The EA follows the format given below:

- Purpose and need for the proposed action;
- Alternatives;
- Affected environment;
- Environmental consequences;
- Other issues;
- Agencies and organizations consulted;
- References.

The EA is used to determine whether the release of the phytophagous biological control agent is expected to result in significant impacts to the human environment. The completed EA is used for consultation with all potentially affected parties including other federal departments, state governments and tribal councils. If a finding of no significant impact (FONSI) is reached, the EA is posted to the US Federal Register for a mandatory 30-day comment period by the general public. All comments received are considered and the EA is revised accordingly. Following this action, the US NPPO will issue a PPQ 526 permit to the petitioner. If a finding of significant impact is reached, the US NPPO advises the petitioner that an Environmental Impact Statement (EIS) needs to be prepared (and evaluated) by the NPPO. Based on the results of the EIS evaluation the petitioner may receive a permit or may discontinue the effort.

Entomophagous biological control agents

The processes for importation and interstate movement of entomophagous biological control agents are essentially the same as those outlined for phytophagous agents. However, the process and guidelines for environmental release of entomophagous agents in the United States are less well developed. The principal difference relates to how host specificity (or test) lists are developed and evaluated.

Host test lists for entomophagous agents, unless they include threatened or endangered species or other beneficial organisms, have historically been less scrutinized, as unintended impacts on non-target arthropods raise fewer concerns than unintended impacts on non-target plants. On the other hand, phytophagous agents are treated as unwanted plant pests until their target specificity and beneficial properties are defined – some would call this approach 'guilty until proven innocent'. Because the perceived risks are higher and host testing is relatively easier (both in containment facilities and in the field where potential agents are being collected), the host test lists for phytophagous agents are much more structured, conservative and extensive.

In addition, many potential non-target hosts of entomophagous agents are considered plant pests in their own right. Furthermore, the arthropod fauna is much larger and less understood (taxonomically and from a community structure standpoint) than the plant fauna and, for these reasons, it is not always obvious what is important to include on host test lists for entomophagous agents. This is further complicated by the fact that it can be challenging to collect, culture and test several potential arthropod non-target hosts in a containment facility. As a result, evaluation of the risks posed by entomophagous agents rely more on literature, host records and comparison to closely related species. However, due to the influence of host searching and acceptance behaviours by entomophagous agents (by stimuli and cues in the environment) in the field, there may be a need to develop an alternate process for evaluating the risks posed by these agents.

Historically, entomophagous biological control agents were not considered as pests by the US NPPO. Prior to the enactment of the PPA, the US NPPO deferred to each state government concerning the importation and release of entomophagous biological control agents. Currently, the US NPPO requires petitioners to follow the NAPPO guidelines provided in RSPM No. 12 when filing a petition to release entomophagous agents. Petition review and evaluation are accomplished with the assistance of NAPPO collaborators from Canada and Mexico, who consolidate comments

and provide feedback, similar to the activities performed by the TAG for phytophagous agents. Consultation and preparation of EAs and their requirements are the same for entomophagous and phytophagous beneficial organisms.

Sterile insects

As noted above, for purposes of this chapter and as defined under ISPM No. 3 and No. 5, sterile insects are considered biological control agents delivered to the environment through repeated inundative releases. Because colony contaminants and the reliability of the sterilization process pose some risks, a PPQ 526 permit is required for the interstate/international movement and release of sterile insects. Although an EA is required for first-time release of sterile insects, the fact that the organisms are sterile and, thus, will not establish in the pest risk analysis area greatly simplifies the review and approval process.

17.7. Summary

Beneficial organisms contribute to managing important pests – but are not without risks of their own. The IPPC provides guidance (in ISPM No. 3 and ISPM No. 11) on analysing the risks associated with beneficial organisms, and for their safe movement between countries. Some of the risks associated with beneficial organisms include non-target effects, attacking relatives of the target pest, movement from intended areas, attacks on rare species, competition and displacement, or interbreeding between beneficial and target species). Different countries have developed analytical processes to address beneficial organisms – this includes conducting pest risk analyses and petitions for release of beneficial organisms. Prior to releasing beneficial organisms, information on the organism's biology and that of the target organism (including host range, life history, taxonomy, etc.) should be assessed to determine if the release of the beneficial organism could have unintended, negative consequences.

References

Albajes, R., Castañe, C., Gabarra, R. and Alomar, O. (2006) Risks of plant damage caused by natural enemies introduced for arthropod biological control, In: Bigler, F., Babendreier, D. and Kuhlmann, U. (eds) *Environmental Impact of Invertebrates for Biological Control of Arthropods – Methods and Risk Assessment*. CABI Publishing, Wallingford, pp. 132–144.

Andow, D.A., Lane, C.P. and Olson, D.M. (1995) Use of *Trichogramma* in maize – estimating environmental risks. In: Hokkanen, H.M.T. and Lynch, J.M. (eds) *Biological Control: Benefits and Risks*. Cambridge University Press, Cambridge, pp. 101–118.

Barbosa, P. (1998) *Conservation Biological Control*. Academic Press, San Diego, California.

Bellows, T.S. and Fischer, T.W. (1999) *Handbook of Biological Control*. Academic Press, San Diego, California.

Bernays, E.A. (2000) Neural limitations in phytophagous insects: implication for diet breadth and evolution of host affiliation. *Annual Review of Entomology* 46, 703–727.

Bigler, F., Bale, J.S., Cock, M.J.W., Dreyer, H., Greatrex, R., Kuhlmann, U., Loomans, A.J.M. and Van Lenteren, J.C. (2005) Guidelines on information requirements for import and release of invertebrate biological control agents in European countries. *CAB Reviews – Perspectives in Agriculture, Veterinary Sciences, Nutrition and Natural Resources* 1–10.

Carpenter, J.E., Bloem, K.A. and Bloem, S. (2001) Applications of F1 sterility for research and management of *Cactoblastis cactorum* (Lepdoptera: Pyralidae). *Florida Entomologist* 84, 531–536.

Carpenter, J.E., Bloem, S. and Marec, F. (2005) Inherited sterility in insects. In: Dyck, V.A., Hendrichs, J. and Robinson, A.S. (eds) *Sterile Insect Technique – Principles and Practice in Area-wide Integrated Pest Management*. Springer, Dordrecht, The Netherlands, pp. 115–146.

Davis, J.C. (1996a) *Comparing Environmental Risks: Tools for Setting Government Priorities*. Resources for the Future, Washington, DC.

Davis, J.C. (1996b) Comparative risk analysis in the 1990s: the state of the art. In: Davies, J.C. (ed.) *Comparing Environmental Risks: Tools for Setting Government Priorities*. Resources for the Future, Washington, DC.

De Clerck-Floate, R.A., Mason, P.J., Parker, D.J., Gillespie, D.R., Broadbent, A.B. and Boivin, G. (2006) *Guide for the Importation and Release of Arthropod Biological Control Agents in Canada*. Agriculture and Agri-Food Canada, Ottawa, Ontario, Canada.

Delfosse, E.S. (2005) Risk and ethics in biological control. *Biological Control* 35, 319–329.

Dyck, V.A., Hendrichs, J. and Robinson, A.S. (2005) *Sterile Insect Technique – Principles and Practice in Area-wide Integrated Pest Management*. Springer, Dordrecht, The Netherlands.

Ehler, L.E. (1990) Introduction strategies in biological control of insects. In: Mackauer, M., Ehler, L.E. and Roland, J. (eds) *Critical Issues in Biological Control*. Intercept, Andover, pp. 111–134.

EPPO (2010) Import and release of non-indigenous biological control agents (2010/ PM 6/2 (2)). European and Mediterranean Plant Protection Organization. *EPPO Bulletin* 40, 3, 335–344. Available online at: http://www.eppo.org/PUBLICATIONS/bulletin/bulletin.htm, accessed 31 March 2011.

Hopper, K.R., Britch, S.C. and Wajnberg, E. (2006) Risks of interbreeding between species used in biological control and native species, and methods for evaluating their occurrence and impact. In: Bigler, F., Babendreier, D. and Kuhlmann, U. (eds) *Environmental Impact of Invertebrates for Biological Control of Arthropods – Methods and Risk Assessment*. CABI Publishing, Wallingford, pp. 78–97.

Howarth, F.G. (1983) Classical biological control: Panacea or Pandora's Box? *Proceedings of the Hawaiian Entomological Society* 24, 239–244.

Howarth, F.G. (1991) Environmental impacts of classical biological control. *Annual Review of Entomolology* 36, 485–509.

Hunt, E.J., Kuhlmann, U., Sheppard, A., Qin, T.-K., Barratt, B.I.P., Harrison, L., Mason, P.G., Parker, D., Flanders, R.V. and Goolsby, J.A. (2008) Review of invertebrate biological control agent regulation in Australia, New Zealand, Canada and the USA: recommendations for a harmonized European system. *Journal of Applied Entomology* 132, 89–123.

IPPC (2004a) International Standards for Phytosanitary Measures, Publication No. 11: *Pest Risk Analysis for Quarantine Pests Including Analysis of Environmental Risks and Living Modified Organisms*. Secretariat of the International Plant Protection Convention (IPPC), Food and Agriculture Organization of the United Nations, Rome.

IPPC (2004b) International Standards for Phytosanitary Measures, Publication No. 20: *Guidelines for a Phytosanitary Regulatory Import System*. Secretariat of the International Plant Protection Convention (IPPC), Food and Agriculture Organization of the United Nations, Rome.

IPPC (2005) International Standards for Phytosanitary Measures, Publication No. 3: *Guidelines for the Export, Shipment, Import and Release of Biological Control Agents and Other Beneficial Organisms*. Secretariat of the International Plant Protection Convention (IPPC), Food and Agriculture Organization of the United Nations, Rome.

IPPC (2007) International Standards for Phytosanitary Measures, Publication No. 2: *Framework for Pest Risk Analysis*. Secretariat of the International Plant Protection Convention (IPPC), Food and Agriculture Organization of the United Nations, Rome.

IPPC (2010) International Standards for Phytosanitary Measures, Publication No. 5: *Glossary of Phytosanitary Terms*. Secretariat of the International Plant Protection Convention (IPPC), Food and Agriculture Organization of the United Nations, Rome.

Kuhlmann, U., Mason, P.G., Hinz, H.L., Blossey, B., De Clerck-Floate, R.A., Dosdall, L.M., McCaffrey, J.P., Schwarlaender, M., Olfert, O., Brodeur, J., Gassmann, A., McClay, A.S. and Wiedenmann, R.N. (2006) Avoiding conflicts between insect and weed biological control: selection of non-target species to assess host specificity of cabbage seedpod weevil parasitoids. *Journal of Applied Entomology* 130, 129–141.

Lonsdale, W.M., Briese, D.T. and Cullen, J.M. (2001) Risk analysis and weed biological control. In: Wajnberg, E., Scott, J.K. and Quimby, P.C. (eds) *Evaluating Indirect Ecological Effects of Biological Control*. CABI Publishing, Wallingford, pp. 185–210.

Louda, S.M., Pemberton, R.W., Johnson, M.T. and Follett, P.A. (2003) Nontarget effects – the Achilles' heel of biological control? Retrospective analyses to reduce risk associated with biocontrol introductions. *Annual Review of Entomology* 48, 365–396.

Mason, P.G., Flanders, R.V. and Arrendondo-Bernal, H.A. (2005) How can legislation facilitate the use of biological control of arthropods in North America?' In: Hoddle, M.S. (ed.) *Proceedings of the Second International Symposium on the Biological Control of Arthropods*. Davos, pp. 701–713.

Messing, R.H., Roitberg, B.D., and Brodeur, J. (2006) Measuring and predicting indirect impacts of biological control, competition, displacement and secondary interactions. In: Bigler, F., Babendreier, D. and Kuhlmann, U. (eds) *Environmental Impact of Invertebrates for Biological Control of Arthropods – Methods and Risk Assessment*. CABI Publishing, Wallingford, pp. 64–77.

Moeri, O.E., Cuda, J.P., Overholt, W.A., Bloem, S. and Carpenter, J.E. (2009) F1 sterile insect technique: a novel approach for risk assessment of *Episimus unguiculus* (Lepidoptera: Tortricidae), a candidate biological control agent of *Schinus terebinthifolius* in the continental USA. *Biocontrol Science and Technology* 19, 303–315.

NAPPO (2008a) NAPPO Regional Standards for Phytosanitary Measures (RSPM): *Guidelines for Petition for First Release of Non-indigenous Phytophagous Biological Control Agents* (RSPM No. 7). North American Plant Protection Organization (NAPPO), Ottawa, Ontario.

NAPPO (2008b) NAPPO Regional Standards for Phytosanitary Measures (RSPM): *Guidelines for Petition for First Release of Non-indigenous Entomophagous Biological Control Agents* (RSPM No. 12). North American Plant Protection Organization (NAPPO), Ottawa, Ontario.

NAPPO (2010) NAPPO Regional Standards for Phytosanitary Measures (RSPM): *NAPPO Glossary of Phytosanitary Terms* (RSPM No. 5). North American Plant Protection Organization (NAPPO), Ottawa, Ontario.

Pemberton, R.W. (2000) Predictable risk to native plants in weed biological control. *Oecologia* 125, 489–494.

Pickett, C.H. and Bugg, R.L. (1998) *Enhancing Biological Control: Habitat Management to Promote Natural Enemies of Agricultural Pests*. University of California, Berkeley and Los Angeles, California.

Robinson, A.S. (2005) Genetic basis of the sterile insect technique. In: Dyck, V.A., Hendrichs, J. and Robinson, A.S. (eds) *Sterile Insect Technique – Principles and Practice in Area-wide Integrated Pest Management*. Springer, Dordrecht, pp. 95–114.

Simberloff, D. and Stiling, P. (1996) How risky is biological control? *Ecology* 77, 1965–1974.

Strong, D.R. and Pemberton, R.W. (2001) Foodwebs, risks of alien enemies and reform of biological control. In: Wajnberg, E., Scott, J.K. and Quimby, P.C. (eds) *Evaluating Indirect Ecological Effects of Biological Control*. CABI Publishing, CABI International, Wallingford, pp. 229–248.

Wapshere, A.J. (1974) A strategy for evaluating the safety of organisms for biological weed control. *Annals of Applied Biology* 77, 201–211.

Wapshere, A.J. (1989) A testing sequence for reducing rejection of potential biological control agents for weeds. *Annals of Applied Biology* 114, 515–526.

18 Special Applications of Pest Risk Analysis – Weed Risk Assessment

Anthony Koop

18.1. Introduction

The IPPC includes plants in its definition of pest (ISPM No. 5; IPPC, 2010). Like pathogens and arthropods, pest plants cause a variety of impacts. They lower crop yield, reduce biodiversity, induce allergies, affect land value, alter ecosystem processes and produce toxins (Elton, 1958). Every year, people spend considerable amounts of time and money controlling these plants in their homes, farmlands, and conservation areas. Some workers consider them to be the most important type of plant pests (Gunn and Ritchie, 1988). Based on a survey of US agriculture, 84% of all pesticides applied to the top ten crops are herbicides. It has been estimated that pest plants cause direct and indirect losses of about $20–34 billion per year in the USA (Gunn and Ritchie, 1988; Westbrooks, 1998; Pimentel et al., 2000). In Australia, they result in losses of AUS$3.5 to $4.5 billion per year, and possibly more (Sinden et al., 2004).

Most people refer to pest plants as weeds. Weeds are unwanted plants (Esler, 1988) because they have some type of negative impact, even if it is simply growing where they should not. Although the term weed is well understood, its usage is highly subjective. What might be a weed to one person may be someone else's favourite plant. Another term that is used to refer to pest plants is 'invasive plant', or more broadly, 'invasive species'. The concept of an invasive species varies widely (Box 18.1), but can be broadly defined as a non-native species capable of spreading across large areas, producing large population sizes quickly, and causing one or more kinds of negative impacts. The terms weed and invasive plant represent slightly different concepts (Rejmánek, 1994), however, in this chapter, we use them interchangeably as they both relate to the IPPC concept of a pest plant (IPPC, 2010).

Many of today's most common and notorious weeds were intentionally introduced for one or more reasons (e.g. for ornament, fodder, food, processing) (Mack, 1991). Other weeds that were not intentionally introduced entered as contaminants of trade goods or via other human pathways (Hulme et al., 2008). Because weeds and invasive plants are relatively difficult to eradicate once established, resource managers advocate that preventing their initial entry should be a high priority (Orton, 1914; Westbrooks and Eplee, 1996; White and Schwarz, 1998; Pimentel et al., 2000). However, not all introduced plants are pests or will become pests. It has been estimated that only about 0.1% to 2% of introduced plants become significant

Box 18.1. Invasive plant terminology.

The term invasive species is commonly used to refer to undesirable non-native species that spread and have some type of negative impact. This term is widely used in government documents, scientific literature and electronic resources. Unfortunately, it has been defined in a variety of ways, leading to confusion (see Occhipinti-Ambrogi and Galil, 2004). Because it is unlikely that managers, policy makers, scientists and lay people will reach a consensus on the definition of this term, it is vital that we be aware of how the term is defined in different works, particularly when conducting risk analysis. Below are different definitions of invasive species from five commonly cited sources. Following, is a short discussion on the underlying conceptual differences among these definitions.

Example Definitions

United States Executive Order on Invasive Species (US Government, 1999) – an alien species whose introduction does or is likely to cause economic or environmental harm or harm to human health.

  National Research Council (National Research Council, 2002) – Non-indigenous species that arrive in a new range, establish, proliferate, spread and cause broadly defined detrimental consequences in the environment.

  Richardson *et al.* (2000) – Naturalized plants [species] that produce reproductive offspring, often in very large numbers, at considerable distances from parent plants, and thus have the potential to spread over a considerable area.

  Heger and Trepl (2003) – A plant invader is any plant species that occurs at a location outside its area of origin; the occurrence of the species must have been prevented in the past by a barrier to dispersal, and not by the conditions in the new habitat.

  CBD (COP 6 Decision VI/23) – An alien [non-native] species whose introduction and/or spread threatens biological diversity.

Stage of Invasion

The process of biological invasion occurs as a series of steps or stages proceeding from entry, to escape, naturalization, and finally, to spread (Heger, 2000; Williamson, 2006). Most concepts of invasive species relate to species that have already established and spread throughout a region, and have begun expressing negative impacts. However, Heger and Trepl's definition of invasive species includes species at earlier stages of the process. They consider a species to be invasive if it colonizes new regions after overcoming barriers to dispersal (e.g. mountain ranges, oceans).

Spread and Impact

The generalized concept of invasive species includes elements of spread and impact. While many definitions consider each of these to a lesser or greater extent, some consider only one. For example, the United States Executive Order on Invasive Species defines invaders as alien species that have negative impacts, without referring to spread. In contrast, Richardson *et al.* define them as naturalized species that reproduce and spread away from their parent populations, without referring to impact. These definitions present profoundly different concepts of the term, and ultimately reflect two perspectives on the issue of biological invasions.

Impacted Systems

The term 'invasive species' has more often been applied to non-native species that have negative impacts in conservation areas, national parks and other wild lands. Some definitions of invasive species restrict their definitions to species impacting the natural environment and biodiversity (e.g. the CBD definition), whereas other definitions include invaders of agricultural and human landscapes (e.g. US Executive Order).

Definition adapted from the definition for invasion.

weeds (Williamson and Fitter, 1996). Given these low rates of invasiveness, how do national plant protection organizations evaluate which plants are safe to import and which contaminants do not require mitigation measures? How do we determine if a plant is going to become weedy or invasive? The answer is through weed risk assessment (WRA).

Similar to pest risk assessments of other types of organisms, a WRA evaluates the likelihood of entry, establishment and spread of a weed, and the potential for direct and indirect impacts. WRAs can be used to not only screen proposed imports, but also to evaluate new weed incursions, and prioritize management efforts (Groves et al., 2001). Because weeds are included in the IPPC's definition of a pest (IPPC, 2010), the IPPC guidelines developed for other types of pests (i.e. ISPM No. 11, IPPC, 2004) are also relevant to weeds.

However, WRAs are generally approached somewhat differently to assessments of other types of species. This is partially an artefact of how the discipline of weed risk assessment developed, but also due to differences in plant policy and regulations. We begin this chapter by taking a brief look at the history of weed risk assessment, followed by a review of the diversity of WRA systems available. We then continue with a short discussion of how WRA processes can be reframed into a traditional pest risk assessment process. Finally, this chapter concludes with some of the challenges associated with WRAs.

18.2. Development and History of WRA

18.2.1. Early Foundations

Interest in characterizing and assessing plants for their weed potential probably started in 1965 when Herbert Baker published a list of weed traits of agricultural and ruderal weeds (Baker, 1965). His list provided a foundation for research in weed ecology and the development of weed risk assessment tools. Later, Baker refined his list somewhat and published it in a paper on 'The evolution of weeds' (Baker, 1974). He categorized weeds as species that can germinate in a wide range of environments, display great longevity of seed, grow quickly to reproductive age, self-pollinate, produce seeds under a wide range of conditions, use multiple dispersal mechanisms and be tolerant if not resilient to disturbance and competition (Box 18.2). Although no single plant will have all of these traits, almost all will have a few of these, particularly if they colonize disturbed environments. Most of Baker's weed traits are evaluated by current WRA systems.

Several other works at the time of Baker's publications also helped lay the foundations for weed risk assessment. For example, Charles Elton in The Ecology of Invasions by Plants and Animals drew attention to biotic invasions of natural communities and started the field of invasion biology (Elton, 1958). Sir Edward Salisbury in Weeds and Aliens provided historical accounts of weed spread and discussed the factors associated with their success (Salisbury, 1961). LeRoy Holm and others in A Geographical Atlas of World Weeds assembled a tremendous amount of information on the distribution and relative importance of agricultural weeds from around the world (Holm et al., 1979).

Prior to pest risk assessment (sensu IPPC), compilations of global or regional weeds (e.g. Reed, 1977; Holm et al., 1979) served as the basis for initial quarantine lists. In the early 1980s in Australia, plant species were identified for prohibition on the basis of their behaviour elsewhere, close relationship to other weeds or past recognition as a weed in Australia (Forcella and Wood, 1984). Panels of weed experts often reviewed the compilations to identify the best candidates. In the USA, a Technical Committee to Evaluate Noxious Weeds recommended the set of 76 species that were initially listed under the Federal Noxious Weed Act of 1974 (Gunn and Ritchie, 1988). Although use of expert opinion or panel determinations to develop weed lists may be simple and quick, some groups are concerned that it is subjective and not transparent (Fox et al., 2003).

The concept and use of weed risk assessments to evaluate weed potential began to crystallize in the 1980s when several workers argued that plants should be evaluated with a systematic science-based process prior to importation (see Forcella and Wood, 1984; Hazard, 1988). In 1988, Hazard (1988) proposed a short scoring system that led to the development of the Australian WRA system (Panetta, 1993; Pheloung, 1995). From there, additional WRA systems were designed and validated, some for small geographic areas (Jefferson et al., 2004; Widrlechner et al., 2004), others for specific taxonomic or ecological groups of plants (Tucker and Richardson, 1995; Rejmánek and Richardson, 1996; Reichard and Hamilton, 1997) and others with a much wider scope (Panetta and Mitchell, 1991; Pheloung et al., 1999; Champion and Clayton, 2000; Lehtonen, 2001).

While researchers and regulatory officials were designing WRAs to support quarantine and other policy decisions, natural resource managers were developing WRA systems to help prioritize weeds and invasive plants already established for management. Motivation for these kinds of assessment systems started when workers realized that they needed an objective and standard way to characterize the wide range of impacts associated with pest plants (Esler, 1988). In addition to evaluating weed impacts, these WRA systems also consider the overall biotic potential of the plant, the conservation value of protected resources, the potential for additional weed spread, and the feasibility of control (Hiebert and Stubbendieck, 1993; Owen et al., 1996; Virtue et al., 2001).

18.2.2. Predicting which species will become weeds

Following Baker's publications, researchers continued studying weed traits in the hope of using them to identify weeds and invasive plants before they are imported to a country. Although technically not a plant trait, researchers believed that one of the best predictors of weediness is a species' behaviour elsewhere,

particularly in the species' introduced range (Daehler and Strong, 1993; Scott and Panetta, 1993; Mack, 1996; Reichard and Hamilton, 1997). This idea probably originated from observations and biogeographic studies showing that different countries and regions have similar weeds, or that weeds in a given area, were also weedy elsewhere (Holm et al., 1977). For example, comparison of Australia's noxious weeds to a list of global weeds showed that 90% of Australia's weeds are considered weeds in one or more other countries (Panetta, 1993). Thus, assuming two regions are climatically similar, it is not unreasonable to assume that a weed of some one region or country is likely to become a weed in another.

Numerous studies since have verified or supported the predictive ability of behaviour elsewhere and have recommended its use in weed risk assessments (Ruesink et al., 1995; Maillet and Lopez-Garcia, 2000; Caley and Kuhnert, 2006; Gordon et al., 2008). Two commonly cited screening tools consider the status of a species elsewhere in some fashion (Reichard and Hamilton, 1997; Pheloung et al., 1999). The Australian WRA considers status elsewhere in four different questions (Pheloung et al., 1999). One question addresses whether the species has naturalized elsewhere, while the other three address whether it is recognized as a weed of natural, agricultural or disturbed habitats. One of the first weed screening tools that was developed used behaviour elsewhere as a question; in fact, scoring a yes on this question alone was enough to support quarantine action (Hazard, 1988).

Another good predictor of weed potential is the breadth or range of a species' native distribution (Forcella and Wood, 1984; Goodwin et al., 1999). This factor is a good predictor because it is correlated with the overall adaptive potential of a species (Ren and Zhang, 2009). Widely distributed species may be better adapted to a wider range of environmental conditions than species with narrow distributions. Adaptive potential not only increases the likelihood that a given species will establish in a new place, but it increases the potential range over which it may express any impacts (Forcella et al., 1986; Forcella and Wood, 1984).

History elsewhere, breadth of native distribution and many other plant traits have been evaluated for their predictive potential. However, the results have not been very consistent (Lodge, 1993). Traits that have been shown to be predictive in some contexts are not predictive in others, or not as important as other traits (Sutherland, 2004). For example, it is commonly believed that vegetative reproduction is associated with weediness because these species can colonize new areas from single individuals and because they may be more resilient to disturbances that break up plants (e.g. tilling) (Baker, 1974; Reichard, 2001; White and Schwarz, 1998). Some studies show that significantly more invasive species reproduce vegetatively than non-invasive species (e.g. Reichard, 2001). One study found no difference in vegetative reproduction between invaders and non-invaders (McClay *et al.*, 2010). And yet another study found that significantly more non-weeds than weeds reproduced vegetatively (Sutherland, 2004).

Inconsistent performance of plant traits as predictors of weed potential, has led some authors to conclude that predictive models and screening systems can never truly predict whether a species will be invasive (Williamson, 2001; Moles *et al.*, 2008). Some researchers maintain that to be successful, predictive WRAs also need to consider traits of the recipient environment, including patterns of disturbance, resource availability and potential biotic interactions (Mack, 1996; White and Schwarz, 1998; Moles *et al.*, 2008). While this is certainly a more comprehensive approach, it will not be practical for most species (Daehler and Strong, 1993), particularly those for which there is little biological information available. For this reason, weed risk assessments that focus on simple biological traits of the species and that can be completed within a couple of days (e.g. the Australian WRA approach) have been popular.

18.3. Diversity of WRAs

Since about the mid-1990s, numerous weed risk assessment systems have been developed

to evaluate weed risk. Because of the diverse needs that led to their development, WRAs vary greatly not only in what elements of risk they evaluate, but also in how they evaluate them and how much they contribute to the final risk score. As a basis for discussion, we have selected ten different WRA models. We list some of their features in Table 18.1, including their format, style and composition. In particular, for each WRA model, we note which elements of risk they evaluate and the extent to which those elements determine the risk score or outcome of the assessment. The ten WRA models were selected to represent the diversity of WRA models available. Some of these are being used to evaluate weeds (e.g. Weber *et al.*, 2009), but a few do not appear to have been applied since their publication.

18.3.1. Pre-border and post-border WRAs

The majority of WRA models can be broadly classified into one of two types based on their goals: pre-border screening tools and post-border characterization/prioritization tools (Table 18.1).[1] Screening tools aim to identify or rank the risk potential of a given species becoming weedy or invasive prior to introduction (Caley and Kuhnert, 2006; Jefferson *et al.*, 2004). Although they are primarily used as predictive tools, they have also been used to evaluate the risk posed by species recently established in the pest risk analysis area. Screening tools typically rely on information about the species' behaviour elsewhere, its inherent traits (e.g. life form, fruit type, dispersal strategy) and its climatic tolerances to evaluate the species' likelihood of establishment, spread and potential impacts in the pest risk analysis area (Table 18.1) (Pheloung *et al.*, 1999). Species obtaining higher risk scores are more likely to be invasive or cause significant impacts, and are recommended for exclusion. Because there are many species for which information is limited, pre-border WRA models often consider whether the taxon has any similar relatives (i.e. congeners) that are also weeds. Species with weedy relatives obtain higher risk scores and are more likely to be rejected from entry into the pest risk analysis area ([e.g. the Reichard and Hamilton model (1997)].

Post-border, characterization and prioritization schemes, in contrast, aim to rank or classify the overall risk potential of a species after it has invaded an area. Focusing more on information from the invaded region, these risk assessments are primarily used to characterize a species' current and potential impact to help develop management priorities (e.g. NatureServe's characterization tool; Morse et al., 2004; Randall et al., 2008). Like pre-border WRAs, characterization and prioritization tools consider factors relating to establishment, reproduction and spread as these affect species invasiveness and management strategies. However, they tend to de-emphasize behaviour elsewhere and climatic suitability, while emphasizing species impacts (Table 18.1; Box 18.2). They also usually consider the current and potential distribution of the species in the country/region, and the factors affecting its feasibility of its control. Typically, species that have only invaded a small portion of their potential range and species that would be relatively easy to manage receive higher risk scores (e.g. The Exotic Species Ranking System; Hiebert and Stubbendieck, 1993). In contrast, species that are widely distributed or difficult to manage would receive lower risk scores. Thus, post-border tools usually prioritize species that can be eradicated, or contained (Downey et al., 2010).

Weed screening and characterization/prioritization tools consider similar kinds of information (Table 18.1), and often ask the same kinds of questions; however, the ways in which they consider that information may differ between them. For example, seed banks contribute to the establishment potential of a species by helping plants avoid unfavourable environmental conditions in time. Screening tools usually consider seed banks as a trait of weedy species and use it as an indicator of potential invasiveness (e.g. the Australian WRA; Pheloung et al., 1999). In contrast, prioritization tools may consider seed banks under feasibility of control (e.g. the Exotic Species Ranking

System; Hiebert and Stubbendieck, 1993) because plants with seed banks are much more difficult to manage and eradicate (e.g. Panetta and Timmins, 2004; Vince, 2011). Several other plant traits are considered by both types of WRAs, but may be used differently in the assessment process; these include allelopathy, vegetative reproduction and dispersal strategies.

Recently, a third type of WRA has been recognized (Downey et al., 2010). These WRAs evaluate the risk of native species extinction or extirpation by widely distributed weeds. In addition to some of the risk factors discussed above, these types of WRAs consider the proximity of invasive weeds to native species, the feasibility of weed control, the recoverability of the native species following management and the degree of urgency for management (Downey et al., 2010). Essentially, they help managers prioritize sites for weed management. Because widely distributed pests do not meet the IPPC definition of a quarantine pest (ISPM No. 5, IPPC, 2010), this chapter does not consider this third of type of WRA any further.

18.3.2. Structure, scoring and conclusions

The WRAs highlighted in Table 18.1 take different approaches to evaluating risk. Some screening systems are designed as decision trees that usually focus on a few factors that are strongly associated with weediness and invasiveness (Panetta, 1993; Reichard and Hamilton, 1997; Widrlechner et al., 2004). Based on the answers to the questions, assessors are then led down a particular path in each tree, ultimately ending in some type of conclusion about the risk potential of the species. Some decision trees may result in a conclusion in as little as two questions. For example, in the Reichard and Hamilton model (Reichard and Hamilton, 1997), species that invade elsewhere outside of North America (question #1) and that have relatives that are already strongly invasive in North America (question #2) are recommended for rejection (i.e. quarantine pest status).

Table 18.1. Variation in the content and structure of ten different weed risk assessment systems.

Question category	Pre-border screening tools					Post-border characterization and prioritization tools				
	Panetta System for Australia	Reichard and Hamilton N.A. Model	Weber and Gut European WRA	PPQ WRA for the US	Australian WRA	NatureServe's Invasive Species Assessment	Exotic Species Ranking System	New Zealand Aquatic Species WRA	Victoria's Noxious Weed Review	Xiamen WRA
Format[a]	DT	DT	MC	Y/N	Y/N	MC	MC	MC	MC	MC
No. of questions	7	7	12	62	49	20	24	35	39	19
Considers uncertainty	No	No	No	Yes	No	Yes	No	No	Yes	No
Contribution to final score[b]										
Climate match	**	*	*	Yes[c]	*	*	*	**		**
Likelihood of entry				Yes[c]						**
Behaviour elsewhere	**	*	*	**	**			**		*
Congeners	**	*	*	*	*					
Establishment	**	**	**	*****	****	**	*	**	*	**
Invasiveness (repro&spread)	***	***	***	***	***	**	**	***	*	**
Impact	**	****	****	**	**	****	****	****	******	****
Current or potential distribution						***			****	*
Feasibility of control						**	*****	**		**

[a]Format: Decision tree (DT); questionnaire of primarily Yes/No (Y/N) questions; primarily multiple choice (MC) questions.
[b]WRA questions were categorized into these nine broad categories. The number of asterisks represents the approximate percentage of points each category contributes to the final weed risk assessment score in increments of 10%. For example, 1–10% (*), 11–20% (**), 21–30% (***), etc.
[c]The PPQ WRA considers climatic compatibility and likelihood of entry and likelihood of establishment and likelihood of entry as two separate risk elements. They do not contribute to the risk score that categorizes the species' overall weedy and invasive potential.

Citations for WRA systems from left to right: Panetta System for Australia (Panetta, 1993); Reichard and Hamilton North American model (Reichard and Hamilton, 1997); Weber and Gut European WRA (Weber and Gut, 2004); Plant Protection and Quarantine (PPQ) WRA (Koop et al., in review); Australian WRA (Pheloung et al., 1999); NatureServe's Invasive Species Assessment (Morse et al., 2004); the Exotic Species Ranking System (Hiebert and Stubbendieck, 1993); The New Zealand aquatic species WRA (Champion and Clayton, 2001); Victoria's Noxious Weed Review (DPI, 2008); Xiamen WRA (Ou et al., 2008).

Box 18.2. Examples of questions from pre-border and post-border WRA tools. Questions selected were those that are more characteristic of that type of WRA.

Pre-border

- Is the species a free-floating (surface or submerged) aquatic or can it survive, grow and reproduce as a free-floating aquatic? (yes or no) (Panetta, 1993).
- Is it in a family or genus with species that are already strongly invasive in North America? (yes or no) (Reichard and Hamilton, 1997).
- What is the size of the global range (native and introduced)? (A) Range is small, species is restricted to a small area within one continent; or (B) range is large, extending over more than 15° latitude or longitude in one continent, or covers more than one continent (Weber and Gut, 2004).
- Invasiveness elsewhere. Select one: Species (A) introduced elsewhere long ago (>75 years) but not escaped; (B) introduced recently (<75 years) but not escaped; (C) never introduced elsewhere; (D) escaped/casual; (E) naturalized; or (F) invasive. (Koop et al., in review).
- Species suited to Australian climates (0 – low; 1 – intermediate; 2 – high) (Pheloung et al., 1999).

Post-border

- Proportion of potential range currently occupied: (A) <10%; (B) 10–30%; (C) 30–90%; (D) >90%; or (U) unknown (Morse et al., 2004).
- Level of effort required for control: (A) Repeated chemical or mechanical control measures; (B) one or two chemical or mechanical treatments required; (C) can be controlled with one chemical treatment; or (D) effective control can be achieved with mechanical treatments (Hiebert and Stubbendieck, 1993).
- Dispersal outside catchment by natural agents, e.g. birds, wind. 0–5 points maximum if propagules well adapted for distribution by birds/wind, 1 if propagule could be spread in bird crop (Champion and Clayton, 2001).
- Impact on water flow within watercourses or water bodies? (yes or no) (DPI, 2008).
- Proportion of current range where the species causes negative impact: (A) <5%; (B) 5–20%; (C) 21–50%; (D) impacts occur in >50% of the species' current generalized range (Ou et al., 2008).

Decision trees are usually relatively short, consisting of only a few questions (Fig. 18.1). The two that are listed in Table 18.1 each consider seven different questions, but others have included as few as four (Caley and Kuhnert, 2006). While this structure favours a quick response to new and emerging threats, their simplicity may limit their widespread applicability (Křivánek and Pyšek, 2006). Relative to the other types of WRAs, a disadvantage of decisions trees is that they do not produce a risk score or other standardized measure of risk. This precludes any kind of comparison of the relative risks of assessed species. Another disadvantage lies in what to do when a particular question cannot be answered. Under these circumstances, the path down the tree is essentially halted. To address this, decision trees can be made more complex by adding surrogate ques-

tions (questions that are correlated with the original question, and that would produce similar responses; Caley and Kuhnert, 2006).

The other WRAs shown in Table 18.1 consist of a series of yes–no and multiple choice questions that are evaluated for all species. They usually consider many more factors than decision tree models. The answers to the questions are scored and combined in some fashion to produce a final score. Because these models rate all species with the same questions, they permit users to compare species risk scores. Screening tools that use these approaches usually relate species risk scores to decision thresholds to evaluate the appropriate risk management action (Fig. 18.2a). Characterization and prioritization tools, in contrast, may or may not apply decision thresholds to reach a conclusion. For these, the primary goal is to rank and compare species

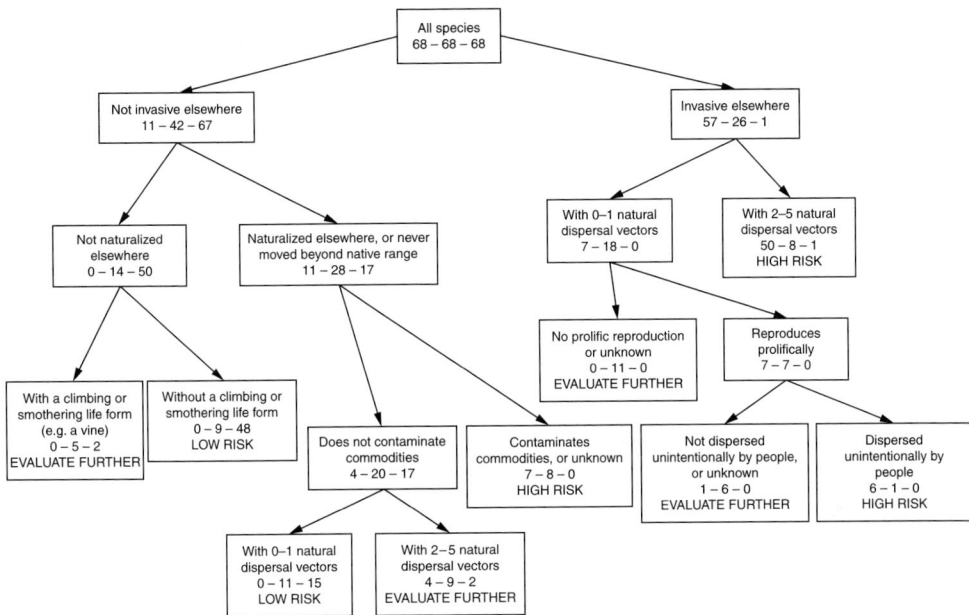

Fig. 18.1. This decision tree was created using data for 204 species whose invasive status in the USA is already known. The number of major-invaders (left), minor-invaders (middle) and non-invaders (right) is shown at each node. The tree was created using Jump's (v. 8) recursive partitioning technique. At each point or node in the tree, Jump selected the best question from among 11 that partitioned the species according to the *a priori* groups.

risk scores. Species with the highest scores are considered the riskiest and are usually targeted for management.

The ways in which question-based WRAs combine question scores differ. Some simply add all scores together to produce a final risk score (e.g. Weber and Gut, 2004), while others first sum scores for similar questions into elements of risk (e.g. establishment, spread, impact). These elements may then be summed or multiplied together to obtain a final risk score (e.g. Parker *et al.*, 2007). Other types of scoring systems are also used. For example, The Exotic Species Ranking System, which was developed to prioritize species for management, graphically 'combines' its two primary risk elements (degree of impact and feasibility of control). It plots the paired risk scores on a graph to evaluate which species would be good candidates for management (Fig. 18.2b; Hiebert and Stubbendieck, 1993). Species appearing in the upper right quadrant of the graph are usually prioritized for management (greater threat and easily controlled).

The PPQ WRA system, which was recently developed as a predictive screening tool (Koop *et al.*, in review), uses its two primary risk scores (establishment/spread potential and impact potential) in a logistic regression model to estimate the probabilities that a given species will be a major-, minor- and non-invader. For any given species, all three probabilities sum to one. This system then uses decision thresholds to determine if the species represents a low- or high-risk potential, or if they should be evaluated further using another tool (Fig. 18.2c).

WRA models with quantitative or semi-quantitative risk scores usually use threshold values to divide the risk space into two or more regions. Models with two risk regions classify species as low or high risk. Those with a third region often classify species with intermediate risk scores as requiring further evaluation (Fig. 18.2). Species with intermediate risk scores may have a mixture of weed and non-weed traits and truly represent moderate risk, or, they may be low-risk

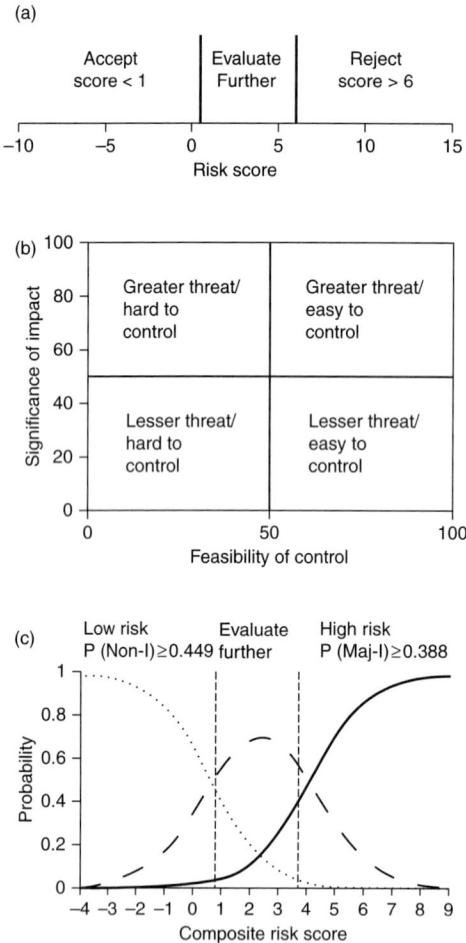

Fig. 18.2. Examples of three different methods for determining WRA outcome. (a) Through an additive model, the Australian WRA (Pheloung et al., 1999) generates one risk score that is evaluated against two decision thresholds. (b) In the Exotic Species Ranking Tool (Hiebert and Stubbendieck, 1993) risk scores for the two risk elements are graphically 'combined' to help managers set priorities. (c) In the PPQ WRA (Koop, in review), the risk scores for establishment/spread potential and impact potential are combined in a logistic regression model to determine the probabilities the species is a major-invader (solid line), minor-invader (dashed line) and non-invader (dotted line). Decision thresholds for 'low risk', 'evaluate further' and 'high risk' categories are shown as two vertical lines.

or high-risk species whose true risk potential could not be resolved due to high uncertainty. While creating three categories of

risk potential is reasonable, it presents a challenge to risk managers who need to determine the appropriate response. Unfortunately, for plants that are imported for planting, there are usually only two management options: allow or deny plant entry.

In some WRA models, the percentage of species classified as intermediate risk can be as high as 30%. In such cases, a secondary screening tool can be used to further evaluate their risk potential (Daehler et al., 2004). These tools focus the evaluation on traits that are strongly associated with weeds, with the goal of classifying the species as either low- or high risk. Daehler et al. (2004) developed one for the Australian WRA that reduced the number of species requiring further evaluation from 30% to 10%. Secondary screening tools developed for the Australian WRA and the PPQ WRA (Daehler et al., 2004; Koop et al., in review) rely on questions that are already evaluated in the risk assessment. However, it is entirely possible to design a tool that relies on a different set of criteria and methodologies (e.g. greenhouse or field studies, Mack, 2005). Assuming that risk analysts and managers cannot resolve the 'evaluate further' species with additional research, what would be the best action for these species?

18.3.3 Uncertainty

Uncertainty and risk are intrinsically linked because knowledge about the factors that affect risk is never perfect or complete. Many factors contribute to uncertainty, including missing and incomplete information, inconsistent or conflicting data, dated or erroneous information, natural variability of biological systems, subjectivity of analysis and sampling error (IPPC, 2004). Some types of uncertainty can be reduced with further investigation (e.g. missing information), but other sources cannot (e.g. natural variability). In the *Framework for Pest Risk Analysis* (ISPM No. 2, IPPC, 2007), the IPPC clarifies that risk analyses should document the nature and degree of uncertainty involved in the analysis. This is not only critical for a proper evaluation of risk, but is also important for transparency.

Despite the fundamental importance of uncertainty in risk analysis, very few WRA systems consider it, either implicitly or explicitly (Table 18.1). This is likely to be an artefact of the historical development of weed risk assessment outside the field of risk analysis. Of the ten weed risk assessment systems listed in Table 18.1, only three consider uncertainty to some extent. For example, in their multiple choice assessment tool, NatureServe allows users to enter letter ranges (e.g. AB, ABC, BC) when only one or two answer choices can be eliminated (Morse et al., 2004). When scored, these answers can be given a score range that is carried through until the final risk score. The Victorian WRA process (DPI, 2008) records a measure of confidence in document quality for each question. Peer reviewed articles receive the highest score possible (1), and 'no data or reference material' receives the lowest score (0). While this confidence measure documents an important component of uncertainty, it does not document other aspects of it such as variation in data, or conflicts in the literature.

Of the WRAs listed in Table 18.1, the PPQ WRA process provides the most rigorous documentation and evaluation of uncertainty. In that system, the answer to each question is accompanied by a qualitative measure of uncertainty that ranges from negligible to maximum. The uncertainty measure includes all factors that can potentially affect our confidence in the answer, including literature quality, and data consistency. Questions that are not answered due to missing information receive the maximum level of uncertainty (maximum). These levels are used to calculate an index of uncertainty for each risk element. Question weighting is factored into the index such that questions that contribute relatively more to the risk score, contribute more to the index. The final index is standardized to range from 0 to 1.

The PPQ WRA process, also evaluates which other risk scores are likely to be possible based on the uncertainties. Specifically, it asks the question: What would the final risk score have been if the questions had been answered somewhat differently? It does this through a Monte Carlo simulation process where the answers to the questions are randomly generated based on the level of uncertainty (see Chapter 10 for a more detailed explanation of the Monte Carlo process). Answers with higher uncertainty levels are more likely to change than those with lower uncertainty levels. For questions that are not answered (i.e. the unknowns), all answer choices are equally likely. The Monte Carlo simulation is run 5000 times, and all simulated risk scores are plotted in an Establishment/Spread vs. Impact risk space. Assessments with higher levels of uncertainty will have a larger spread of simulated risk scores (represented by boxes) around the observed result, than assessments with lower levels of uncertainty (Fig. 18.3). The PPQ uncertainty simulation is a valuable tool because it helps risk analysts evaluate the robustness of the WRA's conclusion. An assessment's conclusion is robust when the majority of the simulated risk scores result in the same conclusion as the analyst-derived risk score.

18.4. Reframing Weed Risk Assessments into an IPPC Risk Analysis Context

Although not framed or organized like the pest risk model described in ISPM No. 11, most WRAs are consistent with the principles of risk analysis and the concepts and risk factors addressed in ISPM Nos 2 and 11 (Downey et al., 2010). The key to reconciling the difference is in recognizing the risk that is being evaluated, and understanding how the risk factors of the WRAs address the likelihood and consequences of the hazard. Some WRAs may build upon prior knowledge of weed risk, and instead focus on evaluating whether risk management is necessary or what risk management strategies may be effective.

Once the risk being evaluated is known, it should be relatively easy to restructure the assessment into a more traditional risk model. Some WRA systems organize their questions (i.e. risk factors) into likelihood and consequence risk elements, or elements that are

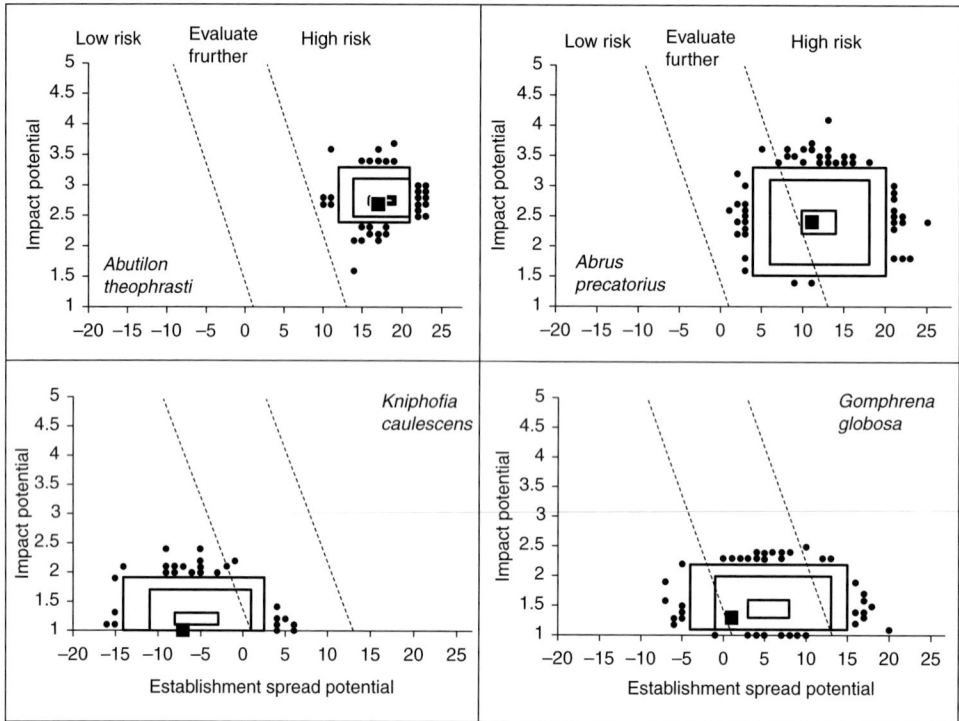

Fig. 18.3. Uncertainty simulations for four species assessed with the PPQ WRA. (Figures show the distribution of 5000 simulated risk scores (points and boxes) produced using a Monte Carlo process. Boxes contain 99%, 95% and 50% of the simulated risk scores. All simulated scores outside of the 99th percentile box are shown as points. The observed risk score (analyst-derived) produced by the WRA model is represented by the square. Of the four assessments shown, the conclusion of High Risk for *A. theophrasti* is the most robust because all simulated risk scores result in the same conclusion. Next are the conclusions for *A. precatorious* and *K. caulescens* because the majority of the simulated risk scores result in the same conclusion as the observed risk score. The conclusion for *G. globosa* is the least robust because based on the simulated risk scores result in all three types of conclusions.)

clear components of these. In other systems, the questions are organized into categories that do not necessarily parallel the components of pest risk assessment (*sensu* ISPM No. 11). However, regardless of how they are organized and how the question scores are combined, most WRAs address aspects of the likelihood and consequences of the hazard being evaluated (Table 18.1). Daehler and Virtue (2010) reorganized the 49 questions in the Australian WRA in this fashion and showed that it improved the predictive accuracy of the model.

Pre-border weed risk assessments differ from other types of pest risk assessments in that most are done on species that are or will be intentionally introduced and cultivated. Consequently, they do not need to consider the likelihood of species introduction (i.e. entry) (Table 18.1). For intentionally introduced species, the IPPC recommends that the analysis focuses on the likelihood of spread from the intended habitat of introduction (i.e. gardens, farmlands, etc.) to unintended habitats such as conservation lands, farmlands, recreational sites, etc. (IPPC, 2004). Almost all pre-border WRA models consider factors that contribute to the establishment and spread potential of a species (Table 18.1) and are thus consistent with ISPM No. 11 (IPPC, 2004). Although none of them specifies or distinguishes between intended and unintended habitats, they assume the risk concerns spread from intended habitats to unintended ones.

18.5. Challenges in Weed Risk Analysis

All groups of organisms pose their own unique set of challenges for risk analysts. Pathologists, for example, struggle with the complexities of fungal taxonomy, and the peculiarities of pathogen–host interactions. Entomologists must consider insect reproductive strategies and the effect of insect predation on colonizing species. It is not any different for the assessment of weeds. Plants differ from other potential groups of pests because people intentionally introduce, promote, breed and cultivate plants at levels far exceeding that of any other taxonomic group. This has resulted in some unique challenges for risk analysts and managers that we briefly explore below.

Plants serve vital functions in our society. They are used for food, ornament, construction, manufacturing, biochemicals, shipping, cultural ceremonies and, more recently, biofuels (Simpson and Conner-Ogorzaly, 1986). There are approximately 250,000–300,000 vascular plants on the planet. A significant portion of these have already been introduced to areas outside of their native distribution. For example, Australia has about 20,000 native species of vascular plants (see the Australian Plant Name Index: http://www.cpbr.gov.au/apni/index.html). An additional 26,000 species have been introduced in the last 200 years, with about 25,000 species currently in cultivation (Randall, 2007). This represents about 8.7–10% of the global flora.

The rate and volume at which people have spread plant species greatly exceeds any natural means of spread and is truly astounding. The high demand for plants challenges port inspectors who must deal with smuggling and improperly labelled consignments. The demand for novel taxa also challenges risk analysts faced with assessing those species. This has probably resulted in the proliferation of rapid WRA

systems in the last 20 years (Randall *et al.*, 2008; Fox and Gordon, 2009; Downey *et al.*, 2010). Because plants are grown in garden centres, sold by thousands of online companies, and traded by plant enthusiasts, it is probably easier to determine whether a given taxon is in the pest risk analysis area than it is for a pathogen or arthropod. However, because plants are privately grown, it may be more difficult to determine their distribution than it is for any other type of pest. Knowledge of the distribution and abundance of a taxon is necessary when evaluating the quarantine status of the organism.

Many of the plant species that our society has found useful have been modified to some extent to better suit our particular needs. We have created tens of thousands of new varieties and cultivars through plant breeding programmes (e.g. 20,000 cultivars of *Rhododendron*; Salley and Greer, 1992; OSU, 2006). While the vast majority of cultivars are not pest plants, some may be or may become so because (i) they are descended from weedy or invasive species; (ii) they were bred for traits associated with weediness (e.g. increased flower production); or (iii) their particular genotype results in hybrid vigour. Hybrid vigour or heterosis, is the increased performance or vigour associated with hybrids. It is the opposite of inbreeding depression, and results when a hybrid is more fit or aggressive than either parent. As more countries scrutinize the invasive potential of plant imports, particularly plants for propagation, there will probably be a need for assessment of the weed potential of plant cultivars. This will be particularly important for cultivars of invasive species that are bred for reduced weed potential (e.g. Japanese barberry; Lehrer and Brand, 2003). The IPPC supports the assessment of taxa at any taxonomic level, as long as it is scientifically justified (IPPC, 2004).

Analysis of the risk potential of plant cultivars presents several challenges for risk analysts. The most obvious concerns the availability of information on the cultivar. Although analysts are tasked with assessing the cultivar, in reality, there will be relatively little scientific information available on it, forcing analysts to fill in data gaps with information about the parental taxa. But at what point does the WRA stop being an assessment of the cultivar, and become an assessment of the parents? Furthermore, if the parents are themselves cultivars that are poorly known, how far back in the genealogy should analysts go? Figure 18.4 shows that most plant breeding histories are incomplete and very complex. Many cultivars are produced by hybridization of two relatively different taxa. Sometimes the unique combination of genetic material from both properties results in an invasive taxon, when neither parent was invasive (Schierenbeck and Ellstrand, 2009). How can we predict or assess this risk?

Risk analysts may also be interested in the risk that the cultivar's form is not 'stable' or that the cultivar may readily cross with other related taxa to produce offspring that are invasive. While these risks may be difficult to evaluate, particularly for new taxa, they are important to consider and reflect another challenge in WRA. For example,

many plant breeders develop variegated forms of commonly cultivated plants. Most variegated plants are not as robust (hardy) as their solid green parental types. But these species sometimes revert back to their original form. Cogongrass (*Imperata cylindrica*) is a highly invasive perennial grass that excludes most native species and changes the fire regime of ecosystems (Dozier *et al.*, 1998). This plant is regulated by the USA as a Federal Noxious Weed. Some states allow the sale of a cultivar of cogongrass known as 'Red Baron'. The cultivar is smaller, has red coloured leaves and is sterile. Recently, researchers have documented the reversion of 'Red Baron' to the original parental form (Kyde, 2010). The reverted plants resemble the invasive form of the species in that they are larger, more aggressive, green and reproductive.

Another good example of cultivar stability involves the popular landscape tree 'Bradford Pear', a cultivar of *Pyrus calleryana*. Being self-sterile, this taxon requires pollen from a different cultivar to produce

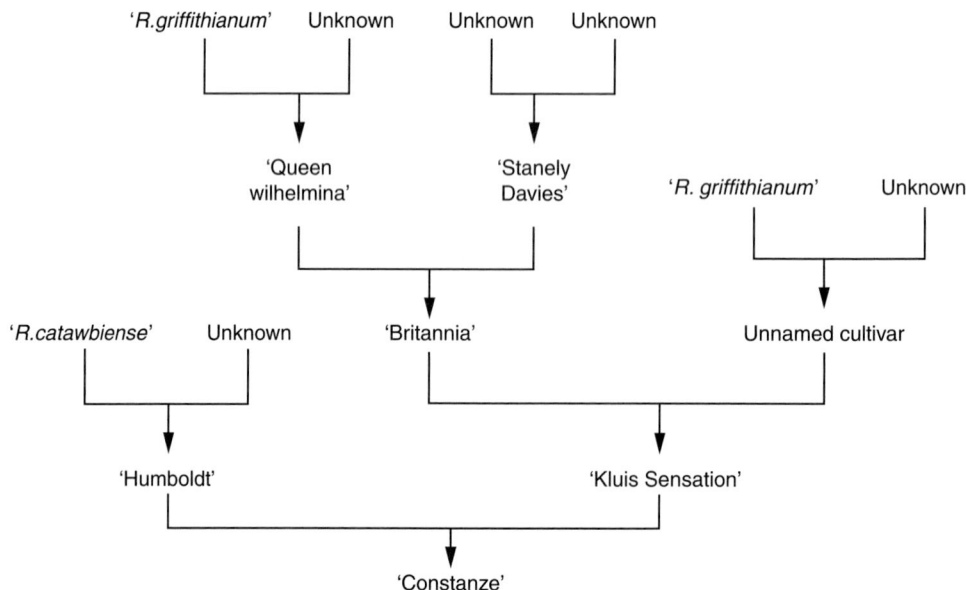

Fig. 18.4. The breeding history of *Rhododendron* 'Constanze'. (This genealogy shows how many cultivars are developed through interspecific crosses and how incomplete plant breeding histories can be. Names in italics represent species and not cultivars. Tree created from data in Salley and Greer (1992).)

fruit. However, as this cultivar and other varieties of *P. calleryana* have been planted more extensively in the USA, there have been more opportunities for cross-pollination. In recent years, this cultivar has established beyond horticultural settings and become invasive (Culley and Hardiman, 2009).

Beyond the scope of most WRAs is the risk of the evolution of weediness and invasiveness in the plants we import. Unlike pathogens and arthropods, we intentionally introduce plants from a variety of locations in their native range and grow these side by side in common gardens. This is a regular part of plant improvement and breeding programmes. However, admixing of previously separate genotypes may increase the evolutionary potential of a population by creating novel genotypes that may be more fit in the new environment than either parent (e.g. Ellstrand and Schierenbeck., 2000; Lavergne and Molofsky, 2007).

18.6. Summary

As discussed in this chapter, the discipline of weed risk assessment is relatively new. Yet because of the strong need for risk assessment tools, dozens of different WRA tools have been developed over the last 10 to 20 years around the world. As a result of diverse needs, these WRAs vary greatly in purpose, structure and content. Prior to the implementation, a WRA should be validated for the purpose and geographic area it is intended. Though not discussed in detail in this chapter, validation is a relatively simple procedure where the accuracy and error of a model is evaluated. If a given WRA is not entirely suitable it can be easily modified to meet most needs. Weed risk assessments can be used to identify the invasive potential of species before they are introduced, and prioritize resources for eradication and containment. They represent one of many useful tools for protecting agricultural and natural resources.

Note

[1] A third type of WRA that focuses on widely distributed species has been recently recognized, however, this chapter does not discuss this it. Please see Downey *et al.* (2010) for a discussion.

References

Baker, H.G. (1965) Characteristics and modes of origin of weeds. In: Baker H.G. and Stebbins, G.L. (eds) *The Genetics of Colonizing Species*. Academic Press, New York, pp. 147–172.

Baker, H.G. (1974) The evolution of weeds. *Annual Review of Ecology and Systematics* 5, 1–24.

Caley, P. and Kuhnert, P.M. (2006) Application and evaluation of classification trees for screening unwanted plants. *Austral Ecology* 31, 5, 647–655.

Champion, P.D. and Clayton, J.S. (2000) Border control for potential aquatic weeds. *Science for Conservation* 141, Department of Conservation, Wellington.

Champion, P.D. and Clayton, J.S. (2001) A weed risk assessment for aquatic weeds in New Zealand. In: Groves, R.H., Panetta, F.D. and Virtue, J.G. (eds) *Weed Risk Assessment*. CSIRO, Collingwood, pp. 194–202.

Culley, T. and Hardiman, N. (2009) The role of intraspecific hybridization in the evolution of invasiveness: a case study of the ornamental pear tree, *Pyrus calleryana*. *Biological Invasions* 11, 5, 1107–1119.

Daehler, C. and Virtue, J.G. (2010) Likelihood and consequences: reframing the Australian weed risk assessment to reflect a standard model of risk. *Plant Protection Quarterly* 25, 2, 52–55.

Daehler, C.C. and Strong, D.R. (1993) Prediction and biological invasions. *Trends in Ecology & Evolution* 8, 380.

Daehler, C.C., Denslow, J.S., Ansari, S. and Kuo, H.C. (2004) A risk-assessment system for screening out invasive pest plants from Hawaii and other Pacific Islands. *Conservation Biology* 18, 2, 360–368.

Downey, P.O., Johnson, S.B., Virtue, J.G. and Williams, P.A. (2010) Assessing risk across the spectrum of weed management. *CAB Reviews: Perspectives in Agriculture, Veterinary Science, Nutrition and Natural Resources* 5, 38, 1–15.

Dozier, H., Gaffney, J.F., McDonald, S.K., Johnson, E.R.R.L. and Shilling, D.G. (1998) Cogongrass in the United States: history, ecology, impacts, and management. *Weed Technology* 12, 4, 737–743.

DPI (2008) Victoria weed risk assessment (WRA) method. Department of Primary Industries (DPI), Victoria, Australia.

Ellstrand, N.C. and Schierenbeck, K.A. (2000) Hybridization as a stimulus for the evolution of invasiveness in plants. *Proceedings of the National Academy of Sciences of the United States of America* 97, 7043–7050.

Elton, C.S. (1958) *The Ecology of Invasions by Plants and Animals*. Chapman & Hall, London.

Esler, A.E. (1988) The naturalisation of plants in urban Auckland, New Zealand 6. Alien plants as weeds. *New Zealand Journal of Botany* 26, 585–619.

Forcella, F. and Wood, J.T. (1984) Colonisation potentials of alien weeds are related to their 'native' distributions: implications for plant quarantine. *Journal of the Australian Institute of Agricultural Science* 50, 35–41.

Forcella, F., Wood, J.T. and Dillon, S.P. (1986) Characteristics distinguishing invasive weeds within *Echium* (Bugloss). *Weed Research* 26, 5, 351–364.

Fox, A.M. and Gordon, D.R. (2009) Approaches for assessing the status of nonnative plants: a comparative analysis. *Invasive Plant Science and Management* 2, 166–184.

Fox, A.M., Gordon, D.R. and Stocker, R.K. (2003) Challenges of reaching consensus on assessing which non-native plants are invasive in natural areas. *HortScience* 38, 1, 11–13.

Goodwin, B.J., McAllister, A.J. and Fahrig, L. (1999) Predicting invasiveness of plant species based on biological information. *Conservation Biology* 13, 2, 422–426.

Gordon, D.R., Onderdonk, D.A., Fox, A.M., Stocker, R.K. and Gantz, C. (2008) Predicting invasive plants in Florida using the Australian weed risk assessment. *Invasive Plant Science and Management* 1, 178–195.

Groves, R.H., Panetta, F.D. and Virtue, J.G. (2001) *Weed Risk Assessment*. CSIRO Publishing, Collingwood.

Gunn, C.R. and Ritchie, C.A. (1988) Identification of disseminules listed in the Federal Noxious Weed Act (Technical Bulletin Number 1719). United States Department of Agriculture, Agricultural Research Service, USA.

Hazard, W.H.L. (1988) Introducing crop, pasture and ornamental species into Australia – the risk of introducing new weeds. *Australian Plant Introduction Review* 19, 19–36.

Heger, T. (2000) A model for interpreting the process of invasion: crucial situations favouring special characteristics of invasive species. In: Brundu, G., Brock, J.H., Camarda, I., Child, L.E. and Wade, P. M. (eds) *Plant Invasions: Species Ecology and Management*. Backhuys Publishers, Leiden, pp. 3–10.

Heger, T. and Trepl, L. (2003) Predicting biological invasions. *Biological Invasions* 5, 313–321.

Hiebert, R.D. and Stubbendieck, J. (1993) *Handbook for Ranking Exotic Plants for Management and Control*. US Department of Interior, National Park Service, Denver, Colorado.

Holm, L.G., Plucknett, D.L., Pancho, J.V. and Herberger, J.P. (1977) *The World's Worst Weeds: Distribution and Biology*. Krieger Publishing Company, Malabar, Florida.

Holm, L.G., Pancho, J.V., Herberger, J.P. and Plucknett, D.L. (1979) *A Geographical Atlas of World Weeds*. Krieger Publishing Company, Malabar, Florida.

Hulme, P.E., Bacher, S., Kenis, M., Klotz, S., Kühn, I., Minchin, D., Nentwig, W., Olenin, S., Panov, V., Pergl, J., Pyšek, P., Roques, A., Sol, D., Solarz, W. and Vilà, M. (2008) Grasping at the routes of biological invasions: a framework for integrating pathways into policy. *Journal of Applied Ecology* 45, 2, 403–414.

IPPC (2004) International Standards for Phytosanitary Measures, Publication No. 11: *Pest Risk Analysis for Quarantine Pests Including Analysis of Environmental Risks and Living Modified Organisms*. Secretariat of the International Plant Protection Convention (IPPC), Food and Agriculture Organization of the United Nations, Rome, Italy.

IPPC (2010) International Standards for Phytosanitary Measures, Publication No. 5: *Glossary of Phytosanitary Terms*. Secretariat of the International Plant Protection Convention (IPPC), Food and Agriculture Organization of the United Nations, Rome, Italy.

Jefferson, L., Havens, K. and Ault, J. (2004) Implementing invasive screening procedures: the Chicago Botanic Garden model. *Weed Technology* 18, 5, 1434–1440.

Křivánek, M. and Pyšek, P. (2006) Predicting invasions by woody species in a temperate zone: a test of three risk assessment schemes in the Czech Republic (Central Europe). *Diversity and Distributions* 12, 3, 319–327.

Kyde, K.L. (2010) Invader of the month: *Imperata cylindrica*. Maryland Invasive Species Council. Available online at: http://www.mdinvasivesp.org/archived_invaders/, accessed 8 February 2011.

Lavergne, S. and Molofsky, J. (2007) Increased genetic variation and evolutionary potential drive the success of an invasive grass. *Proceedings of the National Academy of Sciences of the United States of America* 104, 10, 3883–3888.

Lehrer, J. and Brand, M. (2003) Horticultural strategies to counter invasive Japanese barberry (*Berberis thunbergii*). *HortScience* 38, 5, 757–758.

Lehtonen, P.P. (2001) Pest risk assessment in the United States: guidelines for qualitative assessments for weeds. In: Groves, R.H., Panetta, F.D. and Virtue, J.G. (eds) *Weed Risk Assessment*. CSIRO, Collingwood, pp.117–123.

Lodge, D.M. (1993) Biological invasions: lessons for ecology. *Trends in Ecology & Evolution* 8, 4, 133–137.

Mack, R.N. (1991) The commercial seed trade: an early disperser of weeds in the United States. *Economic Botany* 45, 257–273.

Mack, R.N. (1996) Predicting the identity and fate of plant invaders: emergent and emerging approaches. *Biological Conservation* 78, 107–121.

Mack, R.N. (2005) Predicting the identity of plant invaders: future contributions from horticulture. *HortScience* 40(5), 1168–1174.

Maillet, J. and Lopez-Garcia, C. (2000) What criteria are relevant for predicting the invasive capacity of a new agricultural weed? The case of invasive American species in France. *Weed Research* 40, 1, 11–26.

McClay, A., Sissons, A., Wilson, C. and Davis, S. (2010) Evaluation of the Australian weed risk assessment system for the prediction of plant invasiveness in Canada. *Biological Invasions* 12, 4085–4098.

Moles, A.T., Gruber, M.A.M. and Bonser, S.P. (2008) A new framework for predicting invasive plant species. *Journal of Ecology* 96, 1, 13–17.

Morse, L.E., Randall, J.M., Benton, N., Hiebert, R. and Lu, S. (2004) An invasive species assessment protocol: evaluating non-native plants for their impact on biodiversity (v. 1). NatureServe, Arlington, VA, USA.

National Research Council (2002) *Predicting Invasions of Nonindigenous Plants and Plant Pests*. National Academy Press, Washington DC.

Occhipinti-Ambrogi, A. and Galil, B.S. (2004) A uniform terminology on bioinvasions: a chimera or an operative tool? *Marine Pollution Bulletin* 49, 688–694.

Orton, W.A. (1914) Plant quarantine problems. *Journal of Economic Entomology* 7, 109–116.

OSU (2006) Landscape plants, images, identification, and information: *Rhododendron* (Ericaceae). Oregon State University, Corvallis, OR. Available online at: http://oregonstate.edu/dept/ldplants/rhody.htm, accessed 10 August 2006.

Ou, J., Lu, C. and O'Toole, D.K. (2008) A risk assessment system for alien plant bio-invasion in Xiamen, China. *Journal of Environmental Sciences* 20, 989–997.

Owen, S.J., Timmins, S.M. and West, C.J. (1996) Scoring the weediness of New Zealand's ecological weeds. In: Shepherd, R.C.H. (ed.) *11th Australian Weeds Conference*. Weed Science Society of Victoria, Melbourne, pp. 529–531.

Panetta, F.D. (1993) A system of assessing proposed plant introductions for weed potential. *Plant Protection Quarterly* 8, 10–14.

Panetta, F.D. and Mitchell, N.D. (1991) Homoclime analysis and the prediction of weediness. *Weed Research* 31, 5, 273–284.

Panetta, F.D. and Timmins, S.M. (2004) Evaluating the feasibility of eradication for terrestrial weed incursions. *Plant Protection Quarterly* 19, 1, 5–11.

Parker, C., Caton, B.P. and Fowler, L. (2007) Ranking nonindigenous weed species by their potential to invade the United States. *Weed Science* 55, 386–397.

Pheloung, P. (1995) Determining the weed potential of new plant introductions to Australia. Agriculture Protection Board, Western Australia.

Pheloung, P.C., Williams, P.A. and Halloy, S.R. (1999) A weed risk assessment model for use as a biosecurity tool evaluating plant introductions. *Journal of Environmental Management* 57, 239–251.

Pimentel, D., Lach, L., Zuniga, R. and Morrison, D. (2000) Environmental and economic costs of nonindigenous species in the United States. *BioScience* 50, 1, 53–65.

Randall, J.M. (2007) *The Introduced Flora of Australia and its Weed Status*. CRC for Australian Weed Management, Department of Agriculture and Food, Western Australia.

Randall, J.M., Morse, L.E., Benton,., Hiebert, R., Lu, S. and Killeffer, T. (2008) The invasive species assessment protocol: a tool for creating regional and national lists of invasive nonnative plants that negatively impact biodiversity. *Invasive Plant Science and Management* 1, 1, 36–49.

Reed, C.F. (1977) *Economically Important Foreign Weeds*. Agricultural Research Service, United States Department of Agriculture, USA.

Reichard, S. (2001) The search for patterns that enable prediction of invasion. In: Groves, R.H., Panetta, F.D. and Virtue, J.G. (eds) *Weed Risk Assessment*. CSIRO, Collingwood, pp. 10–19.

Reichard, S.H. and Hamilton, C.W. (1997) Predicting invasions of woody plants introduced into North America. *Conservation Biology* 11, 1, 193–203.

Rejmánek, M. (1994) What makes a species invasive? In: Pyšek, P., Prach, K., Rejmánek, M. and Wade, P.M. (eds) *Plant Invasions*. SPB Academic Publishing, Amsterdam, pp. 3–13.

Rejmánek, M. and Richardson, D.M. (1996) What attributes make some plant species more invasive? *Ecology* 77, 6, 1655–1661.

Ren, M.X. and Zhang, Q.G. (2009) The relative generality of plant invasion mechanisms and predicting future invasive plants. *Weed Research* 49, 5, 449–460.

Richardson, D.M., Pyšek, P., Rejmanek, M., Barbour, M.G., Panetta, F.D. and West, C.J. (2000) Naturalization and invasion of alien plants: concepts and definitions. *Diversity and Distributions* 6, 93–107.

Ruesink, J.L., Parker, I., Groom, M. and Kareiva, P. (1995) Reducing the risks of nonindigenous species introductions. *BioScience* 45, 465–477.

Salisbury, E. (1961) *Weeds and Aliens*. Collins, London.

Salley, H.E. and Greer, H.E. (1992) *Rhododendron Hybrids* (2nd edn). Timber Press, Portland, Oregon.

Schierenbeck, K.A. and Ellstrand, N.C. (2009) Hybridization and the evolution of invasiveness in plants and other organisms. *Biological Invasions* 11, 5, 1093–1105.

Scott, J.K. and Panetta, F.D. (1993) Predicting the Australian weed status of southern African plants. *Journal of Biogeography* 20, 1, 87–93.

Simpson, B.B. and Conner-Ogorzaly, M. (1986) *Economic Botany: Plants in our World*. McGraw-Hill Publishing Company, New York.

Sinden, J., Hester, R.J.S., Odom, D., Kalisch, C., James, R. and Cacho, O. (2004) *The Economic Impact of Weeds in Australia*. CRC for Australian Weed Management, Adelaide.

Sutherland, S. (2004) What makes a weed a weed: life history traits of native and exotic plants in the USA. *Oecologia* 141, 1, 24–39.

Tucker, K.C. and Richardson, D.M. (1995) An expert system for screening potentially invasive alien plants in South African fynbos. *Journal of Environmental Management* 44, 309–338.

US Government (1999) Executive order 13112 on invasive species. *Federal Register* 64(25), 6183–6186.

Vince, G. (2011) Embracing invasives. *Science* 331, 1383–1384.

Virtue, J.G., Groves, R.H. and Panetta, F.D. (2001) Towards a system to determine the national significance of weeds in Australia. In: Groves, R.H., Panetta, F.D. and Virtue, J.G. (eds) *Weed Risk Assessment*. CSIRO, Collingwood, pp. 124–150.

Vogelmann, J.E., Howard, S.M., Yang, L., Larson, C.R., Wylie, B.K. and Van Drel, J.N. (2001) Completion of the 1990s national land cover data set for the conterminous United States. *Photogrammetric Engineering and Remote Sensing* 67, 650–662.

Vose, D. (2000) *Risk Analysis: a Quantitative Guide* (2nd edn). John Wiley and Sons, Ltd, Chichester.

Wade, T. and Sommer, S. (eds) (2006) *A to Z GIS*. ESRI Press, Redlands, California.

Wallner, W.E., Cardé, R.T., Xu, C., Weseloh, R.M., Sun, X., Yan, J. and Schaefer, P.W. (1984) Gypsy moth (*Lymantria dispar* L.) attraction to disparlure enantiomers and the olefin precursor in the People's Republic of China. *Journal of Chemical Ecology* 10, 5, 753–757.

Wearing, C.H., Thomas, W.P., Dugdale, J.P. and Danthanarayana, D. (1991) Tortricid pests of pome and stone fruits, Australian and New Zealand species. In: van der Geest, L.P.S. and Evenhuis, H.H. (eds) *Tortricid Pests: their Biology, Natural Enemies and Control. World Crop Pests*. Elsevier, Amsterdam, pp. 453–472.

Weber, E., and Gut, D. (2004) Assessing the risk of potentially invasive plant species in central Europe. *Journal for Nature Conservation* 12, 171–179.

Weber, J., Panetta, F.D., Virtue, J. and Pheloung, P. (2009) An analysis of assessment outcomes from eight years' operation of the Australian border weed risk assessment system. *Journal of Environmental Management* 90, 2, 798–807.

Werres, S., Marwitz, R., Man Int Veld, W.A., de Cock, A.W., Bonants, P.J., De Weerdt, M., Themann, K., Ilieva, E. and Baayen, R.P. (2001) *Phytophthora ramorum* sp. nov., a new pathogen on Rhododendron and Viburnum. *Mycological Research* 105, 1155–1165.

Westbrooks, R.G. (1998) *Invasive Plants: Changing the Landscape of America*. Federal Interagency Committee for the Management of Noxious and Exotic Weeds, Washington, DC.

Westbrooks, R.G. and Eplee, R.E. (1996) Regulatory exclusion of harmful non-indigenous plants from the United States by USDA APHIS PPQ. *Castanea* 61, 3, 305–312.

White, P.S. and Schwarz, A.E. (1998) Where do we go from here? The challenges of risk assessment for invasive plants. *Weed Technology* 12, 4, 744–751.

Widrlechner, M.P., Thompson, J.R., Iles, J.K. and Dixon, P.M. (2004) Models for predicting the risk of naturalization of non-native woody plants in Iowa. *Journal of Environmental Horticulture* 22, 1, 23–31.

Williamson, M. (2001) Can the impacts of invasive species be predicted? In Groves, R.H., Panetta, F.D. and Virtue, J.G. (eds) *Weed Risk Assessment*. CSIRO, Collingwood, pp. 20–33.

Williamson, M. (2006) Explaining and predicting the success of invading species at different stages of invasion. *Biological Invasions* 8, 7, 1561–1568.

Williamson, M. and Fitter, A. (1996) The varying success of invaders. *Ecology* 77, 6, 1661–1666.

WTO (1994) *Agreement on the Application of Sanitary and Phytosanitary Measures*. World Trade Organization, Geneva.

19 Appropriate Level of Protection, Precaution and Jurisprudence

Robert Griffin

19.1. Acceptable Level of Risk and the Appropriate Level of Protection

The SPS Agreement refers to the *appropriate level of protection* and notes that some WTO members refer to this concept as the *acceptable level of risk* (WTO, 1994). In phytosanitary jargon, terms such as *negligible pest risk* and *quarantine security* may also be used. Other terms, such as *insignificant risk, no significant risk, de minimus risk* and *safe* are also encountered occasionally in documents and discussions related to the same or similar concepts.

Any given hazard, such as damage from the introduction of a harmful pest, may be characterized in complementary terms extracted from the concepts of risk and safety. When risk is low, safety is high; when risk is high, safety is low. The acceptable level of risk (ALR) and the appropriate level of protection (ALP or ALOP) address different sides of the same concept.

The sovereign right of an importing country to establish its appropriate level of [phytosanitary] protection is a central tenet of the SPS Agreement (WTO, 1994). In practical terms, this is determined by identifying the level of acceptable risk following a risk assessment and evaluating the options available to mitigate the risk to an acceptable level. This is the practical application of the concept of ALP because the elements of risk are more tangible or measurable than the elements of protection or safety, which are more abstract and subjective.

Terms such as *insignificant, not significant* and *negligible* can be misleading because they imply agreement concerning the significance of the risk without recognizing the dynamic and subjective nature of the judgements involved. The term *acceptable risk* may be most appropriate for pest risk analysis purposes because it implies that the risk level is selected from a range or judged on a scale relative to other risks, consistent with the principle of *managed risk* and disciplined by the principle of *consistency*.

In general terms, the level of acceptable risk (or the complementary level of protection), are separated by a point which conceptually represents the area where:

- The degree of risk[1] being accepted is commensurate with the benefits and costs of an alternative, i.e. the risks could be significant, but the benefits are greater or the risk mitigation costs are affordable.
- The risk is below what is considered normal or allowable compared to existing risks that are being accepted.
- The risk is unchangeable (must be accepted).

The acceptable level of risk is not necessarily a 'bright-line' concept and should not be expected to be static, nor will it always be identical across similar risks. The key to understanding and challenging the determination of the acceptable level of risk is the *transparency* and *consistency* of the criteria used to define it and the decision making that follows it. The strength of the measures applied in response to the risk should be linked to sound and explicit criteria and may be subject to the principle of *equivalency*. Further, the measures should be consistent, to the extent possible, with the strength of measures for similar situations.

19.2. The Strength of Measures and Their Rational Relationship to Risk

At the point in the pest risk analysis process where risk management is being considered, a judgement must be made concerning whether the pest risk identified through risk assessment is acceptable or unacceptable, i.e. whether it requires management or not. In the case of a quarantine pest that is not established in the endangered area, the decision that specific risk management procedures are *not* necessary does not change the technical status of the pest as a quarantine pest even though phytosanitary measures will not be required or applied (IPPC, 2004). The situation is therefore characterized in terms of *quarantine action*, providing the opportunity for quarantine pests to be further categorized as *actionable* or *non-actionable* as determined by the level of risk being managed in a particular situation. The levels of action which are deemed necessary then become the focus.

Imagine for example that a particular virus meets the defining criteria for a quarantine pest (IPPC, 2010) but it can only be introduced via grafts from the host (plants for planting). The virus may then be present in a fruit of the host and, although clearly a quarantine pest, would not be subject to phytosanitary measures because it has no potential for establishment. The same would not be true for an infected plant. The same quarantine pest is present in both instances, but one would require measures (quarantine action) and the other would not.

Every government has sovereign authority to determine whether phytosanitary measures are required and the level of action necessary for appropriate protection (WTO, 1994). Where risk management is deemed to be necessary, there will nearly always be a few to several risk management options that can be evaluated for their efficacy and may be challenged for their rigour, or *the strength of measures*.

The strength of measures may be metaphorically represented by a sliding scale where increasing risk corresponds to a comparable increase in the strength of measures applied to manage the risk at a specific level. The concept presumes that highly restrictive measures to address low risk are not justified, and vice versa. Begging clarity are the criteria and processes used for comparing and aligning these elements outside the odd instance where very similar situations may be compared for consistency.

We gain significant insight into the strength of measures by applying the principles of consistency and equivalence. In addition, jurisprudence on this point makes it clear that there must be a *rational relationship* (term used by the WTO Dispute Settlement Body) of measures to risk. This is interpreted in two ways. First, that the measures actually have a cause and effect relationship to the risk; second, that the strength of the measures is consistent with the level of risk.

To demonstrate the concept of a rational relationship, imagine requiring a probit 9 treatment for low risk weed seed contaminants when the treatment has no known effect on the weed seeds and the requirement for treatment is based on inspection. Such a requirement would be a challengeable measure because (i) it does nothing to mitigate the risk and (ii) the high risk implied by a probit 9 treatment is inconsistent with the low risk implied by using inspection as the means to determine whether treatment is required. In other words, the measure has no rational relationship to the risk.

The metaphor of a sliding scale is useful for visualizing the concept of strength of measures but in any particular regulatory situation, there may not be such an extensive range of risk management options available to allow for a precise match to the risk level. This means that in some cases, a risk that is found to be low may require a measure designed for higher risk because there is no other measure which is feasible. In a case such as this, it is useful to consider the measure to be provisional while pursuing less rigorous measures.

19.3. Precautionary Approach/ Precautionary Principle

A key point of confusion and controversy for risk analysis is the role and use of precaution in the regulation of risks to plant health and the environment. A concept known as the precautionary approach can be a contentious issue where the concept is not consistently expressed or understood in the same terms. The lack of clarity in this regard is often mistaken for a lack of precaution or a lack of concern for the importance of precaution.

19.3.1. History of the precautionary approach

The precautionary approach was first institutionalized as a distinct concept in an international instrument when it was adopted as Principle 15 of the Rio Declaration (RioDeclaration, 1992). Principle 15 states that:

> in order to protect the environment, the precautionary approach shall be widely applied by States according to their capability. Where there are threats of serious or irreversible damage, lack of full scientific certainty shall not be used as a reason for postponing cost-effective measures to prevent environmental degradation.

The precautionary approach as described in the Rio Declaration in 1992 and reaffirmed in the CBD in 1993 and again in the Cartagena Protocol (2000) is widely held to be both valid and desirable. It has not been found to be technically, legally or otherwise inconsistent with the objectives or obligations of the IPPC or the SPS Agreement. Indeed, in the case of the IPPC, the concept existed and was practised for many years before being labelled by the environmental community. It is however misunderstood and often mischaracterized.

By virtue of its association with the Principles of the Rio Declaration, the precautionary approach has also been referred to as a principle. From the standpoint of risk analysis, it is difficult to envisage how the concept could be considered either an approach or a principle. It is rather a process of accounting for uncertainty in decision making based on a systematic evaluation of the evidence – a process that is integral to risk analysis.

No international agreement or standard explicitly associates the application of the precautionary approach with the 'failure' of risk analysis. Indeed they are all explicit about risk analysis as the basis for evaluating the available scientific information. The question is really a policy judgement as to whether a determination regarding the adequacy of information is made before risk analysis is done, or whether the risk analysis is completed and then becomes the basis for identifying uncertainty, which may lead to precautionary judgements.

The evolution of the precautionary approach is punctuated by significant rhetoric from non-governmental organizations and diverse interpretations by governments claiming to understand the application of the concept in practice and promoting regulatory systems that may not be consistent with existing mechanisms and international obligations. In particular, there is an impression that the precautionary approach is an alternative basis for regulatory decision making to be used where there is judged to be insufficient information to undertake a risk-based approach. The implication is that the process of risk analysis would not be undertaken or completed if there were some level of uncertainty that was deemed to be unacceptable.

19.3.2. Precaution and Uncertainty

This interpretation indicates a fundamental misunderstanding regarding risk analysis and its relationship to regulatory decision making. In particular, it ignores the role of uncertainty in risk analysis. The adoption of such an approach may also prove inconsistent with the IPPC and the SPS Agreement (FAO, 1997; WTO, 1994). These conflicts may be avoided with the proper understanding and practice of risk analysis.

The role of scientific principles and evidence in risk analysis has historically been given greater prominence than the role of uncertainty. This has resulted in a strong focus on 'sufficient scientific evidence' (WTO, 1994) without fully recognizing that uncertainty is a part of *any* scientific evidence and proper risk analysis accounts for uncertainty. Equal emphasis needs to be given to raising the level of awareness regarding the role of uncertainty in risk analysis and the criteria used by governments to account for uncertainty as the justification for precaution in decision making.

It must be understood at the outset that precaution results from uncertainty. Uncertainty is inherent in risk. There is no risk without uncertainty associated with a hazard. Caution is not needed where there is perfect knowledge. When cause and effect are known, predictable and controllable, necessary measures are simply taken to mitigate hazards with full confidence that the measures will be effective and the hazard will cease to exist or will have no impact. It is the unknown, or rather the fear of the unknown that is the basis for precaution. So, uncertainty and precaution have a direct relationship, i.e. high uncertainty creates the need for greater precaution. Therefore the measurement of uncertainty is also useful for estimating the need for precaution. This is fundamental to understanding the role of uncertainty in decision making and the application of the precautionary approach.

19.3.3. Precaution and Risk Analysis in the IPPC-SPS Framework

The IPPC and its standards make no mention of the precautionary approach as a distinct concept. The standards however are based on the premise that phytosanitary measures are inherently precautionary in nature and therefore the decision regarding the strength of measures to be applied should be a function of the risk, taking account of the uncertainty.

At this stage, most governments are aware of the need to demonstrate a scientific basis for their phytosanitary measures, and risk analysis is increasingly being understood and practised as the means to meet this objective. Due to the strong emphasis that has been placed on the use of scientific principles and evidence, and particularly 'sufficient scientific evidence' (WTO, 1994) that is required to justify phytosanitary measures, there has developed a lack of understanding regarding the role of uncertainty in science, risk analysis and regulatory decision making.

The precautionary approach as a concept, whether considered an approach, a principle or simply an integral part of risk analysis, has a direct relationship to uncertainty associated with information or the lack of information about hazards and the risk they pose. Risk analysis as described by the IPPC (IPPC, 2004) is the means to gather and evaluate information to estimate the risk and the strength of phytosanitary measures to apply for mitigating risk associated with protecting plant health from harmful pests. IPPC standards point to uncertainty as an element of risk analysis, but provide little guidance on the role of uncertainty in decision making.

Where the precautionary approach is interpreted as an alternative to risk-based decision making, a judgement must be made that available information is insufficient for undertaking or completing a risk analysis. This is inconsistent with the purpose and proper practice of risk analysis that is undertaken *because* of insufficient information and is designed to account for uncertainty.

Taken to the extreme, it is possible that no risk analysis would ever be undertaken because scientific evidence is always incomplete and uncertain.

At a minimum, the adoption of the precautionary approach as an alternative to risk-based decision making would create enormous potential for abuse in the form of unjustified barriers to trade. The main reason for this prediction is that arbitrary decisions could be taken regarding the amount and quality of information required to overcome the 'fear factor' that is the basis for precaution. In many instances, this fear may not be satisfied by any amount of science. There is the further danger of being caught in the impossible position of 'proving a negative', which can require enormous resources to satisfy concerns for hazards without evidence they exist.

Finally, it must be recognized that in the IPPC-SPS framework, the burden of proof is distributed between the risk creator and the risk acceptor (except in the case of provisional measures) (WTO, 1994; IPPC, 2004). The risk creator has the obligation to demonstrate that the risk is reduced to meet the appropriate level of protection for the risk acceptor. The risk acceptor has the obligation to demonstrate that their appropriate level of protection is justified based on scientific principles and evidence (technically justified), and also meets the principles of transparency, equivalence, non-discrimination and minimal impact (WTO, 1994; IPPC, 2004).

Most important is that decisions regarding phytosanitary measures are consistent with decisions taken for similar risks. In the case where the precautionary approach may be used as an alternative to risk-based decision making, the burden of proof is shifted almost entirely to the risk creator (assuming that the risk acceptor has at minimum the obligation to specify the type, quantity and quality of information required to proceed with risk analysis).

In summary, it is difficult to understand why or how the precautionary approach could be used as an alternative to a proper and complete risk analysis as the basis for deciding phytosanitary measures. However, the IPPC and other organizations providing governments with guidance on the application of risk analysis for these purposes have not been explicit enough about the role of uncertainty in risk analysis and decision making to be clear about how precaution is included and accounted for in the process. Additional effort in this regard will be helpful in providing a better understanding of risk analysis and to avoid the confrontational situation that develops when the precautionary approach is interpreted as an alternative to risk analysis.

19.3.4. Notes on provisional, precautionary, and emergency measures

Emergency measures are not explicit in the SPS Agreement but extend from Annex B paragraph 6 (urgent problems) and the resulting Emergency Notification format adopted by the SPS Committee (G/SPS/7 Rev 1) (WTO, 1994). Article VII.6 of the IPPC is explicit about 'emergency action' based only on the detection of a pest, indicating that such action will be evaluated (implying a pest risk analysis) as soon as possible to ensure that it is justified (FAO, 1997).

The IPPC's Principle 14 (ISPM No. 1) refers to emergency actions for new or unexpected phytosanitary situations based on a preliminary pest risk analysis and indicating that such measures 'shall' be temporary and the subject of a detailed pest risk analysis as soon as possible. Reference to provisional measures is found in Article 5.7 of the SPS Agreement (WTO, 1994). Based on the text of the SPS Agreement and relevant jurisprudence to date, it is clear that such measures have the following characteristics:

- They are taken in the absence of sufficient scientific evidence.
- They are based on the available pertinent information (i.e. Members must search for and consider available evidence), including information provided by relevant international organizations (e.g. the IPPC), and information about measures applied by others.
- They require that the Member imposing the measure actively pursue the

information required for a more objective assessment of the risk and review of the measure within a reasonable period of time.

The term 'precautionary measures' is not explicitly used or described in either the IPPC or the SPS Agreement although SPS jurisprudence indicates that provisional measures 'reflect precaution' (see Section 19.4). It may be argued however that phytosanitary measures are by their nature more or less precautionary depending on the influence of uncertainty in the judgement regarding acceptable risk. The concept of precaution based on uncertainty is therefore implicit in the application of proper risk analysis.

The SPS is not explicit about emergency or precautionary measures, but it is clear that provisional measures are a different concept with heavy obligations attached (WTO, 1994). Provisional measures are designed to facilitate trade by allowing the application what would probably be overly restrictive measures until more or better information or experience is available to decide on permanent measures. The country establishing provisional measures accepts the responsibility for pursuing the information need to reevaluate provisional measures. Although no timeframe is specified for this purpose, SPS jurisprudence makes it clear that a good faith effort is necessary for a provisional measure to be maintained.

Provisional measures need not be emergency measures, i.e. not necessarily in response to an immediate threat. In fact, most provisional measures are not emergency measures but deliberate results of bilateral negotiations to facilitate trade. Likewise, emergency measures need not be provisional, i.e. require additional information and reconsideration. In many cases where emergency measures are implemented, the pest is well known and requires no additional information to justify action (WTO, 1994).

The IPPC, although explicit about emergency *actions*, mixes the concepts of provisional and emergency measures in a single principle, limits it to the detection of a pest,

and also adds mandatory detailed pest risk analysis (FAO, 1997). In the case of so-called precautionary measures, it is important to understand that all phytosanitary measures are more or less precautionary based on the relationship of uncertainty to the judgement of the acceptability of risk as determined by pest risk analysis.

19.4. PRA-related Jurisprudence in the WTO

Six SPS-related disputes are summarized below for their relevance to risk analysis. These cases are not selected to reflect poorly on the policies or processes of any particular country but rather to highlight issues that are useful to understand for risk analysis purposes. The outcomes of these challenges can provide important insight into the practical application of risk analysis in the IPPC-SPS framework and the legal interpretation of the concepts associated with the SPS Agreement. Practitioners of risk analysis can benefit greatly from understanding where there are differences and how these are resolved as a means to avoid shortcomings in their own work that could result in challenges.

Note that the summaries refer to *risk assessment* rather than risk analysis to be consistent with the terminology used by the WTO. Full documentation on these and other WTO disputes, including extensive background on the process is available at http://www.wto.org/english/tratop_e/dispu_e/dispu_e.htm.

19.4.1. Australia – salmon

The Australian salmon dispute is one of the early WTO disputes invoking the SPS Agreement. The Canadian government requested the establishment of a Panel in 1997 to evaluate the Australian import ban on fresh, frozen or chilled salmon products from Canada. The USA, Norway and India reserved third-party rights with Canada. Canada's main issues with the measures were the validity of the risk assessment on which the measure was based, as well as alleging that Australia was maintaining uneven levels of protection between salmon and similar products such as herring and finfish. The Panel found in 1998 that the measure was inconsistent with Article 5.1 of the SPS Agreement (WTO, 1994), because it was not based on a risk assessment, stating that Australia had conducted a risk assessment for ocean-caught Pacific salmon but not for the other kinds of salmon affected by the import restriction (WTO, 1998b).

This finding implied that the import measure was also in violation of Article 2.2 (WTO, 1994), which requires the measure to be based on scientific evidence. The Panel also found that Article 5.5 was violated because the level of protection afforded to salmon by the import ban was higher than for the comparable products of finfish and herring, and could thus be considered arbitrary and unjustifiable. By implication, the measure was also inconsistent with Article 2.3 (WTO, 1994) it created a 'discrimination or disguised restriction on trade'. Finally, the Panel concluded that Article 5.6 was also violated because the import restriction on fresh, frozen or chilled Canadian salmon was more trade-restrictive than necessary to achieve an appropriate level of protection.

Australia appealed against certain findings of the Panel, and the Appellate Body (AB) circulated its report in the autumn of 1998 (WTO, 1998c). The AB reversed some of the Panel findings, but upheld others and still ultimately found the measure to be in violation of several articles of the SPS Agreement. The AB found that although the Panel had examined the wrong measure (the heat-treatment requirement), the correct measure (the import restriction) was still inconsistent with Article 5.1 (WTO, 1994) (and thus Article 2.2 (WTO, 1994) as well) because it was not based on a proper risk assessment pursuant to Australia's obligations under the SPS Agreement. With regards to Article 5.5 (WTO, 1994), the AB agreed with the Panel's finding that the measure resulted in uneven levels of protection between similar products, leading to a disguised restriction or discrimination. As Atik points out, this

was the first case in which a violation of Article 5.5 (WTO, 1994) had been found (Atik, 2004). By implication, the measure was then also in violation of Article 2.3 (WTO, 1994), because it resulted in an unjustifiable or arbitrary discrimination. Finally, the AB reversed the Panel's decision with regards to Article 5.6 (WTO, 1994), citing that the Panel had based its finding on the heat-treatment requirement rather than the import ban itself, which was the true measure at issue.

The main implications of this dispute for risk assessment include:

- A risk assessment can be either quantitative or qualitative. This is important because there is no minimum amount of information required in order for a risk assessment to be conducted; the ability to make a qualitative determination of risk may make it easier for Members to use a risk-based approach to construct import measures.
- Australia had failed to evaluate the likelihood of entry, establishment or spread of pests or diseases that affect salmon, and that they had also failed to evaluate these agents according to the SPS measures that might be applied. The AB expanded this finding to say that if a measure was not based on a risk assessment, then it can be presumed that the measure did not have a scientific basis or that it was maintained without sufficient scientific evidence.
- Members may set their own acceptable level of risk, which may be 'zero'.

19.4.2. Japan – variety testing

The US requested the establishment of a Panel in 1997 to examine Japan's requirements that each variety of certain fruit be tested for efficacy of quarantine treatment for codling moth, even if the same fumigation treatments for different varieties of the same product had proven successful. The USA was specifically concerned with eight varieties of fruits originating in the USA.

The USA alleged that the treatment require-ment was more trade restrictive than neces-sary, and that it was in violation of Articles 2, 5, and 8 of the SPS Agreement (WTO, 1994); Article XI of GATT 1994, and Article 4 of the Agreement on Agriculture. The Panel report, circulated in October 1998, concluded that Japan's measure violated Article 2.2 of the SPS Agreement, finding that the measure was not applied only to the extent necessary to protect animal, plant, and human life and health, and was main-tained without sufficient scientific evidence (WTO, 1998b).

The Panel also found that by implica-tion there was also a violation of Article 5.6 (WTO, 1994), which states that Members must ensure that their SPS measures are not more restrictive than necessary to achieve their appropriate level of protection. The Panel also found the measure to be in viola-tion of Article 5.7 (WTO, 1994), because although Japan claimed that the testing requirement was provisional, the circum-stances of the measure did not meet the cri-teria set out by the SPS Agreement for adoption of a provisional measure. In order for a measure to be provisional, the Member implementing the measure must actively seek additional information in order to con-duct a risk assessment, and the measure must be reviewed within a reasonable period of time. Additionally, the Panel also found that Japan's testing requirement was incon-sistent with both Annex B and Article 7 of the SPS Agreement, which are concerned with transparency of SPS measures.

The Appellate Body upheld the Panel's findings with regards to Article 2.2 (WTO, 1994), and agreed that the measure was maintained in the absence of sufficient sci-entific evidence. Thus there was no rational relationship between existing scientific evi-dence and the import restriction, as required by the SPS Agreement. The AB also upheld the Panel findings that the varietal testing measure could not be classified as a provi-sional measure, and was therefore inconsist-ent with Article 5.7 (WTO, 1994). With regard to Article 5.6 (WTO, 1994), although the USA had failed to establish that Japan's measure was inconsistent with and more restrictive than existing alternative measures,

the AB ultimately found that the measure violated Article 5.6 (WTO, 1994) because it was more restrictive than necessary to attain Japan's appropriate level of protection. Finally, the AB concluded that the measure was also inconsistent with Article 5.1 (WTO, 1994) for the four products not examined by the Panel (apricots, pears, plums and quince) because it was not based on a valid risk assessment and was not applied only to the extent necessary (WTO, 1999).

The main implications of this dispute for risk assessment include:

- Japan was unable to demonstrate a rational relationship between the scien-tific evidence and the measure.
- The use of provisional measures where scientific information is insufficient was not valid because Japan did not actively seek to obtain additional infor-mation in order to review its measure within a reasonable period of time.
- The variety testing requirement was more trade restrictive than required to achieve Japan's appropriate level of protection.
- The variety testing requirement should have been published, although the requirement was not mandatory.

19.4.3. Japan – fire blight

In March 2002 the USA requested Panel con-sultations regarding Japan's import restric-tions on apples based on concerns of fire blight bacterium transmission. The measures included the prohibition of apples originat-ing from orchards in which fire blight had occurred, as well as orchard inspections three times a year. In addition, any orchard within 500 m of a fire blight occurrence was disqualified as a potential exporter.

The USA alleged that these measures were more trade restrictive than necessary, given the amount of relative scientific evi-dence surrounding fire blight transmission. The USA asserted that Japan's import meas-ure violated Articles 2.2, 2.3, 5.1, 5.2, 5.3, 5.6, 6.1, 6.2, 7 and Annex B of the SPS Agreement; Article XI of GATT 1994; and Article 14 of the Agreement on Agriculture.

The Panel released the report of its findings in 2003, and found that Japan's import restrictions were inconsistent with SPS Articles 2.2, 5.7 and 5.1 (WTO, 1994).

Article 2.2 obligates Members to ensure that their SPS measures are not maintained without sufficient scientific evidence, and that they are only applied to the extent necessary to protect animal, plant and human life and health (WTO, 1994). The Panel ruled that the measure was maintained without sufficient scientific evidence, and that no rational relationship existed between the measure and the level of risk that could be determined by available evidence. The scientific evidence established the level of risk as 'negligible', and therefore the measure was stricter than necessary (WTO, 2003a).

The Panel also found that Article 5.7 (WTO, 1994), regarding provisional measures, had been violated as well. The Japanese measure was not a provisional measure as specified by the SPS Agreement, because there has to be a lack of sufficient evidence in order for a measure to be provisionally adopted. The Panel findings determined that 'a large quantity of high-quality scientific evidence' was present, which would allow for a proper evaluation of risk associated with apple importation. Because of the wealth of relevant scientific information available, the measure could not qualify as provisional (WTO, 2003a, 2003b).

The main implications of this dispute for risk assessment include:

- The risk assessment did not qualify as an appropriate risk assessment under the SPS Agreement. Japan's 1999 risk assessment did not 'evaluate (i) the likelihood of entry, establishment, or spread of fire blight specifically through the importation of apple fruit; and (ii) the likelihood of entry 'according to SPS measures that might be applied''. To this end, the measure was not based on an appropriate risk assessment, but rather a general summary of the fire blight disease itself.
- The risk assessment failed to evaluate the risk of fire blight according to the SPS measures that might be applied, not the current measures.

- The measure could not have been provisional as Japan claimed because a large body of scientific evidence existed to evaluate the likelihood of fire blight transmission.
- Japan's measure was maintained without sufficient scientific evidence that apple fruit could serve as a pathway for the entry, establishment or spread of fire blight.
- Japan's risk assessment failed to evaluate the risk according to the SPS measures that might be applied, but rather considered only the existing measures.

19.4.4. EC – hormones

In 1996 the USA requested the establishment of a Panel to review the European Communities' (EC) ban on the import of meat and meat products (specifically beef) that been treated with hormones for growth. The USA claimed that the measures violated the EC's obligations under SPS Articles 2, 3, and 5 (WTO, 1994); TBT Article 2; and GATT Articles III or XI (WTO). The Panel Report, released in 1997, found that the EC measures violated Articles 3.1, 3.3, 5.1 and 5.5 of the SPS Agreement. Article 3.1 states that Members shall base their measures on international standards, if they exist, in order to promote harmonization of SPS measures. Article 5.1 (WTO, 1994) requires measures to be based on an appropriate risk assessment, and Article 5.5 (WTO, 1994) prohibits arbitrary or unjust distinctions in protection levels for similar products. The EC appealed the Panel findings and requested that the AB review the Panel conclusions (WTO, 1997).

In 1998 the AB produced its own report, reversing the Panel's initial conclusions that the import ban violated Articles 3.1 and 5.5 (WTO, 1994). The AB stated that Articles 3.1, 3.2, and 3.3 apply together, yet each addresses a different and separate situation. The AB also said that although Article 3.1 requires the basis for measures to be international standards and guidelines, it does not require a measure to 'conform' to these standards. The AB upheld the Panel's finding that the measure was inconsistent with Article 5.1 (WTO, 1994), and by implication Article 3.3, because it lacked a basis in an appropriate risk assessment (WTO, 1998a).

The Panel findings with respect to Article 5.5 (WTO, 1994), that the EC had restricted trade through arbitrary or unjustifiable distinctions, were reversed by the AB, who said that genuine safety concerns did exist regarding hormone-treated beef (versus pork products treated with the same hormones). Ultimately, the AB upheld most of the Panel findings, with the main exception being the Panel's interpretation regarding the burden of proof under the SPS Agreement. The Panel found that the SPS obligations required the burden of proof to rest on the Member 'imposing an SPS measure'.

The main implications of this dispute for risk assessment include:

- The EC measure reflected a higher level of protection than was described in international standards and was not justified by a risk assessment.
- The EC ban was not based on a risk assessment. The EC's scientific studies on five of the hormones did not support the ban on hormone-treated meat.
- The EC did not use provisional measures but rather invoked the 'precautionary principle' in general. The Panel found that invoking the 'precautionary principle' did not override a country's obligations under the SPS Agreement.
- The EC incorrectly linked their measure to the perception of risk, saying that '"scientific" factors are not the only relevant factors' that come into play when considering import.
- The EC failed to establish a rational relationship between the measure and the scientific evidence submitted on five of the hormones, and found that there was no risk assessment at all for the sixth hormone.

19.4.5 EC – biotech products (GMOs)

In 2003 the USA requested the establishment of a Panel to examine the EC's 1998 alleged moratorium on the approval of biotech products, which the US claimed was restricting trade on food and agricultural products and was inconsistent with the SPS Agreement's Articles 2, 5, 7 and 8 (WTO, 1994). The USA alleged that there existed a moratorium on the approval of biotech products, which since 1998 had limited the number of products receiving approval in EC states. Additionally, so-called 'safeguard measures' were in place in six EC Member states, prohibiting certain biotech products that had already been approved at the EC level. The Panel report was published in 2006.

The Panel first had to examine the allegations of a moratorium, and found that a *de facto* moratorium did in fact exist. However, the Panel determined that because the moratorium was a procedural issue and not intended to achieve appropriate levels of protection, it was not a SPS measure and therefore not subject to Articles 5.1 and 2.2 (WTO, 1994). However, the Panel also found that the existence of a general moratorium created an 'undue delay' in approval of biotech products, and was thus inconsistent with Annex C of the SPS Agreement (WTO, 1994), which requires that 'procedures are undertaken and completed without undue delay and in no less favorable manner for imported products than for like domestic products'. Any violation of Annex C is by implication also a violation of Article 8, which obligates Members to observe Annex C regarding control, inspection and approval procedures (WTO, 2006).

Regarding the Member State safeguard measures, the Panel found that there was no indication that there was insufficient evidence to conduct a risk assessment for the relevant products. Because of this, the Panel determined that the safeguard measures were in violation of both Articles 5.1 and 2.2 (WTO, 1994). The measures were therefore not valid SPS measures because they were not based on an appropriate risk assessment as required by the SPS Agreement.

The main implications of this dispute for risk assessment include:

- Control, inspection and approval procedures covered by Annex C include risk assessment processes, and although failure to initiate a risk assessment where required for approval procedures is not a measure, it creates a *de facto* moratorium, which is a violation of the SPS Agreement.

- 'Safeguard measures' are not a substitute for provisional measures.

19.4.6. Australia – apples

Australia banned the importation of New Zealand apples in 1921 following the introduction of fire blight disease to New Zealand in 1919. Subsequent market access requests by New Zealand in 1986, 1989 and 1995 were rejected. Australia initiated a risk assessment following a new request by New Zealand in 1999. The risk assessment, which was completed by Australia in 2006, was the focus of the dispute. Unlike previous SPS disputes, which included elements associated with risk assessment more broadly, this dispute had a very strong focus on the detail related to the methodologies and data.

The Panel was established in January 2008. New Zealand charged that Australia's measures, both individually and as a whole, were inconsistent with the obligations under Articles 2.2 and 2.3, Articles 5.1, 5.2, 5.5 and 5.6, Article 8, and Annex C(1)(a) (WTO, 1994). The EC and USA joined the consultations. Chile, the EC, Japan, Pakistan, Taiwan and the USA reserved their third-party rights to join the dispute.

The measures imposed by Australia fell into two categories: those of general application and those specific to each of three pests identified in the risk assessment. New Zealand argued that Australia's measures were maintained without sufficient scientific evidence and there was no rational or objective relationship between the measures and scientific evidence. Australia argued that New Zealand did not recognize that risk assessments must confront scientific uncertainty, which may only be resolved or accommodated through the exercise of expert judgement in accordance with the specific requirements of each case. New Zealand suggested and the Panel agreed to seek expert advice on specific scientific and technical aspects of Australia's arguments including the use of semi-quantitative methodologies.

The Panel identified key methodological flaws in Australia's risk assessment that were designed to overestimate the probability of introduction and spread of the pests at issue. One methodological flaw was the choice of an inflated maximum value for the probability of events with a negligible likelihood of occurring. Another important flaw was the inappropriate use of a uniform distribution to model the likelihood of events, particularly those with a 'negligible' likelihood of occurring. New Zealand also argued that Australia overestimated the projected volume of trade in New Zealand apples, but the Panel accepted Australia's argument on this point (WTO, 2010).

Based on inputs from the expert review, the Panel found that Australia had not provided a proper risk assessment, and therefore Australia's measures were not based on a risk assessment. A critical distinction was made between Australia's claim that its risk assessment was consistent with the SPS because it was objective and coherent, versus New Zealand's position that the mechanics and data associated with the risk assessment were designed to exaggerate the risk, and assumptions were not linked to scientific evidence but rather expert opinion. The Panel's opinion favoured New Zealand on this point (WTO, 2010).

There are many lessons to be learned from this dispute, but the main implications for risk assessment include:

- The possibility for the Panel to judge the suitability of a risk assessment by using expert review to examine the methodologies and data.
- Considering that an inappropriate risk assessment is equivalent to no risk assessment.
- Further clarification of the difference between a possibility and a probability, and the need to establish the rational relationship between the probability and the evidence which must be based on more than just expert opinion.
- The proper selection of probability distributions.

- Avoid addressing uncertainty by deliberately exaggerating the risk with worst case assumptions/opinions and selected data extremes.
- Expert opinion must have some basis in evidence.

19.5. Summary

Observers will notice several recurring themes in both the challenges and results. Above all, the importance of scientific evidence and a proper risk assessment are highlighted. The theme of a rational relationship between evidence and risk/measures has a very high profile in several disputes. Likewise, the proper understanding of provisional measures is a central issue that is repeated. Accounting for uncertainty by either substituting the precautionary principle or exaggerating the risk or implementing 'safeguard measures' is not acceptable. Biasing the analysis with expert opinion that is not based on evidence is considered inappropriate, as is the use of negligible events that may be possible but not demonstrated to be probable based on evidence.

Note

[1] Risk in this context refers to the probability of pest introduction with unacceptable consequences.

References

Atik, J. (2004) The weakest link: demonstrating the inconsistency of 'Appropriate levels of protection' in Australia-salmon. *Risk Analysis* 24, 483–490.

CBD (1992) *Convention on Biological Diversity. Text and Annexes*. Secretariat of the Convention on Biological Diversity/UNEP, Montreal.

CBD (2000) *Cartagena Protocol on Biosafety*. Secretariat of the Convention on Biological Diversity/UNEP, Montreal.

IPPC (2004) International Standards for Phytosanitary Measures, Publication No. 11: *Pest Risk Analysis for Quarantine Pests Including Analysis of Environmental Risks and Living Modified Organisms*. Secretariat of the International Plant Protection Convention (IPPC), Food and Agriculture Organization of the United Nations, Rome.

IPPC (2007) International Standards for Phytosanitary Measures, Publication No. 2: *Framework for Pest Risk Analysis*. Secretariat of the International Plant Protection Convention (IPPC), Food and Agriculture Organization of the United Nations, Rome, Italy.

IPPC (2010) International Standards for Phytosanitary Measures, Publication No. 1: *Phytosanitary Principles for the Protection of Plants and the Application of Phytosanitary Measures in International Trade*. Secretariat of the International Plant Protection Convention (IPPC), Food and Agriculture Organization of the United Nations, Rome.

Rio Declaration (1992) *Rio Declaration on Environment and Development*. Report of the United Nations Conference on Environment And Development, Rio de Janeiro, 3–14 June 1992.

WTO (1994) *The WTO Agreement on the Application of Sanitary and Phytosanitary Measures*. World Trade Organization, Geneva.

WTO (1997) *European Communities – Measures Concerning Meat and Meat Products*. Panel Report. WT/DS26/R/USA.

WTO. (1998a) *European Communities – Measures Concerning Meat and Meat Products*. Appellate Body Report. WT/DS26/AB/R & WT/DS48/AB/R.

WTO (1998b) *Australia – Measures Affecting Importation of Salmon*. Panel Report. WT/DS18/AB/R.

WTO (1998c) *Australia – Measures Affecting Importation of Salmon*. Appellate Body Report. WT/DS18/AB/R.

WTO (1998d) *Japan – Measures Affecting Agricultural Products*. Panel Report. WT/DS76/R.

WTO (1999) *Japan – Measures Affecting Agricultural Products*. Appellate Body Report. WT/DS76/AB/R.

WTO (2003a) *Japan – Measures Affecting the Importation of Apples*. Panel Report. WT/DS245/R.

WTO (2003b) *Japan – Measures Affecting the Importation of Apples*. Appellate Body Report. WT/DS245/AB/R.

WTO (2006) European Communities – *Measures Affecting the Approval and Marketing of Biotech Products*. Panel Report. WT/DS291/R; WT/DS292/R and WT/DS293/R.

WTO (2010) *Australia – Measure Affecting the Importation of Apples from New Zealand*. Panel Report. WT/DS367/R.

20 Invasive Species and Living Modified Organisms

Christina Devorshak

20.1. Introduction

Throughout this book we've highlighted different types of organisms for which we might want to analyse risk – this includes conventional 'pests' like various species of arthropods or pathogens that attack both cultivated and wild flora. We also learned about other organisms that have the potential to have effects on plant health, such as weeds (Chapter 18) or beneficial organisms (Chapter 17). Invasive alien species and living modified organisms are two more groups of organisms that warrant some discussion with respect to pest risk analysis.

The purpose of this chapter is to describe key concepts around both invasive alien species and LMOs[1] and the respective roles of the IPPC and other relevant international agreements for these kinds of organisms. Note that terminology used in this chapter is consistent with IPPC definitions (see Chapter 5 in particular to review key definitions related to the CBD and invasive species).

This textbook has addressed pest risk analysis from the perspective of the IPPC and the SPS Agreement. However, the CBD and its supplementary agreement, the Cartagena Protocol on Biosafety, also play a role with regard to invasive alien species and LMOs, respectively. The CBD is an international treaty concerned with protecting biological diversity and has three main objectives:

- The conservation of biological diversity.
- The sustainable use of the components of biological diversity.
- The fair and equitable sharing of the benefits arising out of the utilization of genetic resources.

The CBD Conference of Parties (COP) adopted a supplementary agreement known as the Cartagena Protocol on Biosafety (CP) that entered into force in 2003. The purpose of the CP is to establish means to manage risks associated with LMOs (e.g. seeds for planting rather than products for immediate consumption) resulting from biotechnology that are likely to have environmental impacts affecting biological diversity. The CP establishes an advance informed agreement approach to ensure that countries are provided with the information necessary to make informed decisions prior to importation of LMOs. In addition, it refers to a precautionary approach (see Chapter 19 for more on precautionary approach and risk analysis) and reaffirms the precautionary language of Principle 15 of the Rio Declaration on Environment and Development (1992) regarding action to prevent environmental degradation in the absence of full scientific certainty. Both of these agreements are concerned with *risks* to the environment – those risks (relevant to this text) potentially arising from either invasive alien species or from LMOs. Table 20.1 provides an explanation of the organisms addressed and purpose of each of the agreements.

20.2. Relationship Between the International Agreements

According to the IPPC, the term 'pest' is inclusive of all organisms that harm plant health (e.g. arthropods, weeds, pathogens, etc.) and 'plants' is inclusive of all plants, whether they are cultivated or plants growing naturally in the environment. Protecting plants is not limited to trade, but extends to any means by which pests can be introduced and spread. This fundamental concept is important for us to understand when we look at organisms such as those called 'invasive species' (or alien species, alien invasive species, invasive alien species, etc.) or LMOs.

Figure 20.1 shows the relationship between the IPPC, CBD, SPS and CP. Recall from Chapter 4 that the SPS Agreement is principally a *trade* agreement that makes provisions for protecting health – in this case, plant health specifically, while the IPPC is a *plant protection* agreement that makes provision for trade. The SPS Agreement identifies the IPPC as the standard setting body for plant health, thus creating a clear relationship between these two agreements – that is, promoting the safe trade of plants and plant products. At the same time, the CBD and CP are concerned with risks to biodiversity, and in the case of the CP, the safe transboundary movement of LMOs (primarily through trade, but not exclusively). Like the IPPC, the SPS Agreement intersects with the CBD where SPS measures (applied to protect human, animal and plant life or health) contribute to the aim of protecting biodiversity. The CBD and

Table 20.1. Comparison of organisms addressed and purpose of the IPPC, CBD and CP. (From FAO, 1997; CBD, 1992; CBD, 2000.)

	IPPC	CBD	CP
Organisms	Pests (*sensu* IPPC)	Invasive alien species	Living modified organisms
Purpose	Securing common and effective action to prevent the spread and introduction of pests of plants and plant products, and to promote appropriate measures for their control	Conservation of biological diversity, the sustainable use of its components and the fair and equitable sharing of the benefits arising out of the utilization of genetic resources; and specifically under Article 8 (In situ Conservation): Article 8(h), which states that contracting parties shall as far as possible and as appropriate 'Prevent the introduction of, control or eradicate those alien species which threaten ecosystems, habitats or species'.	To contribute to ensuring an adequate level of protection in the field of the safe transfer, handling and use of living modified organisms resulting from modern biotechnology that may have adverse effects on the conservation and sustainable use of biological diversity, taking also into account risks to human health, and specifically focusing on transboundary movements

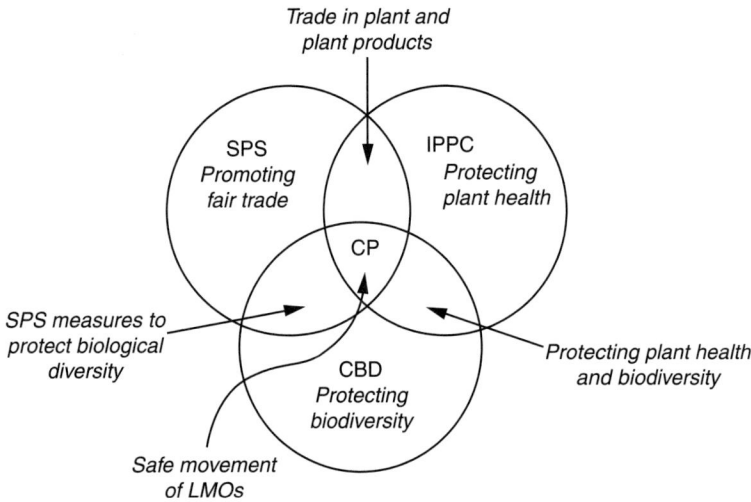

Fig. 20.1. Schematic representation of the overlaps in scope between IPPC, SPS and CBD (and CP). (Adapted from IPPC, http://www.ippc.int, and Macleod *et al.*, 2010.)

CP overlap with the IPPC (concerned with protecting plants and plant health) where:

- Protecting plant health contributes to the aims of the CBD in protecting biological diversity;
- 'Invasive alien species' behave as plant pests (*sensu* IPPC);
- LMOs behave as plant pests (*sensu* IPPC).

20.2.1. Development of cooperation between the CBD and IPPC

This relationship between the IPPC and the CBD was not always clear however. Beginning in the late 1990s, countries that were signatory to both agreements recognized that there were overlaps between the two agreements (Shine, 2007; Schrader *et al.*, 2010). In order to clarify the role of each agreement with respect to organisms that are harmful to plants, a series of joint consultations was organized by the IPPC.

Pests and invasive alien species

The first meeting that addressed overlaps between the IPPC and CBD was the 'ICPM

Exploratory Open-ended Working Group on the Phytosanitary Aspects of GMOs, Biosafety and Invasive Species' (or OEWG), which took place in June 2000. The meeting focused mostly on issues related to Article 8h (alien species), but subsequent consultations went on to address overlaps between the IPPC and CP as well, specifically with regard to LMOs (see following section of this chapter). The purpose of the meeting was to identify the respective roles of the relevant international agreements, identify how ISPMs provide support and guidance to dealing with CBD issues (e.g. alien species), and in particular clarification of how the concepts of invasive alien species and quarantine pests overlap.

One of the major outcomes of the meeting was providing greater explanation of the scope of the IPPC. The meeting noted in particular that 'pests' addressed by the IPPC extends beyond pests directly affecting cultivated plants. The meeting specified that the scope of the IPPC extends to organisms that:

- Directly affect uncultivated/unmanaged plants;
- Indirectly affect plants (e.g. weeds indirectly affect plant health through competition with other plants);

- Indirectly affect plants through effects on other organisms (e.g. parasites of beneficial organisms).

This clarification of the scope created a direct relationship between the IPPC and the CBD by establishing the role of the IPPC in protecting biological diversity, by virtue of protecting plant health.

The OEWG in June 2000 outlined the provisions of the IPPC (see Chapter 4 for a review of key IPPC provisions) that are directly relevant to implementing Article 8h of the CBD with regard to managing risks associated with invasive alien species:

- Providing legal and regulatory frameworks;
- Assessing and managing potential plant pest risks;
- Applying measures to prevent unintentional introduction of plant pests;
- Detecting, controlling and eradicating plant pests in both areas under cultivation and wild flora;
- Protecting areas that may be threatened by plant pests;
- Assessing and managing the intentional introduction of organisms that may be pests of plants and biological control agents;
- Certifying that risk management procedures have been applied for exports;
- Exchanging of scientific and regulatory information relevant to plant pests;
- Cooperating between countries to minimize the impact of plant pests;
- Building capacity and technical assistance for developing countries.

This meant that the structures and processes in place under the IPPC (e.g. legislative frameworks, NPPOs, the conduct of pest risk analysis, etc.) are also directly relevant to protecting biological diversity through preventing the introduction and spread of pests (FAO, 2001).

In addition, the OEWG specifically noted the role of ISPMs in providing guidance to countries on how to address risks associated with alien species; however there were gaps in the existing guidance at that time that needed to be addressed. This included clarifying the terms 'economic

importance' and 'environmental impact' in standards and how these terms relate to threats to ecosystems, habitats and species. Subsequently, additional information was added to different standards, in particular ISPMs No. 2 and 11 for pest risk analysis specifically, and ISPM No. 5 for clarification of terms related to CBD issues. These additions are covered in Section 20.4 of this chapter.

Pests and LMOs

As stated above, initial efforts of the first OEWG in June 2000 were focused on the relationship between invasive alien species and quarantine pest (and how the IPPC and Article 8h of the CBD are related). The issue of LMOs was more fully addressed later when the 'ICPM Open-ended Working Group on Specifications for an International Standard for Phytosanitary Measures for Living Modified Organisms/Products of Modern Biotechnology' was held in September 2001. The terms of reference for the meeting were for the OEWG to:

1. Identify plant pest risks associated with LMOs/products of modern biotechnology.
2. Identify elements relevant to the assessment of these plant pest risks.
3. Consider existing international regulatory frameworks and guidelines.
4. Identify areas within pest risk analysis (PRA) standards and other ISPMs that are relevant to the phytosanitary aspects of LMOs/products of modern biotechnology.
5. Identify plant pest risks associated with LMOs/products of modern biotechnology that are not adequately addressed by existing ISPMs.

This meeting examined the available guidance already included in ISPMs and identified what standards were relevant to assessing and managing risks associated with LMOs, stating that:

- Plant pest concerns that may be presented by LMOs fall within the scope of the IPPC.
- IPPC systems and procedures (including pest risk analysis) apply to analysing and managing the risks posed by LMOs as they relate to the protection of plant health.

- Existing national mechanisms and structures for phytosanitary systems may form a basis or a model for developing other practical approaches to managing risks associated with LMOs.

This simply re-iterated that the scope of the IPPC addresses any pests that have the potential to affect plant health, regardless of the source of those pests; and that environmental impacts also fall within the scope of the IPPC (FAO, 2002). It then went on to identify gaps that could be addressed through the development of any additional guidance (in the form of a standard or a supplement to a standard) that specifically addressed risks associated with LMOs. The gaps were addressed by developing supplemental information to ISPM Nos 2 and 11 with regard to pest risk analysis for LMOs, and ISPM No. 5 with respect to terminology. These additions are covered in Section 20.4 of this chapter.

Current status of cooperation between the IPPC and CBD

Currently, the Secretariats of the IPPC and the CBD continue to have a joint work programme, and have identified areas for cooperation and collaboration, including:

1. Mechanisms of collaboration between the IPPC and the CBD;
2. Common issues related to plant pests, invasive alien species and living modified organisms;
3. Development of standards, guidance and materials on risk analysis (including risk assessment and risk management) of plant pests and invasive alien species;
4. Terminology;
5. Capacity-building;
6. Information sharing;
7. Collaboration through an invasive alien species liaison group.

20.3. Role of risk analysis in the CBD and CP

We've already learned that both the IPPC and the SPS Agreement require risk analysis (technical justification) as the basis for

measures to prevent the introduction and spread of pests. The CBD is not explicit about risk analysis with respect to invasive alien species, but managing risk is implied in Article 8h (In Situ Conservation). Furthermore, in 2002, the Conference of the Parties to the CBD adopted the *Guiding Principles for the Prevention, Introduction and Mitigation of Impacts of Alien Species that Threaten Ecosystems, Habitats or Species.* The Guiding Principles include reference to risk analysis with regard to analysing the risks associated with the introduction of invasive alien species. However, the Guiding Principles stop short of providing specific guidance as to how to conduct risk analysis, and refer instead to guidance provided by other organizations.

Unlike the CBD, the CP explicitly includes risk assessment and risk management in Article 15 (Risk Assessment) and Annex III (Risk Assessment), which addresses general principles and methods for risk assessment.

According to Annex III, and as further elaborated by the draft guidance document:

- Risk assessment for LMOs should be carried out in a scientifically sound and transparent manner.
- Lack of scientific knowledge or scientific consensus should not necessarily be interpreted as indicating a particular level of risk, an absence of risk or an acceptable risk.
- Risks should be considered in the context of risks posed by the non-modified recipients or parental organisms.
- Risks should be assessed on a case-by-case basis.

It further describes general methodology, consistent with most risk analysis models: namely that risk assessment begins with the identification of a hazard, and that risk is a combination of likelihood and consequences of adverse events. Finally, the CP hosts the Biosafety Clearinghouse (bch.cbd.int) – a mechanism for the exchange of information related to the CP and its requirements, including information on risk assessment for LMOs. More specific information on the relevant types of guidance for risk assessment for LMOs is provided in the next sections of this chapter.

20.4. Incorporating and Clarifying CBD and CP issues into ISPMs

The results of the OEWGs, and the establishment of the joint programme of work between the IPPC and CBD clearly identified that the IPPC, and its processes, play a role in protecting biodiversity, where protecting plant health contributes to the aims of the CBD and CP. This includes the provisions of the IPPC identified in Section 20.2 above, and the applicability of ISPMs to invasive alien species and LMOs. However, the IPPC recognized that there were gaps in guidance relating specifically to environmental risks and LMOs, and introduced several amendments to existing standards – in the form of supplements and annexes – to address these gaps. Table 20.2 shows the amendments added to ISPM No. 5 and ISPM No. 11.

20.4.1. Amendments to ISPM No. 5

One of the gaps in the ISPMs with respect to environmental concerns that needed clarification was in measuring the potential for economic harm to the environment. Recall that under the IPPC, the potential to cause economic harm is a requirement for an organism to meet the definition of a quarantine pest (see Chapters 5 and 7 for more information on economic harm). Thus, a major step to incorporating CBD and CP issues into ISPMs was the development of the supplement to ISPM No. 5, which clarifies how 'potential economic importance' (from the definition of quarantine pest) also applies to environmental concerns. A second step to addressing environmental concerns was the development of Appendix 1 to the *Glossary*, which relates key IPPC terms to terms used under the CBD. Chapter 5 of this text also includes a discussion of these terms.

20.4.2. Amendments to ISPM No. 11

It may be generally agreed that harmful effects to ecosystems, uncultivated flora or endangered species from pest introductions are undesirable. Likewise, it may be relatively simple to demonstrate that the introduction of a pest could have deleterious effects on wild flora, an ecosystem or the environment. The difficulty, however, lies in demonstrating economic consequences to situations that are not easily quantified in economic terms, or that have no apparent market value (e.g. the introduction of a pest that affects wild flora). Amendments to ISPM No. 11 further clarified that a number of methods exist to model economic impacts of activities that could have negative effects on the environment (see Table 20.2). In most cases, these methods have been used to assess economic effects resulting from pollution, development or other human-related activities that may be environmentally harmful. These methods can be adapted to assess potential economic impacts that arise from environmental harm as a result of a pest introduction.

Annex 1 of ISPM No. 11 clarifies the types of environmental risks that fall within the IPPC plant protection mandate to include: (i) risks from pests that directly affect uncultivated or unmanaged plants, as in the case of forest disease and pests [e.g. Dutch Elm Disease, (*Ophiostoma novo-ulmi* (Brasier)]; (ii) risks from pests that indirectly affect plants, as in the case of weeds that could affect cultivated plants or wild flora through processes like competition [e.g. Purple loosestrife, (*Lythrum salicaria* Linnaeus) that competes in natural and semi-natural habitats]; and (iii) risks from pests that indirectly affect plants through effects on other organisms, as in the case of parasites of beneficial organisms. In addition, Annex 1 states that in order to protect the environment and biological diversity without creating disguised barriers to trade, environmental risks and risks to biological diversity should be analysed in a pest risk analysis.

The scope of the IPPC in regard to risks from LMOs and the conduct of pest risk analysis for LMOs is described in Annexes 2 and 3 of ISPM No. 11. ISPM No. 11 relates to pest risk analysis and it is acknowledged in Annex 2 (see Box 20.1) that assessment of risks beyond the scope of the IPPC (i.e.

Table 20.2. Amendments to IPPC Standards (ISPMs) relating to environmental, invasive species and CBD considerations.

ISPM No. 5 (2010)	*Glossary of Phytosanitary Terms* Supplement 2: 2002. Guidelines on the understanding of potential economic importance and related terms including reference to environmental considerations Appendix 1: 2009. Terminology of the Convention on Biological Diversity in relation to the *Glossary of Phytosanitary Terms*
ISPM No. 11 (2004)	*Guidelines for Pest Risk Analysis for Quarantine Pests, Including Analysis of Environmental Risks and Living Modified Organisms* Annex 1 of ISPM No. 11, Comments on the scope of the IPPC in regard to environmental risks Annex 2 of ISPM No. 11, Comments on the scope of the IPPC in regard to pest risk analysis for living modified organisms Annex 3 of ISPM No. 11, Determining the potential for a living modified organism to be a pest Annex 4 of ISPM No. 11 *currently under development*, Pest risk analysis for intentionally imported plants as quarantine pests

human, animal health or environmental) may be required for LMOs (see Table 20.2 and Box 20.1 for the excerpted annex from ISPM No. 11). Annex 3 of ISPM No. 11 identifies potential phytosanitary risks for LMOs. This includes changes in adaptive characteristics that may increase the potential for introduction and spread, adverse effects of gene follow or gene transfer, adverse effects on non-target organisms, genotypic and phenotypic instability. It also includes factors that may lead to further consideration in a risk assessment (including insufficient data on the behaviour of the LMO in environments similar to the pest risk analysis area) and factors that may lead to the conclusion that an LMO is not a potential pest (including experience from research trials or experience in other countries). Annex 4 is currently under development (as of 2012) but will address risks associated with the intentional import of plants (e.g. analysing whether a plant behaves as a pest or 'weed' – see Chapter 18).

20.4.3 Revision of ISPM No. 2

ISPM No. 2 [originally *Guidelines for Pest Risk Analysis* (1996), currently *Framework for Pest Risk Analysis* (2007)] was written before the most recent revision of the IPPC (1997) was completed. It represented the earliest efforts at harmonizing concepts around pest risk analysis, based on experience of countries at that time. However, since its original adoption by the FAO Conference in 1996, NPPOs have gained more experience in the field of pest risk analysis, and many points needed to be updated and revised. The standard was revised in 2007, and the revision also incorporated concepts relevant to the CBD and CP.

The revision re-focused ISPM No. 2 to provide greater guidance on the initiation stage (stage 1) of pest risk analysis, since it was felt that ISPM No. 11 already provided detailed guidance on pest risk assessment (stage 2) and pest risk management (stage 3). In other words, the standard provides more information on what kinds of organisms may be considered pests, and how to decide whether or not they should be analysed in a pest risk analysis. It describes evaluating organisms such as beneficial organisms and biological control agents, plants as pests, or organisms new to science – these three cases having particular relevance to concerns identified in Article 8h of the CBD. It also describes how to evaluate an LMO in the initiation stage to determine whether a pest risk analysis is necessary.

Box 20.1. Comments on the scope of the IPPC in regard to pest risk analysis for living modified organisms. (Excerpted from ISPM No. 11, Annex 2, IPPC, 2004.)

Phytosanitary risks that may be associated with a living modified organism (LMO) are within the scope of the International Plant Protection Convention (IPPC) and should be considered using pest risk analysis (PRA) to make decisions regarding pest risk management.

The analysis of LMOs includes consideration of the following:

- Some LMOs may present a phytosanitary risk and therefore warrant a PRA. However other LMOs will not present a phytosanitary risks beyond those posed by related non-LMOs and therefore will not warrant a complete PRA. For example, modifications to change the physiological characteristics of a plant (e.g. ripening time, storage life) may not present any phytosanitary risk. The pest risk that may be posed by an LMO is dependent on a combination of factors, including the characteristics of the donor and recipient organisms, the genetic alteration, and the specific new trait or traits. Therefore, part of the supplementary text (see Annex 3) provides guidance on how to determine if an LMO is a potential pest.
- PRA may constitute only a portion of the overall risk analysis for import and release of an LMO. For example, countries may require the assessment of risks to human or animal health, or to the environment, beyond that covered by the IPPC. This standard only relates to the assessment and management of phytosanitary risks. As with other organisms or pathways assessed by an NPPO, LMOs may present other risks not falling within the scope of the IPPC. When an NPPO discovers potential for risks that are not of phytosanitary concern it may be appropriate to notify the relevant authorities.
- Phytosanitary risks from LMOs may result from certain traits introduced into the organism, such as those that increase the potential for establishment and spread, or from inserted gene sequences that do not alter the pest characteristics of the organism, but that might act independently of the organism or have unintended consequences.
- In cases of phytosanitary risks related to gene flow, the LMO is acting more as a potential vector or pathway for introduction of a genetic construct of phytosanitary concern rather than as a pest in and of itself. Therefore, the term 'pest' should be understood to include the potential of an LMO to act as a vector or pathway for introduction of a gene presenting a potential phytosanitary risk.
- The risk analysis procedures of the IPPC are generally concerned with phenotypic characteristics rather than genotypic characteristics. However, genotypic characteristics may need to be considered when assessing the phytosanitary risks of LMOs.
- Potential phytosanitary risks that may be associated with LMOs could also be associated with non-LMOs. It may be useful to consider risks associated with LMOs in the context of risks posed by the non-modified recipient or parental organisms, or similar organisms, in the PRA area.

20.5. Guidance on Risk Analysis for LMOs and Invasive Alien Species in ISPMs

As mentioned above, the CBD does not itself produce guidance on risk analysis for invasive alien species, but instead has identified (and cooperates with) other organizations that have competence in risk analysis. In the case of invasive alien species that are pests of plants, the IPPC is the organization which is responsible for providing guidance for pest risk analysis. This guidance comes in the form of ISPMs – in particular ISPM No. 2 and ISPM No. 11, although other ISPMs are

relevant to managing pest risk as well (refer to Table 13.2, Chapter 13, for ISPMs with particular relevance to managing risk).

By contrast, considerable efforts have been made by the CP to elaborate on Article 15 and Annex III and most recently, a draft[2] *Guidance on Risk Assessment of Living Modified Organisms* has been published to provide greater guidance on risk assessment under the CP. The guidance applies to LMOs as defined by the CP – such LMOs may or may not also be plant pests. For instance, genetically modified salmon are a type of LMO that would not be considered to meet the criteria of a plant pest (and therefore would not be subject to IPPC procedures), while a genetically modified strain of corn could meet the criteria for being a plant pest, depending on the genetic modification.

This may seem to create a possible divergence between the CP and guidance provided in ISPMs. However, the CP also recognizes guidance produced by the IPPC as having direct relevance to risk assessment required under the CP, and the IPPC recognizes that certain types of LMOs can potentially act as plant pests and are therefore subject to IPPC procedures. Box 20.1 provides an excerpt from ISPM No. 11, defining the scope of the IPPC with regard to LMOs. Furthermore, the guidance produced for risk assessment under the CP is voluntary – in other words, while the agreement itself has legal status and obligations for signatories, the guidance produced under the CP does not. This is in contrast to ISPMs, which have legal status by virtue of being requirements in the SPS Agreement, where Member countries are obliged to use standards (or base measures on risk analysis).

20.5.1. Specific guidance on risk analysis for invasive alien species in ISPMs

Initiation stage of pest risk analysis for invasive alien species

ISPM No. 2 provides the strongest guidance for initiating pest risk analyses for organisms that may have negative effects on the environment. These organisms include plants as pests, biological control agents and other beneficial organisms, identification of new pests, organisms that have not been reported as pests previously (including organisms new to science). It provides specific criteria that should be considered that are indicative that an organism will exhibit pest characteristics, including:

- Previous history of successful establishment in new areas;
- Phytopathogenic characteristics;
- Phytophagous characteristics;
- Presence detected in connection with observations of injury to plants, beneficial

organisms, etc. before any clear causal link has been established;
- Belonging to taxa (family or genus) commonly containing known pests;
- Capability of acting as a vector for known pests;
- Adverse effects on non-target organisms beneficial to plants (such as pollinators or predators of plant pests).

The guidance for initiating pest risk analyses for plants for planting is particularly important with respect to invasive alien species – due largely to the importance of weeds (see also Chapter 18) and their impact on natural environments. Several characteristics are identified specifically for plants, including adaptability, rate of propagation, competiveness, ability to persist, mobility of seeds and ability to hybridize.

Pest risk assessment – likelihood of introduction for invasive alien species

The likelihood of introduction for invasive alien species is assessed in largely the same way as for organisms we conventionally view as 'pests'. As such the existing guidance in ISPM No. 11 was viewed to be sufficient guidance for assessing this element of risk for invasive alien species. Clarifications were added, however, to clarify points related to plant for planting. This is because plants for planting (which might have the potential to impact the environment if they behave as weeds) are intentionally introduced and planted; therefore the likelihood of introduction (into an area) is virtually certain. The question then becomes whether or not the plants for planting will move from an intended area (where they are intentionally planted) to an unintended area (in particular a natural area) and exhibit weedy traits.

Pest risk assessment – consequences of introduction for invasive alien species

One of the main gaps in available guidance for environmental concerns was in assessing impacts to the environment. Thus, amendments to ISPM No. 11 give these points special consideration. The amendments describe the types of impacts on the environment that can be considered in a pest risk analysis, and then goes on to describe how those impacts can be described in economic terms. Particular attention is given to indirect effects on the environment, since those are often the most difficult to predict, describe and measure. Such impacts include:

- Significant effects on plant communities;
- Significant effects on designated environmentally sensitive or protected areas;
- Significant change in ecological processes and the structure, stability or processes of an ecosystem (including further effects on plant species, erosion, water table changes, increased fire hazard, nutrient cycling, etc.);
- Effects on human use (e.g. water quality, recreational uses, tourism, animal grazing, hunting, fishing);
- Costs of environmental restoration.

We know that a range of methods can be used to determine economic impacts (see Chapters 7 and 11) – and that the IPPC requires we demonstrate the potential for economic harm for a pest to be considered a quarantine pest. There was a misconception that 'economic harm' excluded impacts to the environment. However, the supplement to ISPM No. 5 (described above) clarified that economic impacts include environmental impacts, and that environmental impacts can be valued in economic terms. ISPM No. 11 then goes on to provide more specific guidance on this point.

ISPM No. 11 states that the environment can be valued in economic terms according to its 'use' and 'non-use' values. 'Use' values are those that commonly arise from consumption or utilization of a good, and use values include non-consumptive uses of the environment. For example, the use of a national park for recreation would be a non-consumptive use of an environmental good. Non-use values are often derived from the knowledge that something exists, such as knowing that certain species are protected and conserved for future generations. Non-use values may be further divided into option values (a benefit of the environment not derived from direct use), existence values (deriving benefit from an environmental good from knowing that it exists) and bequest values (valuing an environmental good for future generations).

20.5.2 Specific guidance on risk analysis for LMOs in ISPMs

Initiation Stage of PRA for LMOs

As with invasive alien species, the greatest guidance on initiating a pest risk analysis for LMOs is provided in ISPM No. 2. It describes the types of LMOs for which a pest risk analysis may be conducted:

- Plants for use in agriculture, horticulture or silviculture, bioremediation of soil, for industrial purposes, or as therapeutic agents (e.g. LMO plants with an enhanced vitamin profile or 'pharming' – growing plants with pharmaceutical properties).
- Biological control agents and other beneficial organisms modified to improve their performance.
- Pests modified to alter their pathogenic characteristics.

ISPM No. 2 goes on to describe the general types of risks an LMO may pose to plant health. These risks include increased potential for establishment and spread, and the potential to act as a vector for entering the inserted gene sequence into wild or domestic relatives, thereby altering those organisms (so they become pests), or otherwise negatively affecting health. ISPM No. 2 also describes the types of information that would be needed to conduct a pest risk analysis for LMOs (this information being different from what would normally be collected for conventional 'pests'). As with any organism, identification of the organism is a key step in the initiation stage (e.g. identification of the hazard in our risk equation). For LMOs, identification requires information regarding the taxonomic status of the recipient *and* the donor organism, and description of the vector, the nature of the genetic modification, and the genetic sequence and its insertion site in the recipient genome.

Importantly, ISPM No. 2 specifies that if the LMO does not exhibit traits consistent with the definition of a pest, then a pest risk analysis may not need to be conducted. Furthermore, it states that the LMO should be evaluated in the context of risks posed by the same organisms that have not been genetically modified using techniques of modern biotechnology. This distinction was important, since the same characteristics introduced through specialized techniques (e.g. biotechnology) could in principle also occur naturally, or be introduced through selective plant breeding.

ISPM No. 2 identifies factors that could result in the need for further analysis in a pest risk analysis, including cases where the LMO exhibits characteristics of a pest, data (e.g. field research or other information) indicates the LMO may behave as a pest, experience in other countries with the LMO indicates it could act as a pest or where a lack of knowledge requires that further analysis be done before any conclusion is reached. Full pest risk analyses may not need to be done if the LMO has previously been analysed, or a similar organism/trait has been analysed, where the LMO will be contained, or evidence suggests that the LMO will not exhibit characteristics of a pest.

Pest risk assessment – likelihood of introduction for LMOs

Like plants for planting, LMOs may be introduced intentionally into a country or an area. In the case of intentional introductions, the likelihood of entry does not require evaluation. However, they may also be unintentionally introduced – through natural spread, or human-mediated means, and thus unintentional entry may need to be assessed in the pest risk analysis. Likewise, even where LMOs are intentionally introduced, they may need to be evaluated for their potential (or their traits) to spread to unintended habitats. This is similar to the evaluations that may need to be done for plants for planting that have the potential to become weeds.

Pest risk assessment – consequences of introduction for LMOs

Consequences associated with the introduction of LMOs are similar to those for any other pest. ISPM No. 2 and ISPM No. 11 acknowledge that there can also be adverse consequences on the environment; however the guidance specified for invasive alien species (and in general for *pests*) sufficiently addresses the adverse consequences that should be evaluated for LMOs.

20.6. Summary

In this chapter we learned how two other international agreements – the Convention on Biological Diversity and the Cartagena Protocol on Biosafety – intersect with the IPPC and the SPS Agreement. Specifically, invasive alien species (under the CBD) and LMOs (under the CP) may fall under the scope of the IPPC when those organisms act as plant pests. In cases where either an invasive alien species or an LMO meet the definition for a pest under the IPPC, then IPPCs processes apply, including the application of ISPMs. In particular, the IPPC is recognized by all of the agreements as the authoritative source for guidance on conducting risk analysis for plant pests. Thus, pest risk analysis methods described in ISPMs No. 2 and 11 (and in this text) can also be used to analyse risks associated with invasive alien species and LMOs. The conduct of risk analysis for these organisms supports the aims of the IPPC (protecting plant health), the CBD (protection biological diversity) and the CP (concerned with the safe transboundary movement of LMOs).

Notes

[1] The definition of 'living modified organism' according to the CP 'means any living organism that possesses a novel combination of genetic material obtained through the use of modern biotechnology'. This term is related to, but distinct from 'genetically modified organism' or 'GMO'. GMOs may or may not be living, and capable of growth, whereas LMOs are living, and therefore capable of growing.

[2] The document is draft as of the writing of this text (2012) and is available online at the CBD website (www.cbd.int).

References

CBD (1992) *Convention on Biological Diversity*. Text and Annexes. Secretariat of the Convention on Biological Diversity/UNEP, Montreal.

CBD (2000) *Cartagena Protocol on Biosafety*. Secretariat of the Convention on Biological Diversity/UNEP, Montreal.

CBD (2002) Decision VI/23: Alien species that threaten ecosystems, habitats or species to which is annexed guiding principles for the prevention, introduction and mitigation of impacts of alien species that threaten ecosystems, habitats or species. Sixth Conference of the Parties, The Hague, the Netherlands, 7–19 April 2002.

CBD (2011) *Guidance on Risk Assessment of Living Modified Organisms*. Convention on Biological Diversity, Montreal. Available online at: http://bch.cbd.int/onlineconferences/ra_guidance_text.shtml, accessed 20 December 2011.

FAO (1997) *New Revised Text of the International Plant Protection Convention*. Food and Agriculture Organization of the United Nations, Rome.

FAO (2001) Report of the Third Interim Commission on Phytosanitary Measures. Appendix XIII: Statements of the ICPM Exploratory Open-ended Working Group on Phytosanitary Aspects of GMOs, Biosafety and Invasive Species, 2–6 April, Rome.

FAO (2002) Adoption of international standards, ICPM 02/9 Annex VI, Report of the ICPM Open-ended Working Group on Specifications for an International Standard for Phytosanitary Measures for Living Modified Organisms/Products of Modern Biotechnology, Fourth Session of the Interim Commission on Phytosanitary Measures, 11–15 March, Rome.

IPPC (2004) International Standards for Phytosanitary Measures, Publication No. 11: *Pest Risk Analysis for Quarantine Pests Including Analysis of Environmental Risks and Living Modified Organisms*. Secretariat of the International Plant Protection Convention (IPPC), Food and Agriculture Organization of the United Nations, Rome.

IPPC (2007) International Standards for Phytosanitary Measures, Publication No. 2: *Framework for Pest Risk Analysis*. Secretariat of the International Plant Protection Convention (IPPC), Food and Agriculture Organization of the United Nations, Rome.

IPPC (2010) International Standards for Phytosanitary Measures, Publication No. 5: *Glossary of Phytosanitary Terms*. Secretariat of the International Plant Protection Convention (IPPC), Food and Agriculture Organization of the United Nations, Rome.

MacLeod, A., Pautasso, M., Jeger, M.J. and Haines-Young, R. (2010) Evolution of the international regulation of plant pests and challenges for future plant health. *Food Security* 2, 49–70.

Rio Declaration (1992) *Rio Declaration 1992*. Report of the United Nations Conference on Environment and Development.

Schrader, G., Unger, J.G. and Starfinger, U. (2010) Invasive alien plants in plant health: a review of the past ten years. *EPPO Bulletin*, 40, 239–247.

Shine, C. (2007) Invasive species in an international context: IPPC, CBD, European strategy on invasive alien species and other legal instruments. *EPPO Bulletin* 37, 1, 103–113.

WTO (1994) *The WTO Agreement on the Application of Sanitary and Phytosanitary Measures*. World Trade Organization, Geneva.

Acronyms

AGM – Asian gypsy moth
AHP – Analytical Hierarchy Process
ALB – Asian longhorn beetle
APHIS – Animal and Plant Health Inspection Service
BBN – Bayesian Belief Network
bmp – bitmap image file
CBD – Convention on Biological Diversity
CFR – code of federal regulations
CMI – Composite Match Index
CP – Cartagena Protocol on Biosafety
CPHST – USDA-APHIS-PPQ Center for Plant Health Science and Technology
CPM – Commission on Phytosanitary Measures
CRA – commodity risk assessment
CSIRO – Commonwealth Scientific and Industrial Research Organization
csv – comma separated values
EA – Environmental Assessment
EI – Ecoclimatic Index
EIS – environmental impact statement
emf – enhanced metafile
EPPO – European and Mediterranean Plant Protection Organization
FAO – Food and Agriculture Organization of the United Nations
FONSI – Finding of No Significant Impact
GATT – General Agreement on Tariffs and Trade
geotiff – geo tagged image file format
gif – graphics interchange format
GIS – Geographic Information Systems
GRASS – Geographic Resources Analysis Support System
IBCAs – invertebrate biological control agents
IPPC – International Plant Protection Convention
ISPM – International Standards for Phytosanitary Measures
jpg – joint photographic experts group (also written as JPEG)
LBAM – light brown apple moth

MCDM – Multi Criteria Decision Making
NAPPFAST – North Carolina State University Animal and Plant Health Inspection Service
    Plant Pest Forecast System
NAPPO – North American Plant Protection Organization
NASS – National Agricultural Statistics Service
NEPA – National Environmental Policy Act
NPPO – National Plant Protection Organization
.pdf – portable document format
PDF – probability distribution function
PERAL – Plant Epidemiology and Risk Analysis Laboratory (CPHST-PPQ-APHIS)
PPA – Plant Protection Act
PPQ – Plant Protection and Quarantine
PRA – pest risk analysis
RNPQ – regulated non-quarantine pest
RPPO – Regional Plant Protection Organization
RSPM – Regional Standards for Phytosanitary Measures
SC – Standards Committee
SIT – sterile insect technique
SPS – Agreement on the Application of Sanitary and Phytosanitary Measures
TAG – technical advisory group
TCD – thousand cankers disease
tiff – tagged image file format
USC – United States Congress
USDA – United States Department of Agriculture
USDA-APHIS – United States Department of Agriculture – Animal and Plant Health
    Inspection Service
WTO – World Trade Organization

Glossary

Area: An officially defined country, part of a country or all or parts of several countries (Chapter 5: IPPC, 2010).

Area of low pest prevalence: An area, whether all of a country, part of a country or all or parts of several countries, as identified by the competent authorities, in which a specific pest occurs at low levels and which is subject to effective surveillance, control or eradication measures (Chapter 5: IPPC, 2010).

Bark-free wood: Wood from which all bark, except ingrown bark around knots and bark pockets between rings of annual growth, has been removed (Chapter 5: IPPC, 2010).

Beta distribution: A distribution that models the probability of success given the number of trials and number of successes (Chapter 12: Vose, 2000).

Binomial distribution: A distribution that models the number of successes given the number of trials and the probability of success (Chapter 12: Vose, 2000).

Biological control agent: A natural enemy, antagonist or competitor, or other organism, used for pest control (Chapter 17: IPPC, 2010).

Bulbs and tubers: A commodity class for dormant underground parts of plants intended for planting (includes corms and rhizomes) (Chapter 5: IPPC, 2010).

Commodity: A type of plant, plant product, or other article being moved for trade or other purpose (Chapter 5: IPPC, 2010).

Commodity class: A category of similar commodities that can be considered together in phytosanitary regulations (Chapter 5: IPPC, 2010).

Composite Match Index: The measure that Climex uses to characterize climate similarity between geographic locations (Chapter 12: Sutherst *et al.*, 2007).

Consignment: A quantity of plants, plant products and/or other articles being moved from one country to another and covered, when required, by a single phytosanitary certificate (a consignment may be composed of one or more commodities or lots) (Chapter 5: IPPC, 2010).

Contaminating pest: A pest that is carried by a commodity and, in the case of plants and plant products, does not infest those plants or plant products (Chapter 5: IPPC, 2010).

Controlled area: A regulated area which an NPPO has determined to be the minimum area necessary to prevent spread of a pest from a quarantine area (Chapter 5: IPPC, 2010).

Cut flowers and branches: A commodity class for fresh parts of plants intended for decorative use and not for planting (Chapter 5: IPPC, 2010).

Debarked wood: Wood that has been subjected to any process that results in the removal of bark. (Debarked wood is not necessarily bark-free wood.) (Chapter 5: IPPC, 2010).

Degree Day: The accumulated heat units above a threshold during a day (Gordh and Headrick, 2001). Degree days relate to the rate of development of organisms like weeds and insects and can be used to predict their phenology (Chapter 12: Gordh and Headrick, 2001; Magarey et al., 2007a).

Diapause: A state of reduced metabolic activity that can assist an organism in surviving unfavorable environmental conditions (Chapter 12: Gordh and Headrick, 2001).

Domestic Programme: A cooperative programme between PPQ, state departments of agriculture and other agencies whose objective is to mitigate the impacts of an invasive plant pest (Chapter 12: USDA-APHIS, 2011).

Dunnage: Wood packaging material used to secure or support a commodity but which does not remain associated with the commodity (Chapter 8: IPPC, 2010).

Ecoclimatic Index: The measure that Climex uses to characterize the environmental suitability of a location for a species to permanently establish based on growth and stress indices (Chapter 12: Sutherst et al., 1995, 2007).

Emergency measure: A phytosanitary measure established as a matter of urgency in a new or unexpected phytosanitary situation. An emergency measure may or may not be a provisional measure (Chapter 4: IPPC, 2010).

Endangered area: An area where ecological factors favour the establishment of a pest whose presence in the area will result in economically important loss (Chapter 5: IPPC, 2010).

Entry (of a pest): Movement of a pest into an area where it is not yet present, or present but not widely distributed and being officially controlled (Chapter 11: IPPC, 2010).

Equivalence (IPPC definition): The situation where, for a specified pest risk, different phytosanitary measures achieve a contracting party's appropriate level of protection (Chapter 4: IPPC, 2010).

Establishment: Perpetuation, for the foreseeable future, of a pest within an area after entry (Chapter 11: IPPC, 2010).

Field: A plot of land with defined boundaries within a place of production on which a commodity is grown (Chapter 5: IPPC, 2010).

Fruits and vegetables: A commodity class for fresh parts of plants intended for consumption or processing and not for planting (Chapter 5: IPPC, 2010).

Geographic Information Systems: Technology used to analyse, store, and visualize geospatial information (Chapter 12: Wade and Sommer, 2006).

Germplasm (IPPC definition): Plants intended for use in breeding or conservation programmes (Chapter 8: IPPC, 2010).

Grain (IPPC definition): A commodity class for seeds intended for processing or consumption and not for planting (Chapter 8: IPPC, 2010).

Growth Index: The measure that Climex uses to characterize the potential of a location to facilitate population development of a species based on abiotic and biotic factors (Chapter 12: Sutherst et al., 2007).

Habitat: Part of an ecosystem with conditions in which an organism naturally occurs or can establish (Chapter 5: IPPC, 2010).

Harmonization: The establishment, recognition and application by different countries of phytosanitary measures based on common standards (Chapter 4: IPPC, 2010).

Innundative release: The release of large numbers of mass-produced biological control agents or beneficial organisms with the expectation of achieving a rapid effect (Chapter 17: IPPC, 2010).

Inspection: Official visual examination of plants, plant products or other regulated articles to determine if pests are present and/or to determine compliance with phytosanitary regulations (Chapter 14: IPPC, 2010).

Interpolation: The process of estimating the values of unknown geospatial points based on surrounding geospatial points with known values (Chapter 12: Wade and Sommer, 2006).

Introduction: The entry of a pest resulting in its establishment (Chapter 11: IPPC, 2010).

Leaf wetness: An important parameter for modelling plant diseases whose biology requires that moisture be present during periods with suitable temperatures in order for infection to occur (Chapter 12: Magarey *et al.*, 2005). Note this can also represent general surface moisture on the plant (Magarey *et al.*, 2005, 2007a).

Legislation: Any act, law, regulation, guideline or other administrative order promulgated by a government (Chapter 5: IPPC, 2010).

Lot: A number of units of a single commodity, identifiable by its homogeneity of composition, origin etc., forming part of a consignment (Chapter 5: IPPC, 2010).

Natural enemy: An organism which lives at the expense of another organism in its area of origin and which may help to limit the population of that organism. This includes parasitoids, parasites, predators, phytophagous organisms and pathogens (Chapter 17: IPPC, 2010).

Official: Established, authorized or performed by a National Plant Protection Organization (Chapter 6: IPPC, 2010).

Organism (IPPC definition): Any biotic entity capable of reproduction or replication in its naturally occurring state (Chapter 5: IPPC, 2010).

Packaging: Material used in supporting, protecting or carrying a commodity (Chapter 5: IPPC, 2010).

Parasite: An organism which lives on or in a larger organism, feeding upon it (Chapter 17: IPPC, 2010).

Parasitoid: An insect parasitic only in its immature stages, killing its host in the process of its development, and free living as an adult (Chapter 17: IPPC, 2010).

Pathway: A means of pest entry or dispersal (Chapter 8: IPPC, 2010).

Pest: Any species, strain or biotype of plant, animal or pathogenic agent injurious to plants or plant products (Chapter 5: IPPC, 2010).

Pest categorization: The process for determining whether a pest has or has not the characteristics of a quarantine pest or those of a regulated non-quarantine pest (Chapter 5: IPPC, 2010).

Pest free area: An area in which a specific pest does not occur as demonstrated by scientific evidence and in which, where appropriate, this condition is being officially maintained (Chapter 5: IPPC, 2010).

Pest free place of production: Place of production in which a specific pest does not occur as demonstrated by scientific evidence and in which, where appropriate, this condition is being officially maintained for a defined period (Chapter 5: IPPC, 2010).

Pest free production site: A defined portion of a place of production in which a specific pest does not occur as demonstrated by scientific evidence and in which, where appropriate, this condition is being officially maintained for a defined period and that is managed as a separate unit in the same way as a pest free place of production (Chapter 5: IPPC, 2010).

Pest risk analysis: The process of evaluating biological or other scientific and economic evidence to determine whether an organism is a pest, whether it should be regulated, and the strength of any phytosanitary measures to be taken against it (Chapter 8: IPPC, 2010).

Pest Risk Assessment: Evaluation of the probability of the introduction and spread of a pest and the magnitude of the associated potential economic consequences (Chapter 11: IPPC, 2010).

Pest Risk Management: Evaluation and selection of options to reduce the risk of introduction and spread of a pest (Chapter 13: IPPC, 2010).

Pest status (in an area): Presence or absence, at the present time, of a pest in an area, including where appropriate its distribution, as officially determined using expert judgement on the basis of current and historical pest records and other information (Chapter 5: IPPC, 2010).

Phytosanitary import requirement: Specific phytosanitary measures established by an importing country concerning *consignments* moving into that country (Chapter 5: IPPC, 2010).

Phytosanitary measure: Any legislation, regulation or official procedure having the purpose to prevent the introduction and/or spread of quarantine pests, or to limit the economic impact of regulated non-quarantine pests (Chapter 5: IPPC, 2010).

Phytosanitary procedure: Any official method for implementing phytosanitary measures including the performance of inspections, tests, surveillance or treatments in connection with regulated pests (Chapter 5: IPPC, 2010).

Phytosanitary regulation: Official rule to prevent the introduction and/or spread of quarantine pests, or to limit the economic impact of regulated non-quarantine pests, including establishment of procedures for phytosanitary certification (Chapter 5: IPPC, 2010).

Place of production: Any premises or collection of fields operated as a single production or farming unit. This may include production sites which are separately managed for phytosanitary purposes (Chapter 5: IPPC, 2010).

Plants: Living plants and parts thereof, including seeds and germplasm (Chapter 5: IPPC, 2010).

Plants for planting: Plants intended to remain planted, to be planted or replanted (Chapter 5: IPPC, 2010).

Plant Hardiness Zones: Thermal bands representing the average extreme minimum temperatures for an area in 10°F increments (Chapter 12: Cathey, 1990; Magarey *et al.*, 2008).

Plants *in vitro*: A commodity class for plants growing in an aseptic medium in a closed container (Chapter 5: IPPC, 2010).

Point of entry: Airport, seaport or land border point officially designated for the importation of consignments, and/or entrance of passengers (Chapter 5: IPPC, 2010).

PRA area: Area in relation to which a pest risk analysis is conducted (Chapter 11: IPPC, 2010).

Predator: A natural enemy that preys and feeds on other animal organisms, more than one of which are killed during its lifetime (Chapter 17: IPPC, 2010).

Processed wood material: Products that are a composite of wood constructed using glue, heat and pressure, or any combination thereof (Chapter 5: IPPC, 2010).

Prohibition: A phytosanitary regulation forbidding the importation or movement of specified pests or commodities (Chapter 14: IPPC, 2010).

Protected area: A regulated area that an NPPO has determined to be the minimum area necessary for the effective protection of an endangered area (Chapter 5: IPPC, 2010).

Provisional measure: A phytosanitary regulation or procedure established without full technical justification owing to current lack of adequate information. A provisional measure is subjected to periodic review and full technical justification as soon as possible (Chapter 4: IPPC, 2010).

Quarantine: Official confinement of regulated articles for observation and research or for further inspection, testing and/or treatment (Chapter 3: IPPC, 2010).

Quarantine area: An area within which a quarantine pest is present and is being officially controlled (Chapter 5: IPPC, 2010).

Quarantine pest: A pest of potential economic importance to the area endangered thereby and not yet present there, or present but not widely distributed and being officially controlled (Chapter 5: IPPC, 2010).

Raster data: Geo-referenced data represented by grid cells (Chapter 12: Wade and Sommer, 2006).

Raw wood: Wood which has not undergone processing or treatment (Chapter 5: IPPC, 2010).

Regulated area: An area into which, within which and/or from which plants, plant products and other regulated articles are subjected to phytosanitary regulations or procedures in order to prevent the introduction and/or spread of quarantine pests or to limit the economic impact of regulated non-quarantine pests (Chapter 5: IPPC, 2010).

Regulated article: Any plant, plant product, storage place, packaging, conveyance, container, soil and any other organism, object or material capable of harbouring or spreading pests, deemed to require phytosanitary measures, particularly where international transportation is involved (Chapter 5: IPPC, 2010).

Regulated non-quarantine pest: A non-quarantine pest whose presence in plants for planting affects the intended use of those plants with an economically unacceptable impact and which is therefore regulated within the territory of the importing contracting party (Chapter 5: IPPC, 2010).

Regulated pest: A quarantine pest or a regulated non-quarantine pest (Chapter 5: IPPC, 2010).

Resolution: The spatial dimensions of raster grid cells, e.g. 10 km (Chapter 12: Wade and Sommer, 2006).

Risk: The likelihood of an adverse event and the magnitude of the consequences (Chapter 2).

Risk Analysis: A systematic way of gathering, evaluating and recording information leading to recommendations for a position or action in response to an identified hazard (Chapter 2).

Round wood: Wood not sawn longitudinally, carrying its natural rounded surface, with or without bark (Chapter 5: IPPC, 2010).

Sawn wood: Wood sawn longitudinally, with or without its natural rounded surface with or without bark (Chapter 5: IPPC, 2010)

Seeds: A commodity class for seeds for planting or intended for planting and not for consumption or processing (see grain) (Chapter 5: IPPC, 2010).

Spread: Expansion of the geographical distribution of a pest within an area (Chapter 5: IPPC, 2010).

Sterile insect technique: Method of pest control using area-wide inundative release of sterile insects to reduce reproduction in a field population of the same species (Chapter 17: IPPC, 2010).

Stored product: Unmanufactured plant product intended for consumption or processing, stored in a dried form (this includes in particular grain and dried fruits and vegetables) (Chapter 5: IPPC, 2010).

Stress Index: The measure that Climex uses to characterize the unsuitability of a location for a species to persist based on environmental stress factors (Chapter 12: Sutherst *et al.*, 2007).

Systems approach: The integration of different risk management measures, at least two of which act independently, and which cumulatively achieve the appropriate level of protection against regulated pests (Chapter 14: IPPC, 2010).

Technically justified: Justified on the basis of conclusions reached by using an appropriate pest risk analysis or, where applicable, another comparable examination and evaluation of available scientific information (Chapter 4: IPPC, 2010).

Transparency: The principle of making available, at the international level, phytosanitary measures and their rationale (Chapter 4: IPPC, 2010).

Vector data: Georeferenced data represented by points, lines and polygons (Chapter 12: Wade and Sommer, 2006).

Wood: A commodity class for round wood, sawn wood, wood chips or dunnage, with or without bark (Chapter 5: IPPC, 2010).

Wood packing material: Wood or wood products (excluding paper products) used in supporting, protecting or carrying a commodity (includes dunnage) (Chapter 8: IPPC, 2010).

Index

.